Lecture Notes in Computer Science   8210

Commenced Publication in 1973
Founding and Former Series Editors:
Gerhard Goos, Juris Hartmanis, and Jan van Leeuwen

Editorial Board

David Hutchison
Lancaster University, UK

Takeo Kanade
Carnegie Mellon University, Pittsburgh, PA, USA

Josef Kittler
University of Surrey, Guildford, UK

Jon M. Kleinberg
Cornell University, Ithaca, NY, USA

Alfred Kobsa
University of California, Irvine, CA, USA

Friedemann Mattern
ETH Zurich, Switzerland

John C. Mitchell
Stanford University, CA, USA

Moni Naor
Weizmann Institute of Science, Rehovot, Israel

Oscar Nierstrasz
University of Bern, Switzerland

C. Pandu Rangan
Indian Institute of Technology, Madras, India

Bernhard Steffen
TU Dortmund University, Germany

Madhu Sudan
Microsoft Research, Cambridge, MA, USA

Demetri Terzopoulos
University of California, Los Angeles, CA, USA

Doug Tygar
University of California, Berkeley, CA, USA

Gerhard Weikum
Max Planck Institute for Informatics, Saarbruecken, Germany

T0212604

Tetsuya Yoshida   Gang Kou
Andrzej Skowron   Jiannong Cao
Hakim Hacid   Ning Zhong (Eds.)

Active Media Technology

9th International Conference, AMT 2013
Maebashi, Japan, October 29-31, 2013
Proceedings

 Springer

Volume Editors

Tetsuya Yoshida
Hokkaido University, Sapporo, Japan
E-mail: yoshida@meme.hokudai.ac.jp

Gang Kou
Southwestern University of Finance and Economics, Chengdu, China
E-mail: kougang@yahoo.com

Andrzej Skowron
University of Warsaw, Poland
E-mail: skowron@mimuw.edu.pl

Jiannong Cao
Hong Kong Polytechnic University, Hong Kong
E-mail: csjcao@comp.polyu.edu.hk

Hakim Hacid
Bell Labs, Paris, France
E-mail: hakim.hacid@gmail.com

Ning Zhong
Maebashi Institute of Technology, Japan
E-mail: zhong@maebashi-it.ac.jp

ISSN 0302-9743   e-ISSN 1611-3349
ISBN 978-3-319-02749-4   e-ISBN 978-3-319-02750-0
DOI 10.1007/978-3-319-02750-0
Springer Cham Heidelberg New York Dordrecht London

Library of Congress Control Number: 2013950264

CR Subject Classification (1998): H.5, H.3, I.2, H.2.8, K.3, H.4, C.2, J.1

LNCS Sublibrary: SL 3 – Information Systems and Application,
incl. Internet/Web and HCI

© Springer International Publishing Switzerland 2013

This work is subject to copyright. All rights are reserved by the Publisher, whether the whole or part of the material is concerned, specifically the rights of translation, reprinting, reuse of illustrations, recitation, broadcasting, reproduction on microfilms or in any other physical way, and transmission or information storage and retrieval, electronic adaptation, computer software, or by similar or dissimilar methodology now known or hereafter developed. Exempted from this legal reservation are brief excerpts in connection with reviews or scholarly analysis or material supplied specifically for the purpose of being entered and executed on a computer system, for exclusive use by the purchaser of the work. Duplication of this publication or parts thereof is permitted only under the provisions of the Copyright Law of the Publisher's location, in its current version, and permission for use must always be obtained from Springer. Permissions for use may be obtained through RightsLink at the Copyright Clearance Center. Violations are liable to prosecution under the respective Copyright Law.

The use of general descriptive names, registered names, trademarks, service marks, etc. in this publication does not imply, even in the absence of a specific statement, that such names are exempt from the relevant protective laws and regulations and therefore free for general use.

While the advice and information in this book are believed to be true and accurate at the date of publication, neither the authors nor the editors nor the publisher can accept any legal responsibility for any errors or omissions that may be made. The publisher makes no warranty, express or implied, with respect to the material contained herein.

Typesetting: Camera-ready by author, data conversion by Scientific Publishing Services, Chennai, India

Printed on acid-free paper

Springer is part of Springer Science+Business Media (www.springer.com)

Preface

This volume contains the papers selected for presentation at *The 2013 International Conference on Active Media Technology* (AMT 2013), held jointly with The 2013 International Conference on Brain and Health Informatics (BHI 2013) at Maebashi Terrsa, Maebashi, Japan, during October 29–31, 2013. Organized by the Web Intelligence Consortium (WIC), by the IEEE Computational Intelligence Society Task Force on Brain Informatics (IEEE TF-BI), and by Maebashi Institute of Technology, this conference constitutes the ninth of the AMT series since the initial conference at Hong Kong Baptist University in 2001 (followed by AMT 2004 in Chongqing, China, AMT 2005 in Kagawa, Japan, AMT 2006 in Brisbane, Australia, AMT 2009 in Beijing, China, AMT 2010 in Toronto, Canada, AMT 2011 in Lanzhou, China, and AMT 2012 in Macau, China).

We are witnessing many rapid scientific and technological developments in human-centered, seamless computing environments, interfaces, devices, and systems with applications ranging from business and communication to entertainment and learning. These developments are collectively best characterized as *active media technology*, a new area of intelligent information technology and computer science that emphasizes the proactive and seamless roles of interfaces and systems as well as new media in all aspects of digital life. An AMT-based system offers services to enable the rapid design, implementation, and support of customized solutions. AMT 2013 aimed at providing a leading international forum to bring together researchers and practitioners from diverse fields, to increase the cross-fertilization of ideas and explore the fundamental roles, interactions as well as practical impacts of intelligent information technology and computer science on the next generation of computing environments, systems, and media.

The AMT conference series aims to explore and present state-of-the-art research in a wide spectrum of realms related to active media technology. We had full-paper submissions from 20 countries comprising of Japan, Bangladesh, Brazil, Canada, China, Egypt, Hong Kong, India, Iran, Italy, Korea, New Zealand, Pakistan, Singapore, Taiwan, Thailand, The Netherlands, Tunisia, UK, and USA. After a rigorous evaluation process, 26 full papers and two short papers were accepted. The topics of these papers centered on the main themes of AMT 2013, which encompass: active computer systems, interactive systems, and application of AMT-based systems; active media machine learning and data mining techniques; AMT for the Semantic Web, social networks, and cognitive foundations. The core issues of AMT have been investigated using various criteria that were carefully identified from both theoretical and practical perspectives to ensure the fruitful incorporation of technologies.

Apart from the 28 high-quality papers accepted by the main conference of AMT 2013, another 16 papers were selected for oral presentations in the workshop and special sessions. The Workshop on Intelligence for Strategic Foresight is based on the success of the Special Session on Technology Intelligence at AMT 2011, and aims at enabling companies to identify the technological opportunities and threats that could affect the future growth and survival of their business. The Special Session on Technologies and Theories of Narrative aims at providing a forum for discussing original and developing ideas about narrative-related information technologies, as well as various sorts of narrative theories for humanities and social sciences as the background of the technologies. The Special Session on Evolutionary Computation and Its Application deals with evolutionary computation (e.g., genetic algorithms, probabilistic approaches) and its application to real-world problems. The Special Session on Intelligent Media Search Techniques provides a venue for scholars to share and exchange their ideas of applying intelligent techniques to media search and management problems.

We would like to express our gratitude to all members of the Conference Committee for their support. AMT 2013 had a very exciting program with a number of features, ranging from keynote talks, technical sessions, workshops, special sessions, and social programs. A program of this kind would not have been possible without the generous dedication of the Program Committee members and the external reviewers in reviewing the papers submitted to AMT 2013. Many thanks also to all the authors who contributed their research results to this volume. We greatly appreciate our keynote speakers, Yuichiro Anzai of the Japan Society for the Promotion of Science, Yuzuru Tanaka of Hokkaido University, Carl K. Chang of Iowa State University, and Andrzej Skowron of Warsaw University. Special thanks to Hanmin Jung, Seungwoo Lee, and Sa-Kwang Song for organizing the Workshop on Intelligence for Strategic Foresight; Takashi Ogata, for the Special Session on Technologies and Theories of Narrative; Kenichi Ida, for the Special Session on Evolutionary Computation and Its Application; Zhiguo Gong, for the Special Session on Intelligent Media Search Techniques, and to the AMT-BHI 2013 Organizing Chairs, Kazuyuki Imamura, and Tetsumi Harakawa. Special thanks are also extended to all the panelists of the AMT-BHI panel on Brain Big Data in the Hyper World, Stephen S. Yau of Arizona State University (chair), Bin Hu of Lanzhou University, China, Guoyin Wang of Chongqing University of Posts and Telecommunications, China, Jianhua Ma of Hosei University, Kazuhiro Oiwa of the National Institute of Information and Communications Technology, Japan, as well as Shinsuke Shimojo and Marcel A. Just.

Finally, we extend our highest appreciation to Springer's *Lecture Notes in Computer Science* (LNCS/LNAI) team for their generous support. We thank Alfred Hofmann and Anna Kramer of Springer for their help in coordinating the publication of this special volume in an emerging and interdisciplinary research field. Also, we are grateful to Tetsumi Harakawa, Juzhen Dong, and Shinich Motomura for their strong support and dedication to AMT 2013. We also would like

Co-chairs/Directors

ong	Maebashi Institute of Technology, Japan
Liu	Hong Kong Baptist University, SAR China

CIS-TFBI Chair

ong — Maebashi Institute of Technology, Japan

Advisory Board

A. Feigenbaum	Stanford University, USA
Ohsuga	University of Tokyo, Japan
in Wah	Chinese University of Hong Kong, SAR China
Yu	University of Illinois, Chicago, USA
adeh	University of California, Berkeley, USA

Technical Committee

Bradshaw	UWF/Institute for Human and Machine Cognition, USA
ercone	York University, Canada
Fensel	University of Innsbruck, Austria
Gottlob	Oxford University, UK
hi Jain	University of South Australia, Australia
ia Ma	Hosei University, Japan
hang Mao	Yahoo! Inc., USA
e Morizet-Mahoudeaux	Compiegne University of Technology, France
hi Motoda	Osaka University, Japan
ki Nishida	Kyoto University, Japan
zej Skowron	Warsaw University, Poland
ong Wu	Okayama University, Japan
ong Wu	University of Vermont, USA
Yao	University of Regina, Canada

gram Committee

Abraham	Norwegian University of Science and Technology, Norway
ia Bordogna	The National Research Council, Italy
nio Chella	The University of Palermo, Italy
ang Chen	Institute of Computing Technology, Chinese Academy of Sciences, China

to thank conference, including Gunma Prefecture Government, Maebashi City Government, Maebashi Convention Bureau, Web Intelligence Lab Inc., Mitsuba Gakki Co. Ltd., GCC Inc., Japan High Comm, Kuribara Medical Instruments, Yamato Inc., etc.

August 2013

Tetsuya Yoshida
Gang Kou
Andrzej Skowron
Jiannong Cao
Hakim Hacid
Ning Zhong

Conference Organizat

Honorary General Chair

Setsuo Ohsuga   University of Tokyo, Ja

Conference General Chairs

Andrzej Skowron   Warsaw University, Pola
Jian-Nong Cao   Hong Kong Polytechnic

WIC

Edward
Setsuo
Benjan
Philip
L.A. Z

Program Chairs

Tetsuya Yoshida   Hokkaido University, Japa
Gang Kou   University of Electronic S
     Technology of China, C

WIC

Jeffrey

Nick (
Dieter
Georg
Lakhr
Jianh
Jianc
Pierr
Hiros
Toyo
Andr
Jingl
Xind
Yiyu

Workshop/Special Session Organizing Chair

Hakim Hacid   Bell Labs, France

AMT-BHI 2013 Organizing Chairs

Kazuyuki Imamura   Maebashi Institute of Techn
Tetsumi Harakawa   Maebashi Institute of Techn
Ning Zhong   Maebashi Institute of Techn

Panel Chair

Stephen S. Yau   Arizona State University, US

Publicity Chairs

Shinichi Motomura   Maebashi Institute of Technolc
Dominik Slezak   Infobright Inc., Canada and U
     Warsaw, Poland
Jian Yang   Beijing University of Technolog

Chin-Wan Chung	Korea Advanced Institute of Science and Technology, Republic of Korea
Adrian Giurca	Brandenburg University of Technology at Cottbus, Germany
William Grosky	University of Michigan, USA
Bin Guo	Northwestern Polytechnical University, China
Hakim Hacid	SideTrade, France
Yoshikatu Haruki	Ferris University, Japan
Enrique Herrera-Viedma	University of Granada, Spain
Masahito Hirakawa	Shimane University, Japan
Wolfgang Huerst	Utrecht University, The Netherlands
Hajime Imura	Hokkaido University, Japan
Hiroki Inoue	Kyoto University, Japan
Hiroshi Ishikawa	Kagawa University, Japan
Hanmin Jung	KISTI, Republic of Korea
Yoshitsugu Kakemoto	The JSOL, Ltd., Japan
Gang Kou	University of Electronic Science and Technology, China
Jing Li	University of Science and Technology of China, China
Wen-bin Li	Shijiazhuang University of Economics, China
Xining Li	University of Guelph, Canada
Xiaohui Liu	Brunel University, UK
Marco Luetzenberger	DAI-Labor, TU-Berlin, Germany
Wenji Mao	Institute of Automation, Chinese Academy of Sciences, China
Kouzou Ohara	Aoyama Gakuin University, Japan
Yoshihiro Okada	Kyushu University, Japan
Yoshiaki Okubo	Hokkaido University, Japan
Gang Pan	Zhejiang University, China
Naoki Saiwaki	Konan University, Japan
Eugene Santos	University of Connecticut, USA
Gerald Schaefer	Loughborough University, UK
Dominik Slezak	University of Warsaw and Infobright Inc., Poland
Tsuyoshi Sugibuchi	Internet Memory Research, France
Kazunari Sugiyama	National University of Singapore, Singapore
Yuqing Sun	Shandong University, China
Akio Takashima	Shohoku College, Japan
Xijin Tang	Academy of Mathematics and Systems Science, Chinese Academy of Sciences, China
Xiaohui Tao	University of Southern Queensland, Australia
Takao Terano	Tokyo Institute of Technology, Japan
Vincent Toubiana	Bell Labs France, France
Athena Vakali	Aristotle University of Thessaloniki, Greece

Egon L. Van den Broek	University of Twente/Karakter University Center, The Netherlands
Natalie van der Wal	VU University Amsterdam, The Netherlands
Neil Y. Yen	The University of Aizu, Japan
Tetsuya Yoshida	Hokkaido University, Japan
Yi Zeng	Institute of Automation, Chinese Academy of Sciences, China
Guoqing Zhang	University of Windsor, Canada
Shichao Zhang	University of Technology, Australia
Zhangbing Zhou	Institut Telecom & Management SudParis, France

External Reviewers

Dongmin Seo
Dandan Zhou
Chu Du

Table of Contents

AMT for Semantic Web, Social Networks, and Cognitive Foundations

Workshop on Intelligence for Strategic Foresight

Special Session on Technologies and Theories of Narrative

Special Session on Evolutionary Computation and Its Application

Special Session on Intelligent Media Search Techniques

Interactive Rough-Granular Computing in Wisdom Technology

Andrzej Jankowski[1], Andrzej Skowron[2], and Roman Swiniarski[3]

[1] Institute of Computer Science, Warsaw University of Technology
Nowowiejska 15/19, 00-665 Warsaw, Poland
a.jankowski@ii.pw.edu.pl
[2] Institute of Mathematics, Warsaw University
Banacha 2, 02-097 Warsaw, Poland
skowron@mimuw.edu.pl
[3] Department of Computer Science, San Diego State University
5500 Campanile Drive San Diego, CA 92182, USA
and
Institute of Computer Science Polish Academy of Sciences
Jana Kazimierza 5, 01-248 Warsaw, Poland
rswiniarski@mail.sdsu.edu

Constructing the physical part of the theory and unifying it
with the mathematical part should be considered as one of
the main goals of statistical learning theory
– Vladimir Vapnik
([24], Epilogue: Inference from sparse data, p. 721)

Abstract. Understanding of interactions is the critical issue of complex systems. Interactions in physical world are represented by information granules. We propose to model complex systems by interactive intelligent systems (IIS) created by societies of agents. Computations in IIS are based on complex granules (c-granules, for short). Adaptive judgement allows us to reason about c-granules and interactive computations performed on them. In adaptive judgement, different kinds of reasoning are involved such as deduction, induction, abduction or reasoning by analogy as well as intuitive judgement. In modeling of mental parts of c-granules, called information granules (infogranules, for short), we use the approach based on the rough set methods in combination with other soft computing approaches. Issues related to interactions among objects in the physical and mental worlds as well as adaptive judgement belong to the fundamental issues in Wisdom Technology (WisTech). In the paper we concentrate on some basic issues related to interactive computations over c-granules. WisTech was developed over years of work on different real-life projects. It can also be treated as a basis in searching for solutions of problems in such areas as Active Media Technology and Wisdom Web of Things.

Keywords: granular computing, rough sets, interactions, information granule, physical object, complex granule, interactive intelligent system, active media technology.

T. Yoshida et al. (Eds.): AMT 2013, LNCS 8210, pp. 1–13, 2013.
© Springer International Publishing Switzerland 2013

1   Introduction

Granular Computing (GC) is now an active area of research (see, e.g., [16]). Objects we are dealing with in GC are *information granules* (or *infogranules*, for short). Such granules are obtained as the result of information granulation [25, 27]: *Information granulation can be viewed as a human way of achieving data compression and it plays a key role in implementation of the strategy of divide-and-conquer in human problem-solving.*

The concept of granulation is rooted in the concept of a linguistic variable introduced by Lotfi Zadeh in 1973. Information granules are constructed starting from some elementary ones. More compound granules are composed of finer granules that are drawn together by indistinguishability, similarity, or functionality [26].

Understanding of interactions of objects on which are performed computations is fundamental for modeling of complex systems [3]. This requirement is fundamental for modeling of complex systems [3]. For example, in [13] this is expressed in the following way: *[...] interaction is a critical issue in the understanding of complex systems of any sorts: as such, it has emerged in several well-established scientific areas other than computer science, like biology, physics, social and organizational sciences.*

Interactive Rough Granular Computing (IRGC) is an approach for modeling interactive computations (see, *e.g.*, [19, 21–23]) based on rough sets in combination with other soft computing approaches such as fuzzy sets or evolutionary computing, and also with machine learning and data mining techniques. The notion of the highly interactive granular system is clarified as the system in which intrastep interactions [4] with the external as well as with the internal environments take place.

We extend the existing approach by introducing *complex granules* (*c-granules*) making it possible to model interactive computations performed by agents operating in the physical world. Any c-granule consists of three components, namely soft_suit, link_suit and hard_suit. These components are making it possible to deal with such abstract objects from soft_suit as infogranules as well as with physical objects from hard_suit. The link_suit of a given c-granule is used as a kind of c-granule interface for expressing interaction between soft_suit and and hard_suit. Any agent operates in a local world of c-granules. The agent control is aiming to control computations performed by c-granules from this local world for achieving the target goals. Actions (sensors or plans) from link_suits of c-granules are used by the agent control in exploration and/or exploitation of the environment on the way to achieve their targets. C-granules are also used by agents for representation of perception of interactions in the physical world. Due to the bounds of the agent perception abilities usually only a partial information about the interactions from physical world may be available for agents. Hence, in particular the results of performed actions by agents can not be predicted with certainty.

The reasoning making it possible to derive relevant information c-granules for solutions of the target tasks is based on interactions between many c-granules and is called *adaptive judgement*. *Adaptive rational judgement* and *adaptive intuitive judgement* are two main kinds of adaptive judgement [8]. In some sense this is some kind of dialogue and/or discussion between c-granules representing different perspectives for the problem solving. For example, any agent classifying a new object typically uses some arguments *for* and *against*. One should make some kind of dialogue between such "arguments". It is a good idea to distinguish dialogue and discussion [2]: *A key difference between a dialogue and an ordinary discussion is that, within the latter, people usually hold relatively fixed positions and argue in favor of their views as they try to convince others to change.* When a given agent has enough time and recourses then this agent can continue such kind of "internal agent dialogue" until acceptable solutions will be achieved. It is a good example of "slow thinking" [9]. However, in practice usually agent does not have "enough time and recourses". Thus agent has to make the decision fast. In many practical cases "lack of decision is the worst decision". In order to "survive" in such situations agent has to make a decision based on agent hierarchy of values representing current agent priorities of needs, habits, intuitions, emotions and others. This way of thinking corresponds in many aspects to "fast thinking" (Kehneman and Tversky [9]) or it corresponds to concept of "discussions" [2]. In some sense this approach to problem solving has roots in the Socrates approach to logic. According to "Dialogues" of Plato, Socrates prefers concept of "dia-logos" instead of logos. In other words, the problem solving process is a result of some kind of interaction processes like "dialogue" or/and "discussion" between agent internal c-granules and/or with other agents. Two agent processes are important for realisation of interactions:

1. Adaptive intuitive judgement - corresponds to making decisions mainly based on agent current hierarchy of values which represents current agent priorities of needs, habits, intuitions, emotions and others. They are computed fast using simple heuristics. One can observe strong relationships of adaptive intuitive judgement with "fast thinking" [9] or "discussion" [2] as well as to some extent with reactive agents. The rules of adaptive intuitive judgement are subject of continuous adaptive change toward improvement of the quality of non perfect judgement (it is result of interactions with environment).

2. Adaptive rational judgement - corresponds to making decisions mainly based on rational (*i.e.*, acceptable and verifiable by agent) inference rules. One can observe strong relationships of adaptive rational judgment with "slow thinking" [9], "dialogue" [2] as well as to some extent with deliberative agents. The rules of adaptive rational judgement are subject of rather rare changes (relative to the dynamic speed of changes of the analyzed phenomena). The changes typically are results of important new discoveries, paradigm shifts (*e.g.*, science revolution according to Thomas Kuhn) and/or language meaning shifts.

Definitely, in the language used by agents for dealing with adaptive judgment some deductive systems known from logic may be applied to closed worlds for reasoning about knowledge relative to them. This may happen, *e.g.*, if the agent languages are based on classical mathematical logic. However, if we move to interactions in open worlds then new specific rules or patterns relative to a given agent or group of agents in such worlds should be discovered. Such rules or patterns are influenced by uncertainty because they are induced by agents under uncertain knowledge about the environment. Hence, considering only the absolute truth becomes unsatisfactory. Deduction and induction as well as abduction or analogy based reasoning may be involved in adaptive judgement. Among the tasks for adaptive judgement are the following ones supporting reasoning toward: searching for relevant approximation spaces, discovery of new features, selection of relevant features, rule induction, discovery of inclusion measures, strategies for conflict resolution, adaptation of measures based on the minimum description length principle, reasoning about changes, perception (action and sensory) attributes selection, adaptation of quality measures over computations relative to agents, adaptation of object structures, discovery of relevant context, strategies for knowledge representation and interaction with knowledge bases, ontology acquisition and approximation, learning in dialogue of inclusion measures between information granules from different languages (*e.g.*, the formal language of the system and the user natural language), strategies for adaptation of existing models, strategies for development and evolution of communication language among agents in distributed environments, strategies for risk management in distributed computational systems.

The question arises about the logical tools relevant for the above mentioned tasks of adaptive judgement on computations performed on c-granules. First let us observe that the satisfiability relations in the IRGC framework can be treated as tools for constructing new c-granules. If fact, for a given satisfiability relation, the semantics of formulas over c-granules relative to this relation is defined. In this way the candidates for new relevant c-granules can be obtained. We would like to emphasize a very important feature. The relevant satisfiability relation for the considered problems is not given but it should be induced (discovered) on the basis of a partial information encoded in information (decision) systems obtained as results of aggregation (*e.g.*, join with constraints [17]) of other information (decision) systems. For real-life problems, it is often necessary to discover a hierarchy of satisfiability relations before we obtain the relevant target level. Information granules constructed at different levels of this hierarchy finally lead to relevant ones for approximation of complex vague concepts expressed in natural language. The rough set approach in combination with other soft computing approaches is used in inducing approximations of complex vague concepts and reasoning schemes over them [20].

Issues related to interactions among objects in the physical and mental worlds as well as to adaptive judgement about interactive computations on c-granules belong to the fundamental issues in Wisdom Technology (WisTech) [6, 7] based on the following meta-equation:

WISDOM = INTERACTIONS+ADAPTIVE JUDGEMENT+KNOWLEDGE.

Wistech can be treated as a basis in searching for solutions of problems related to the goals of Active Media Technology (AMT) as well as of Wisdom Web of Things (W2T) [28, 10].

In Section 2 a general structure of c-granules is described and some illustrative examples are included. Moreover, some preliminary concepts related to agents performing computations on c-granules are discussed. Comments on risk management in IIS are presented in Section 3.

This paper covers some issues of the keynote talk at AMT 2013.

2   Complex Granules and Physical World

In this section we discuss the basic concepts related to c-granule relative to a given agent ag. Let us assume that the agent ag has access to a local clock making it possible to use the local time scale. In this paper we consider discrete linear time scale.

We distinguish several kinds of objects in the environment in which the agent ag operates:

- *physical objects* (called also as *hunks of matter*, or *hunks*, for short) [5] such as physical parts of agents or robots, specific media for transmitting information; we distinguish hunks called as artifacts used for labeling other hunks or stigmergic markers used for indirect coordination between agents or actions [11]; note that hunks may change in time and are perceived by the agent ag as dynamic (systems) processes; any hunk h at the local time t of ag is represented by the state $st_h(t)$; the results of perception of hunk states by agent ag are represented by value vector of relevant attributes (features);
- *complex granules* (c-*granules*, for short) consisting of three parts: *soft_suit*, *link_suit*, and *hard_suit* (see Figure 1); c-granule at the local time t of ag is denoted by G; G receives some inputs and produces some outputs; inputs and outputs of c-granule G are c-granules of the specified admissible types; input admissible type is defined by some input preconditions and the output admissible type is defined by some output postconditions, there are distinguished inputs (outputs) admissible types which receive (send) c-granules from (to) the agent ag control;
  - • G soft_suit consists of
    1. G name, describing the behavioral pattern description of the agent ag corresponding to the name used by agent for identification of the granule,
    2. G type consisting of the types of inputs and outputs of G c-granule,
    3. G status (*e.g.*, active, passive),
    4. G information granules (infogranules, for short) in mental imagination of the agent consisting, in particular of G specification, G implementation and manipulation method(s); any implementation distinguished in infogranule is a description in the agent ag language of transformation of input c-granules of relevant types into output

c-granules of relevant types, *i.e.*, any implementation defines an interactive computation which takes as input c-granules (of some types) and produces some c-granules (of some types); inputs for c-granules can be delivered by the agent ag control (or by other c-granules), we also assume that the outputs produced by a given c-granule depend also on interactions of hunks pointed out by link_suite as well as some other hunks from the environment - in this way the semantics of c-granules is established;

- G link_suit consists of
  1. a representation of configuration of hunks at time t (*e.g.*, mereologies of parts in the physical configurations perceived by the agent ag);
  2. links from different parts of the configuration to hunks;
  3. G links and G representations of elementary actions; using these links the agent ag may perform sensory measurement or/and actions on hunks; in particular, links are pointing to the sensors or actuators in the physical world used by the considered c-granule; using links the agent ag may, *e.g.*, fix some parameters of sensors or/and actions, initiate sensory measurements or/and action performance; we also assume that using these links the agent ag may encode some information about the current states of the observed hunks by relevant information granules;
- G hard_suit is created by the configuration of interacting hunks encoding G soft_suit, G link_suit and implementing G computations;
- soft_suit and link_suit of G are linked by G links for interactions between the G hunk configuration representation and G infogranules;
- link_suit and hard_suit are linked by G links for interactions between the G hunk configuration representation and hunks in the environment.

The interactive processes during transforming inputs of c-granules into outputs of c-granules are influenced by (i) interaction of hunks pointed out by link_suit and (ii) interaction of pointed hunks with relevant parts of configuration in link_suit.

Agent can establish, remember, recognize, process and modify relations between c-granules or/and hunks. Agents and societies of agents may be also represented as (generalized) c-granules [7]. A general structure of c-granules is illustrated in Figure 1. Figure 2 illustrates c-granules corresponding to sensory measurement. Note that in this case, the parameters fixed by the agent control may concern sensor selection, selection of the object under measurement by sensor and selection of sensor parameters. They are interpreted as actions selected from the link_suit. In the perception of hunk configuration of c-granule are distinguished infogranules representing sensor(s), object(s) under measurement and the configuration itself. The links selected by the agent control represent relations between states of hunks and infogranules corresponding to them in the link_suit. Figure 3 illustrates how an interactive information (decision) system is created and updated during running of c-granule implementation according to scenario(s) defined in the soft_suit and related G links. Such information (decision) systems are used for recording information about the computation steps

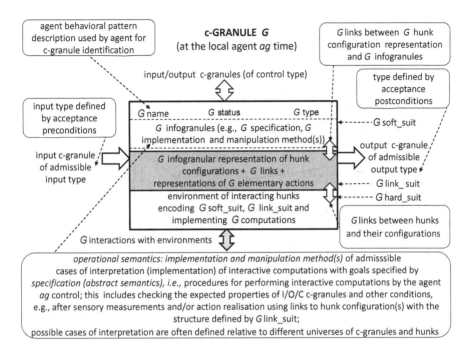

Fig. 1. General structure of c-granules

during the c-granule implementation run. Note that the structure of such information systems is different from the classical definition [14, 15, 20]. In particular, such systems are open due to the fact that links to physical objects as well as interactions are changing (often in unpredictable way) in time. We assume that for any agent *ag* there is distinguished a family of *elementary c-granules* and constructions on c-granules leading to more compound c-granules. The agent *ag* is using the constructed granules for modeling attention and interaction with the environment. Note that for any new construction on elementary granules (such as network of c-granules) should be defined the corresponding c-granule. This c-granule should have appropriate soft_suit, link_suit and hard_suit so that the constructed c-granule will satisfy the conditions of the new c-granule construction specification. Note that one of the constraints on such construction may follow from the interactions which the agent *ag* will have at the disposal in the uncertain environment.In the construction o new c-granules information systems play the fundamental role (see, *e.g.*, [1, 21, 17]).

It is worthwhile mentioning that the interactive computations based on c-granules are different from Turing computations. In IRGC computations are progressing due to controlled by actions interactions among different physical objects and their interactions with inforgranules. One can imagine operations non-computable in the Turing sense which are computable in IRGC. In Figure 4 we illustrate how the abstract definition of operation from soft_link interacts

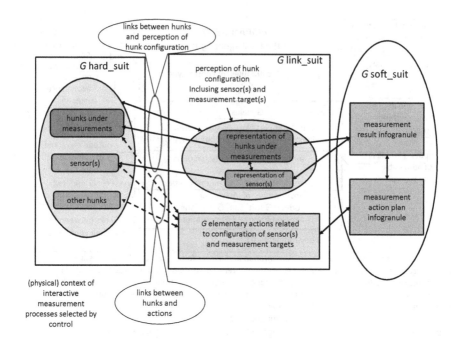

Fig. 2. Interactions caused by sensors

with other suits of c-granule. It is necessary to distinguish two cases. In the first case, the results of operation ⊗ realized by interaction of hunks are consistent with the specification in the suit_link. In the second case, the result specified in the soft_suit can be treated only as an estimation of the real one returned as the result of interactions which may be different from the estimated result due to the unpredictable interactions in the hard_suit.

A transition relation related to a given agent ag is defined between configurations of ag at time t and the time moment next to t. A configuration of ag at time t consists of all configurations of c-granules existing at time t. A configuration of c-granule G at time t consists of G itself as well as all c-granules selected on the basis of links in the link_suit of G at time t. These are, in particular all c-granules pointed by links corresponding to the c-granules stored in the computer memory during the computation process realised by c-granule as well as c-granules corresponding to perception at time t of the configuration of hunks at time t.

Note that the transition relation is making it possible to predict the next possible configurations. However, the real successive configuration may be different from all of the predicted ones due to unpredictable interactions in the environment.

Languages of agents consists of special c-granules called expressions. From one hand side any expression may be used without its 'support' in corresponding link_suit and hard_suit of the c-granule defining the expression, *i.e.*, only

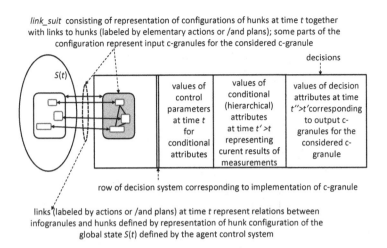

link_suit consisting of representation of configurations of hunks at time *t* together with links to hunks (labeled by elementary actions or /and plans); some parts of the configuration represent input c-granules for the considered c-granule

decisions

| $S(t)$ | values of control parameters at time t for conditional attributes | values of conditional (hierarchical) attributes at time $t' > t$ representing curent results of measurements | values of decision attributes at time $t'' > t'$ corresponding to output c-granules for the considered c-granule |

row of decision system corresponding to implementation of c-granule

links (labeled by actions or /and plans) at time *t* represent relations between infogranules and hunks defined by representation of hunk configuration of the global state *S(t)* defined by the agent control system

Fig. 3. Example: Row of interactive information (decision) system corresponding to registration of computation of c-granule according to implementation scenario

soft_suit may fully characterize expression. This can be done under the assumption that there are fixed reliable coding methods (representing, *e.g.*, reading and storing methods) between soft_suit and hard_suit of the expression by the agent. On the other hand, under the same assumption as before, one may consider only hard_link as an object characterizing the considered expression. Expressions for communication are usually representing classes of hunks rather than single hunks. This follows from the fact that the agents have bounded abilities for discernibility of perceived objects. Different behavioral patterns may be indiscernible relative to the set of attributes used by the agent. Hence, it follows that the agents perceive objects belonging to the same indiscernibility or/and similarity class in the same way. This is an important feature making it possible to use generalization by agents. For example, the situations classified by a given set of characteristic functions of induced classifiers (used as attributes) may be indiscernible. On the other hand, a new situation unseen so far may be classified to indiscernibility classes which allows agents to make generalizations. The new names created by agents are names of new structured objets or their indiscernibility (similarity) classes. Languages of agents consist of partial descriptions of situations (or their indiscernibility or similarity classes) perceived by agents as well as description of approximate reasoning schemes about the situations and their changes by actions and/or plans. The situations may be represented in hierarchical modeling by structured objects (*e.g.*, relational structures over attribute value vectors or/and indiscernibility (similarity classes) of such structures). In reasoning about the situation changes [18] one should take into account that the predicted actions or/and planes may depend not only on the changes recognized in the past situations but also on the performed actions and plans in the past. This is strongly related to the idea of perception pointed out in [12]: *The main*

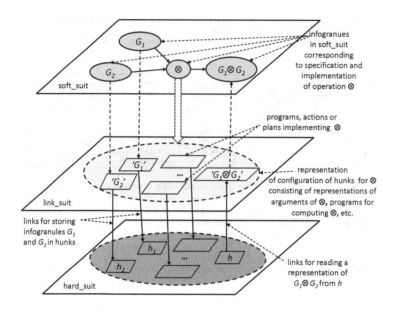

Fig. 4. Computability of operation $\otimes$ in IRGC

idea of this book is that perceiving is a way of acting. It is something we do. Think of a blind person tap-tapping his or her way around a cluttered space, perceiving that space by touch, not all at once, but through time, by skillful probing and movement. This is or ought to be, our paradigm of what perceiving is. Agents should be equipped with adaptation strategies for discovery of new structured objects and their features (attributes). This is the consequence of the fact that the agents are dealing with vague concepts. Hence, the approximations of these concepts represented by the induced classifiers evolve with changes in uncertain data and imperfect knowledge.

3   Risk Managemant in IIS

For the risk management in IIS one of the most important task is to develop strategies for inducing approximations of vague complex concepts which are next used for activation of actions performed by agents on the basis of satisfiability (to a degree) of these concepts in a given situation. A typical example is the statement of the form: "now we do have very risky situation". The development of strategies for inducing approximations of such vague complex concepts are based the activation of actions performed by agents. These vague complex concepts are represented by the agent hierarchy of needs.

In risk management one should consider a variety of complex vague concepts and relations between them as well as reasoning schemes related, *e.g.*, to the

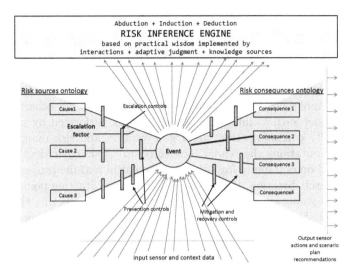

Fig. 5. Bow-tie diagram

bow-tie diagram (see Figure 5). One can consider the mentioned above task of approximation of vague complex concepts initiating actions as the task of discovery of complex games (see Figure 6) from data and domain knowledge which the agents are playing in the environment attempting to achieve their targets. The discovery process can be based on hierarchical learning supported by domain knowledge [7, 1]. It is also worthwhile mentioning that such games are evolving in time when the data and knowledge about the approximated concepts are changing (drifting in time) and the relevant adaptive strategies for such games are required. These adaptive strategies are used to control the behavior of agents toward achieving by them targets. Note that also these strategies should be learned from data and domain knowledge.

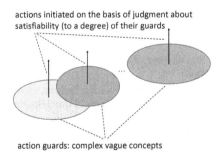

Fig. 6. Games based on complex vague concepts

4   Conclusions and Future Research

The outlined research on the nature of interactive computations is crucial for understanding complex systems. Our approach is based on complex granules (c-granules) performing computations through interaction with physical objects (hunks). Computations of c-granules are controlled by the agent control. More compound c-granules create agents and societies of agents. For other issues outlined in this paper such as interactive computations performed by societies for agents, especially communication language evolution and risk management in interactive computations the reader is referred to [7]. Further development of the IRGC based on c-granules for risk management in real-life projects is one of our future research direction. IRGC seems to be a good starting point for better understanding computations in nature as well as for constructing the physical part of the statistical learning theory and unifying it with the mathematical part of this theory (see motto of his article). Interactive computations based on c-granules create a good background for modeling computations in AMT and W2T. Yet another research direction is related to relationships of interactive computability based on c-granules and Turing computability.

Acknowledgements. This work was supported by the Polish National Science Centre grants 2011/01/B/ ST6/03867, 2011/01/D/ST6/06981, and 2012/05/B/ST6/03215 as well as by the Polish National Centre for Research and Development (NCBiR) under the grant SYNAT No. SP/I/1/77065/10 in frame of the strategic scientific research and experimental development program: "Interdisciplinary System for Interactive Scientific and Scientific-Technical Information" and the grant No. O ROB/0010/ 03/001 in frame of the Defence and Security Programmes and Projects: "Modern engineering tools for decision support for commanders of the State Fire Service of Poland during Fire & Rescue operations in the buildings".

References

1. Bazan, J.: Hierarchical classifiers for complex spatio-temporal concepts. In: Peters, J.F., Skowron, A., Rybiński, H. (eds.) Transactions on Rough Sets IX. LNCS, vol. 5390, pp. 474–750. Springer, Heidelberg (2008)
2. Bohm, D., Peat, F.D.: Science, Order, and Creativity, pp. 240–247. Bantam, New York (1987)
3. Goldin, D., Smolka, S., Wegner, P. (eds.): Interactive Computation: The New Paradigm. Springer (2006)
4. Gurevich, Y.: Interactive algorithms 2005. In: Goldin, et al. (eds.) [3], pp. 165–181
5. Heller, M.: The Ontology of Physical Objects. Four Dimensional Hunks of Matter. Cambridge Studies in Philosophy. Cambridge University Press (1990)
6. Jankowski, A., Skowron, A.: A WisTech paradigm for intelligent systems. In: Peters, J.F., Skowron, A., Düntsch, I., Grzymała-Busse, J.W., Orłowska, E., Polkowski, L. (eds.) Transactions on Rough Sets VI. LNCS, vol. 4374, pp. 94–132. Springer, Heidelberg (2007)
7. Jankowski, A., Skowron, A.: Practical Issues of Complex Systems Engineering: Wisdom Technology Approach. Springer, Heidelberg (2013)

8. Kahneman, D.: Maps of bounded rationality: Psychology for behavioral economics. The American Economic Review 93, 1449–1475 (2002)
9. Kahneman, D.: Thinking, Fast and Slow. Farrar, Straus and Giroux, New York (2011)
10. Liu, J.: Active media technologies (AMT) from the standpoint of the Wisdom Web. In: Li, Y., Looi, M., Zhong, N. (eds.) Advances in Intelligent IT - Active Media Technology 2006, Proceedings of the 4th International Conference on Active Media Technology, AMT 2006, Brisbane, Australia, June 7-9. Frontiers in Artificial Intelligence and Applications, vol. 138. IOS Press, Brisbane (2006)
11. Marsh, L.: Stigmergic epistemology, stigmergic cognition. Journal Cognitive Systems 9, 136–149 (2008)
12. Noë, A.: Action in Perception. MIT Press (2004)
13. Omicini, A., Ricci, A., Viroli, M.: The multidisciplinary patterns of interaction from sciences to computer science. In: Goldin, et al. (eds.) [3], pp. 395–414
14. Pawlak, Z.: Information systems - theoretical foundations. Information Systems 6, 205–218 (1981)
15. Pawlak, Z.: Rough Sets: Theoretical Aspects of Reasoning about Data, System Theory, Knowledge Engineering and Problem Solving, vol. 9. Kluwer Academic Publishers, Dordrecht (1991)
16. Pedrycz, W., Skowron, S., Kreinovich, V. (eds.): Handbook of Granular Computing. John Wiley & Sons, Hoboken (2008)
17. Skowron, A., Stepaniuk, J.: Hierarchical modelling in searching for complex patterns: Constrained sums of information systems. Journal of Experimental and Theoretical Artificial Intelligence 17, 83–102 (2005)
18. Skowron, A., Stepaniuk, J., Jankowski, A., Bazan, J.G., Swiniarski, R.: Rough set based reasoning about changes. Fundamenta Informaticae 119(3-4), 421–437 (2012)
19. Skowron, A., Stepaniuk, J., Swiniarski, R.: Modeling rough granular computing based on approximation spaces. Information Sciences 184, 20–43 (2012)
20. Skowron, A., Suraj, Z. (eds.): Rough Sets and Intelligent Systems. Professor Zdzisław Pawlak in Memoriam. Series Intelligent Systems Reference Library. Springer (2013)
21. Skowron, A., Szczuka, M.: Toward interactive computations: A rough-granular approach. In: Koronacki, J., Raś, Z.W., Wierzchoń, S.T., Kacprzyk, J. (eds.) Advances in Machine Learning II. SCI, vol. 263, pp. 23–42. Springer, Heidelberg (2010)
22. Skowron, A., Wasilewski, P.: Information systems in modeling interactive computations on granules. Theoretical Computer Science 412(42), 5939–5959 (2011)
23. Skowron, A., Wasilewski, P.: Interactive information systems: Toward perception based computing. Theoretical Computer Science 454, 240–260 (2012)
24. Vapnik, V.: Statistical Learning Theory. John Wiley & Sons, New York (1998)
25. Zadeh, L.A.: Fuzzy sets and information granularity. In: Advances in Fuzzy Set Theory and Applications, pp. 3–18. North-Holland, Amsterdam (1979)
26. Zadeh, L.A.: Toward a theory of fuzzy information granulation and its centrality in human reasoning and fuzzy logic. Fuzzy Sets and Systems 90, 111–127 (1997)
27. Zadeh, L.A.: A new direction in AI: Toward a computational theory of perceptions. AI Magazine 22(1), 73–84 (2001)
28. Zhong, N., Ma, J.H., Huang, R., Liu, J., Yao, Y., Zhang, Y.X., Chen, J.: Research challenges and perspectives on Wisdom Web of Things (W2T). The Journal of Supercomputing 64, 862–882 (2013)

Vision-Based User Interface for Mouse and Multi-mouse System

Yuki Onodera[1] and Yasushi Kambayashi[2]

[1] Department of Retail Service Systems
Cube System Inc., 1-2-33 Higashigotanda, Shinagawa-ku, Tokyo 141-0022, Japan
yuki-onodera@cubesystem.co.jp
[2] Department of Computer and Information Engineering
Nippon Institute of Technology, 4-1 Gakuendai, Miyashiro-machi, Saitama 345-8501, Japan
yasushi@nit.ac.jp

Abstract. This paper proposes a vision-based methodology that recognizes the users' fingertips so that the users can perform various mouse operations by gestures as well as implements multi-mouse operations. By using the Ramer-Douglas-Peucker algorithm, the system retrieves the coordinates of the finger from the palm of the hand. The system also recognizes the users' intended operation on the mouse through the movements of recognized fingers. When the system recognizes several palms of hands, it changes its mode to the multi-mouse mode so that several users can coordinate their works on the same screen. The number of mice is the number of recognized palms. In order to implement our proposal, we have employed the Kinect motion capture camera and have used the tracking function of the Kinect to recognize the fingers of users. Operations on the mouse pointers are reflected in the coordinates of the detected fingers. In order to demonstrate the effectiveness of our proposal, we have conducted several user experiments. We have observed that the Kinect is suitable equipment to implement the multi-mouse operations. The users who participated in the experiments quickly learned the multi-mouse environment and performed naturally in front of the Kinect motion capture camera.

Keywords: Kinect, Multi-mouse, Hand Tracking, Skelton Tracking, OpenNI.

1 Introduction

We have witnessed the advent of novel man-machine interfaces in various fields including human computer interaction. In particular, the field of the human computer interaction in mobile phones and games are eye-opening; operations by voice and by gestures become common scenery today. It is advantageous if one can operate apparatuses without touching them. Even in the operation of PCs, there are quite a few situations that one does not want to touch input devices.

In this paper, we propose a vision-based methodology that recognizes the users' fingertips so that the users can perform various mouse operations by gestures as well as implements multi-mouse operations. By using the Ramer-Douglas-Peucker

T. Yoshida et al. (Eds.): AMT 2013, LNCS 8210, pp. 14–23, 2013.
© Springer International Publishing Switzerland 2013

algorithm, the system retrieves the coordinates of the finger from the palm of the hand. The system also recognizes the users' intended operation on the mouse through the movements of recognized fingers. When the system recognizes several palms of hands, it changes its mode to the multi-mouse mode so that several users can coordinate their works on the same screen such as sharing information, transferring files by dragging them. The number of mice is the number of recognized palms.

In order to implement our proposal, we have employed the Kinect motion capture camera and have used the tracking function of the Kinect to recognize the fingers of users. Operations on the mouse pointers are reflected in the coordinates of the detected fingers. Even in the multi-mouse mode, operations such as clicking are detected independently so that multiple users' operations do not confuse each other.

In order to demonstrate the effectiveness of our proposal, we have conducted several user experiments. We have observed that the Kinect is suitable equipment to implement the multi-mouse operations. The users who participated in the experiments quickly learned the multi-mouse environment and performed the vision based interface gestures naturally in front of the Kinect motion capture camera.

We believe that our development opens a new horizon toward a novel human computer interaction that particularly assists brainstorming. Because natural, stress-free interface is the most important factor to make brain storming be success.

The structure of the balance of this paper is as follows. In the second section, we describe the related works. In the third section, we present our methodology. The core of our methods is the Ramer-Douglas-Peucker algorithm to retrieve the coordinates of the fingers from the palm of the hand in an open space. The fourth section describes how the system is implemented. We have employed the Kinect motion capture camera to retrieve the coordinates of the palms of the users, as well as to detect their intentions by gestures. The fifth section describes the user experiments and demonstrates the effectiveness of our system. Finally, in the sixth section, we conclude our discussion and present the direction of future works.

2   Related Works

Versatility and applicability of the non-contact operation are the containers of enormous ideas. There are various situations where the operators of a machine cannot touch any physical devices. In fact, non-contact operations have the possibility to make impossible desires be possible. Pioneers in this field have shown the possibility of non-contact operations. For example, Farzin et al. have demonstrated the non-contact mouse is possible [1].

Ueda et al. have developed a multi-mouse system [2]. Even though they have shown the potential collaborative works through multi-mouse system, all the participants have to have their own physical mice.

Several researchers have conducted studies in real time tracking systems using the Villa-Jones-based cascade classifier [3-4]. Other researchers have shown that cascade classifiers can also be used to recognize hands and various parts of the human body [5-9]. Marcel et al. have detected the hand by the different technique [10].

They have used a variation that tracks hand by the skin color of human body. Yu et al. proposed a technique to detect gestures of hands by image of processing [11]. They have utilized that the bipolarization of the extracted image increases the contrast so that the silhouette and distinct features of the hands are accurately and efficiently extracted from the image. The Gauss-Laplace edge detection approach has been utilized to get the hand edge.

Another notable study and development is accomplished by Chiba et al. [12]. They have successfully substituted the physical mouse by the Kinect [12]. The system is already commercially available from Nichii [13]. Ahn et al. have developed the other system, an interactive presentation system, for slideshow presentations [14].

As we mentioned in the first section, the Kinect has several advantages over the conventional mouse. In the field of medical care and welfare, there are lots of occasions that the users want to avoid hand-contact. Substituting the mouse by the Kinect already exists for medical use [12]. The system, however, is designed just for single user. We adopt a multi-mouse function in our system so that the system can be used for multi-user brainstorming.

For the multi-user brainstorming, the multi-mouse system developed by Ueda et al. has great success. The participants are able to communicate directly though pointing the same screen. At a brainstorming scene, we often experience that exchanging roles and input devices are annoying and interfere the discourse of thought and discussion. Therefore the multi-mouse environment is desirable for brainstorming. However, the conventional multi-mouse system has certain disadvantage. In fact, the multi-muse system using conventional physical mice takes too much space to operate [2]. Our system provides the multi-mouse environment operated entirely by gestures.

3   Methodology

The purpose of this system is to provide a powerful tool that substitutes the traditional input device, i.e. mouse, and realizes the multi-mouse system. This system is aimed to assist the brainstorming of multiple people by using the Kinect motion capture camera. The system sets the mouse pointer coordinate as follows:

1. The Kinect recognizes the hand of a user. The information acquisition of the frame model starts.
2. The system acquires the coordinate of the palm from the frame model that is provided by the Kinect.
3. The system acquires the coordinate of the each finger from the palm. The system can acquire the coordinates of the entire hand in this way. If the system detects (through the Kinect) more than one palms, it acquires the coordinate of the each palm.
4. The system sets the mouse pointer at the coordinate of the finger acquired at the second step.

Thus, when the system recognizes the fingertip of the user, it makes the coordinate of the mouse pointer be the coordinate of the finger. The system displays several mouse pointers depending on the number of the recognized palms, and assigns the click and the mouse wheel functions to each mouse pointer. By simple finger gestures, the user can perform those click and mouse wheel functions in our vision based user interface system. The user can perform all the features that the basic mouse has. For example, in order to perform the mouse wheel function, the user is supposed to move his or her finger as if tracking the touch panel. Table 1 shows the gestures that the user is expected to perform in order to activate the mouse functions.

Table 1. Gesture functions

Number of Finger	Function	The user 's performance
One finger	Mouse cursor	Fingertip indicates the mouse position
Two fingers	Mouse click	Second finger indicates the mouse clicked
Three fingers	Mouse wheel	Scroll the screen by moving fingers

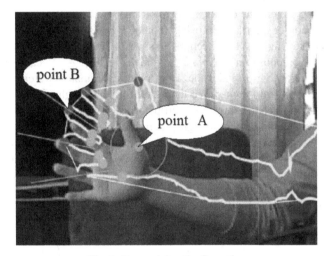

Fig. 1. Recognizing the fingertips

We have chosen the C# as the implementation language for our system [15]. This is the language recommended for programming with the Kinect. We use OpenNI API to recognize the fingers of the user [16]. OpenNI API acquires the skeleton information of a human body (the yellow outline) as shown in Fig. 1. The decision to use OpenNI as the base library is due to NITE [17]. OpenNI uses another set of library programs called NITE, and this supports development of applications using OpenNI, and is middleware to support gesture recognition in particular. We describe the details of gesture recognition in the next implementation section.

We employ Ramer-Douglas-Peucker algorithm to detect fingers [18]. The Ramer-Douglas-Peucker algorithm is intended to reduce the number of points to represent a curved line while preserving the rough drawing figure. This is also called as End-Point-Fit Algorithm. The algorithm takes the first point (point A) and the last point

(point B) (Fig. 1). Point A is the coordinate of the palm of the user. The skeleton tracking feature of the Kinect provides the coordinate of this point, and we can extract the values by using the OpenNI library. Point B is a point of approximately the same depth close to point A, and is the furthest point on the plane of the same depth. Since the depth sensor provides the contour of the hand by bipolarization, we can determine the coordinates of the tip of the finger furthest from the coordinates of the palm (point A). It then looks for a remote point more than the approximate precision (ε) from the line with the first and the last points as end points. This point is the furthest on the curved line from the approximating line between the end points. If the point furthest from the line is greater than the approximate precision (ε), then that point must be kept. The algorithm recursively calls itself until the point furthest from the line is less than the approximate precision (ε). By using this algorithm, our system captures the image of the user's hand as shown in Fig. 1. This algorithm makes the outline of a given hand image be expressed smoothly by minimum number of points as shown in Fig. 1. By the preliminary experiments, i.e. trial and error, on a number of adult male hands, we have found that the condition $\varepsilon = 15$ (pixels) provides the best result.

After obtaining the outline of a hand by the Ramer-Douglas-Peucker algorithm, we identify the position of the fingertip using another algorithm called convex hull algorithm. The convex hull is a collection of straight line connecting the outermost points in the set of points in the plane as shown in Fig. 2. Because the Ramer-Douglas-Pecker algorithm provides the set of point coordinates, we can easily determine the convex hull from them. We then look for acute angles in the convex hull. Because we assume that the only a fingertip forms an acute angle. We define the acute angle in our context be zero to eighty degree.

When the system recognizes a hand, it acquires the coordinates of the palm and constructs the coordinates of fingers from them. When the system recognizes several hands, each coordinate is kept in the data structure as an array. Then each finger of each hand is also kept in the data structure as also an array in a hierarchical way. Fig. 2 shows how the data structure is constructed.

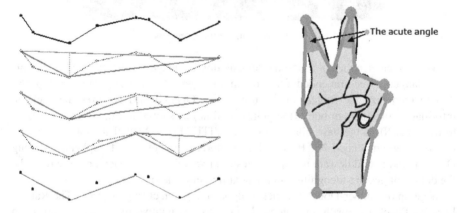

Fig. 2. Left: Douglas-Pecker Algorithm, Right: The Convex hull

4   Implementation

We employed the Kinect motion capture camera as the base technology for our system. The Kinect is the peripheral device for Xbox 360 released by Microsoft Corporation [19]. Kinect can capture the movement of whole body of the user without wearing any physical devices by using a RGB camera, a depth sensor, a multi-array microphone. The RGB camera uses three colors (red, blue, green) of signals with three different wires. The camera is placed to between of two depth sensors. The Kinect adopts a technique to perform the triangulation by its own laser pattern. In fact, we have observed innumerable points are arranged as a laser pattern. Skeleton tracking is to distinguish each part of the body using the image that the RGB camera and the depth sensor of the Kinect acquire. With the Kinect, we can extract the movement of each part of the body in dynamic setting.

In order to develop our system with the Kinect, we select OpenNI for API. OpenNI is the library that is provided by PrimeSence [16]. The notable feature of OpenNI is that the library makes the near-mode available for any types of the Kinect. By using this library, the Kinect can detect an object as close as 405mm. This makes the operation in front of the PC be recognizable. We intensively utilize this function to make the system detect fingertips of users just in front of the Kinect. OpenNI uses another set of library programs called NITE, and this supports development of OpenNI. NITE is the middleware to support gesture recognition in particular [17]. The NITE library can detect various gestures by using its own algorithm, such as circular motion or the vertical motion. In addition, it supports the function for multi-processing using a shared memory and performs the acquisition and the disposal of camera images by a different process. This feature is useful when many users participate in the multi-mouse mode.

When the user uses single finger and moves it, the system reflects the numerical value of the coordinate of the mouse according to the movement. When the user adds another finger, i.e. using two fingers, the system recognizes the user pushes the mouse button. Therefore, if the user display two fingers in a short period, the system recognizes the gesture as the mouse click, and if the user two fingers and moves them, the system recognizes the gesture as the dragging the mouse. When the user displays three fingers, the system recognizes the gesture mouse wheel operation. These gestures and operations are summarized in Table 1.

When the system recognizes more than one palm, the system registers those positions of palms in an array of the data structure as shown in Fig. 3. The system checks the data structure and displays the mouse pointer as many elements as in the array. Once a certain palm is recognized and corresponding mouse pointer is created.

The correspondence of the hand and pointer is never broken. The image of the hand and the mouse pointer are tightly coupled in the data structure. Therefore confusions between multiple mouse operations should not be happen.

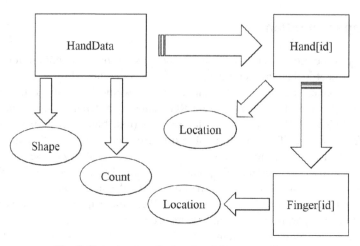

Fig. 3. Data structure for hand and finger coordinates

5   User Experiments

In order to demonstrate the effectiveness of our system, we have conducted some user experiments. In this section, we report the result of the user experiments. Fig. 4 shows the users conducting an experiment using two mice.

Fig. 4. Users operating PC with two mice

We have conducted the usability evaluation by using questionnaires. For the test of the multi-mouse, we made pairs of users. Each pair consists of a teacher and a learner and they perform a role playing game. They do not have previous knowledge about the Kinect. The questionnaire sheet contains five questions. They are: the comprehensibility of the operation, the response of the operation, the novelty of the system, the error rate and the degree of satisfaction. The simplicity of the operation makes the user learn the operation of the system easily. The bigger the value means the easier to learn. The error rate means how the user feels uncomfortable of the system. We recorded the users' impressions and opinions during the experiments and summed up the numerical value on the check sheet as shown in Table 2. The higher the value, the more the user satisfies.

Interviewing with the users of the experiments, we have found that they had experienced uncomfortable phenomena due to misrecognitions of our system. In order to ameliorate the situation, we have investigated and conducted the numerical experiment to reveal how the system misrecognizes the users' gestures. We have observed the system tends to misrecognize the number of the fingers. Also, when the user changes the number of fingers to indicate the system for mouse operations, it takes some time delay to recognize the changes. Since the system is designed to recognize the users' intended operation by the number of fingers, this may cause serious problems. We compiled the results of the experiments in Table 3. We have measured the time period that the system recognizes the change of the number of fingers, and be stable. When the user displays single finger, the system recognizes it immediately, i.e. less than one second. Regardless the user changes the number of fingers one to two or three to two, the time period the system stably recognizes the two fingers, i.e. mouse-down, is one or two second. For the three fingers, it takes three seconds as well. We can observe that the system is fairly stable.

Table 2. User evaluations

Users	Comprehensibility	Response	Error rate	Novelty of System	Satisfaction
A	5	4	4	5	4
B	5	3	4	5	5
C	5	3	3	5	5
D	5	4	4	4	4
E	5	5	3	5	5
F	5	4	3	5	4
Average	**5.0**	**3.8**	**3.5**	**4.8**	**4.5**

Table 3. Time to recognize the fingers

Changes of the number of fingers	Response time in second
Display single finger	Less than 1
Changes to two fingers	1-2
Changes to three fingers	2-3

The user experiments show the system is easy to learn. We have found the users get used to the vision-based interface immediately after starting to use. We have also observed that the Kinect and multi-mouse system have high affinity. The system demonstrates considerably low misrecognitions. The system certainly misrecognizes the number of fingers the users show, but the error rate is in the tolerable range. Erroneous recognition was more likely to occur when:

1.   More than two hands are present in front of the Kinect in close proximity, and
2.   Some objects that resembles to hand or finger including hair, near the hand.

In order to improve the recognition success rate, we are designing to combine the skeleton tracking system of the Kinect and image processing. RGB camera can easily recognize the color of the hand, and should be able to distinguish the fingers from the background. This should augment the depth sensor data.

When the user is about to perform a mouse operation and the other user of the pair perform another mouse operation, their operations do not interfere each other. But while one is scrolling the screen, the other clicks the mouse on an icon, their works interfere. Technically such a situation is no problem, but the users feel nuisance. We have found the source of the irritation is that it is not clear what the company is doing, because there is no visible mouse. In order to ameliorate the situation, we are designing the icons of the mouse pointer that indicate what the users operate on it.

6   Conclusion and Future Direction

We have proposed a vision-based human computer interface that recognizes the users' fingertips so that the users can perform various mouse operations by gestures. We have also implemented multi-mouse features in the vision-based human computer interface system. The system retrieves the coordinates of the finger from the palm of the hand by using the Ramer-Douglas-Peucker algorithm.

We have employed the Kinect motion capture camera and have used the tracking function of Kinect to retrieve the coordinates of the fingers of users. Operations on the mouse pointers are reflected in the coordinates of the detected fingers. Even in the multi-mouse mode, operations such as clicking are detected independently so that multiple users' operations do not confuse each other.

In order to demonstrate the effectiveness of our proposal, we have conducted several user experiments. We have observed that the Kinect is suitable equipment to implement the multi-mouse operations. The users who participated in the experiments quickly learned the multi-mouse environment and performed naturally in front of the Kinect motion capture camera. The system sometimes misrecognizes the number of the users' fingers. We should be able to mitigate the misrecognition due to the inaccurate depth sensor by combing the image processing of the RGB camera data.

Even though there is much room to improve the system, we believe that our development opens a new horizon toward a novel human computer interaction that particularly assists brainstorming. Because natural, stress-free interface is the most important factor to make brain storming be success.

References

1. Farhadi-Niaki, F., Aghaei, R.G., Arya, A.: Empirical Study of a Vision-based Depth Sensitive Human-Computer Interaction System. In: Tenth Asia Pacific Conference in Computer Human Interaction, pp. 101–108. ACM Press (2012)
2. Ueda, M., Takeuchi, I.: Mouse Cursors Surf the Net - Developing Multi-computer Multi-mouse Systems. In: IPSJ Programming Symposium, pp. 25–32 (2007) (in Japanese)
3. Viola, P., Jones, M.: Robust real-time object detection. In: Second International Workshop on Statistical and Computational Theories of Vision (2001)
4. Viola, P., Jones, M.: Rapid object detection using a boosted cascade of simple feature. IEEE Computer Vision and Pattern Recognition 1, 511–518 (2001)
5. Chen, Q., Cordea, M.D., Petriu, E.M., Varkonyi-Kockzy, A.R., Whalen, T.E.: Human-computer interaction for smart environment applications using hand-gesture and facial-expressions. International Journal of Advanced Media and Communication 3(1/2), 95–109 (2009)
6. Kolsch, M., Tuck, M.: Robust hand detection. In: International Confernce on Automatic Face and Gesture Recognition, pp. 614–619 (2004)
7. Kolsch, M., Tuck, M.: Analysis of rotational robustness of hand detection with a Viola-Jones detector. In: IAPR International Conference of Pattern Recognition, vol. 3, pp. 107–110 (2004)
8. Zhang, Q., Chen, F., Liu, X.: Hand gesture detection and segmentation based on difference background image with complex background. In: International Conference on Embedded Software and Systems, pp. 338–343 (2008)
9. Anton-Canalis, L., Sanchez-Nielsen, E., Castrillon-Santana, M.: Hand pose detection for vision-based gesture interfaces. In: Conference on Machine Vision Applications, pp. 506–509 (2005)
10. Marcel, S., Bernier, O., Viallet, J.E., Collobert, D.: Hand gesture recognition using input-output hidden Markov models. In: Conference on Automatic Face and Gesture Recognition, pp. 456–461 (2000)
11. Yu, C., Wang, X., Huang, H., Shen, J., Wu, K.: Vision-based hand gesture recognition using combinational features. In: Sixth International Conference on Intelligent Information Hiding and Multimedia Signal Processing, pp. 543–546 (2010)
12. Chiba, S., Yosimitsu, K., Maruyama, M., Toyama, K., Iseki, H., Muragaki, Y.: Opect: Non-contact image processing system using Kinect. J. Japan Society of Computer Aided Surgery 14(3), 150–151 (2012) (in Japanese)
13. Nichi, Opect: Non-contact image processing system using Kinect, http://www.nichiiweb.jp/medical/category/hospital/opect.html
14. Ahn, S.C., Lee, T., Kim, I., Kwon, Y., Kim, H.: Computer vision-based interactive presentation system. In: Asian Conference for Computer Vision (2004)
15. Wagner, B.: Effective C# (Cover C# 4.0): 50 Specific Ways to Improve Your C#, 2nd edn. Addision-Wesley Professional (2010)
16. OpenNI: The standard framework for 3D sensing, http://www.openni.org/
17. NiTE 2: OpenNI, http://www.openni.org/files/nite/
18. Douglas, D., Peucker, T.: Algorithms for the reduction of the number of points required to represent a digitized line or its caricature. In: The Canadian Cartographer, pp. 112–122 (1973)
19. Kinect for Windows: Voice, Movement & Gesture Recognition Technology, http://www.microsoft.com/en-us/kinectforwindows/

An Automated Musical Scoring System for Tsugaru Shamisen by Multi-agent Method

Juichi Kosakaya

Hachinohe Institute of Tech.
kosakaya@hi-tech.ac.jp

Abstract. Local traditional arts people and the city's traditional music preservation society have eagerly wished a technology to accurately score traditional music, especially Tsugaru Shamisen. This music will be preserved as scores, which avoids relying solely on the oral transmission of this music to the young performers. At this time, we have experimentally produced an "Electronic Shamisen" equipped with a pick-up microphone attached to each string and automatic scoring equipment, which automatically records scores from the sound source with an agent method.

Keywords: pattern recognition, agent, automatic scoring, noise reduction, resonance.

1   Introduction

Aomori prefecture is abundant with traditional arts. Most of them, especially music has been handed down orally and traditionally hardly any scores have been in existence. Given the tradition, classical melodies have not been communicated accurately to students over the years and some of the melody has even been disappearing. Heritage Shamisen players and traditional music preservation societies (the public) have great concern over the issue of preservation of traditional music. This research aims at developing automatic scoring system on the basis of local traditional musical instrument such as Tsugaru and with information technology to solve the issue. Scoring refers to writing the musical composition into scores. This research saves the effort of the scoring specialists in taking down the masters' performance and makes it easy to develop the score unique to the musical instrument.

2   Automatic Scoring System

Automatic scoring system processes audio source entry – analysis – scoring as in the previous studies [1]. This research focus on scoring/replaying of the traditional folk songs played by the Japanese musical instrument such as Tsugaru Shamisen.

T. Yoshida et al. (Eds.): AMT 2013, LNCS 8210, pp. 24–36, 2013.
© Springer International Publishing Switzerland 2013

2.1   Composition of Equipment

The research equipment is composed of "electronic Shamisen" and "automatic scoring system" as shown Fig.1. While electronic Shamisen is that with independent pickup exclusive for strings as shown Photo.1, automatic scoring system composes scores automatically from the audio source. The former is equipped with "pick-up microphone for each string" to avoid resonance of the strings, making it possible to extract single scale sound or frequency. On the other hand, the latter is composed of audio source frequency analysis process; Shamisen scale or Tsubo deciphering/ scoring process into tablature or Western scores; speed designated equipment. First of all, Shamisen sound is entered into the automatic scoring equipment to analyze frequency by the audio frequency analysis process. Then, Shamisen scale deciphering process compares the frequency with the registered scale frequency and if the deviation is within the allowance, it is identified as scale or Tsubo. Scoring process or conversion in the form of the tablature or the Western score then takes place.

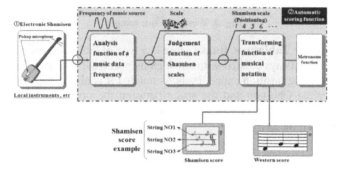

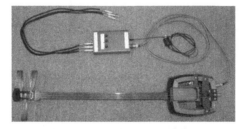

Fig. 1. Automatic scoring system structure

Photo. 1. Electronic Shamisen

2.2   Comparison between the Western Scores and Shamisen Scores (Bunka Tablature)

Traditional Japanese music such as Shamisen has not recorded in scores until Yo Sakagami, later known as Kine-ya Yahichi IV published the staff notation called "Sangen score" in 1910. However, most of the music fanciers of the time did not comprehend the staff notation and as a result the notation was never popularized. Further researches and contribution of Dr. Inazo Nitobe have compilation of "Shamisen Bunka tablature". The Bunka tablature was beginner friendly and as a result contributed in popularizing

Shamisen music. Currently it is in use not only the mainstream of Shamisen music but also Naga-uta, Ko-uta and Ha-uta. The following describes the difference between the Western score and Shamisen score (Bunka tablature).

(1) Western score. Western score is the most widely used score, which is considered as a type of point graph using high-pitch and time. As shown in Fig. 2, the symbols mainly used are clef, key signature, time signature, notes, rest, accidentals, musical symbols as well as slogan in language.

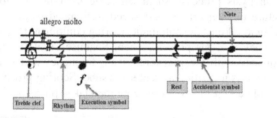

Fig. 2. Western score sample

(2) Bunka score or Tablature. Bunka score is a type of tablature mixed with Shamisen score that had always existed with some elements of the staff notation. Instead of intervals, performing position called Tsubo is notated It is also called Aka-fu. Tuning Shamisen is called Cho-gen and the space between the open strings varies among first, second and third strings. Fig.3 and Table 1 show the tunings types of strings and mutual response of the strings.

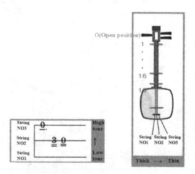

Fig. 3. Shamisen clef & positioning

Table 1. String tunings

	Tone interval between String NO1 and NO2	Tone interval between String NO2 and No3	Ex. String NO1: C (String NO1, String NO2, String NO3)
Honchoshi	Complete 5 degrees	Complete 5 degrees	NO1: C, NO2: F, NO3: C
Niagari	Ditto	Complete 4 degrees	NO1: C, NO2: G, NO3: C
Sansagari	Complete 4 degrees	Ditto	NO1: C, NO2: F, NO3: B

(3) Correspondence to the musical notation. Table 2 shows the partial relationship between the Western musical notation and Bunka score. In Bunka sore, notes are shown as Tsubo numbers that are held and rest are expressed in black dot. Length of notes and rest are expressed by the location of bars.

Table 2. Comparison Western score with Shamisen score sample

※ indicates the positioning number

	Western clef	Shamisen clef		Western clef	Shamisen clef
Full note	◯	※－－－	Full rest	▬	●●●●
Half note	♩ ♩	※－	Half rest	▬	●●
Quarter note	♪ ♪	※	Quarter rest	⌇	●
Eighth note	♪ ♪	※	Eighth rest	⁊	●
Sixteenth note	♪ ♪	※	Sixteenth rest	⁊	●

2.3   Indications of the Musical Instrument (Tsugaru Shamisen)

Symbols unique to Tsugaru Shamisen are on sale in various guides [2]. Tsugaru Shamisens have their own unique finger manipulation and combination of the bachi or plectrum and left hand to create particular ways of performance. Table 3 shows the main indications. They are helpful when the player cannot think of how actual hand motions just by looking at the score, However, as Western scores do not employ such symbols and it can be confusing and obscure at the time of notation.   Verification of notes of "Su", "Ha", tremolo or trill in this research was extremely difficult to decipher. Details are described in Chapter 4.

Table 3. Priority of Shamisen finger handlings

a1～a9 : Parameter

Item	Music symbols of Shamisen unique definitions	Priority Number
1	Su (Up plectrum)	a1
2	Ha (Flip finger)	a2
3	Suri (Slide finger)	a3
4	U (Inside finger)	a4
5	· (Staccato)	a5
6	～～～ (Tremolo)	a6
7	> (Accent)	a7
8	Oshi (Push plectrum)	a8
9	Suberi (Slide plectrum)	a9
N	Definition N	an

3   Electronic Shamisen

Most of the Japanese musical instruments are not made to be electronic. This is mainly due to the loss of the unique sound quality. However, in the process of this research, Shamisen was recorded in acoustic sound and analyzed the source by frequency, it was impossible to block the influence of sound noise; slight high and/or low frequency sound inaudible to human ears. Therefore we have developed an electronic Shamisen so that its acoustic source was directly entered by lineout and scored. Electronic Shamisen is equipped with high performance pickup microphone attached to each string shown in Photo 1, preventing resonance of the strings as much as possible or unnecessary harmonic overtone; created the skeleton in such a way that provides the feel for the real Shamisen. The fundamental is that the string vibration is converted into electronic symbols and after the analog-digital conversion, scoring takes place upon extracting the frequency corresponding to the scale or Tsubo by spectral analysis.

4   Automatic Notation Process

4.1   Automatic Notation Process

The automatic notation equipment is composed mainly of the following process as in Fig. 1.

(1) Audio source frequency analysis process: Frequency range of Shamisen strings (first to third strings) is registered and notes made by the plectrum is analyzed by the acoustic spectrum analysis, delete interference or resonance with the other strings enabling to extract the single string sound.

(2) Shamisen scale decipherment process: The unique tuning of Shamisen such as "Hon-choshi (full scale)", "Ni/ two – agari (two degrees up)", or "San/3 sagari (3 degrees down)" is registered, analyze each frequency unit and analog-digitally convert the scale of each string corresponding to the tuning of respective string (0 to 18 degrees per string).

(3) Notation process: The analog-digital conversion (numerical designation scale) and the registered frequency scale (digital data) are compared. If the result is within the allowance, formal scale of the players ("Tsubo": Scale of Shamisen score is in numbers), the scale numbers is saved in the memory of the score.

4.2   Issues with Notation Particular to Shamisen

In the notation process, following problems have happened which are particular to Shamisen.

(1) In a vibrato on single sound, the sound is recognized as a bundle of multiple scales instead of a single sound.
(2) "Su or Sukui-bachi" and "Ha or Hajiki-yubi" are not differentiated.
(3) In a tremolo or trill sound, recording process of the sound cannot keep up. In tackling the issues, the multi-agent system was utilized to decipher Shamisen scale or Tsubo as follows:

4.3   Shamisen Scale or Tsubo Decipherment Process by Multi-agent System

In order to correctly recognize the input scale frequency, this process employed the multi agent (MA) for decipherment. The term "multi agent" is currently used broadly and is described in detail in 4.3.1. However, in this research we refer to the function of discriminating priority invariables of unique Shamisen sound or estimating array of notes; dynamically generating necessary information under the necessary condition as needed; capability of alignment of notes among the first through the third strings in a ideal manner by negotiation and coordination process. Such composition of the agent has been reported in various fields as in [3]. This particular research covers decipherment of Shamisen scale or Tsubo.

4.3.1   Role of Multi-agent (MA) in an Automatic Scoring Function (ASF)

To implement the above functions by using MA method, we need to resolve the conflicts that occur when Judgment Function of Shamisen Scales (JFSS) processed by different music data frequencies is inconsistent or irregular. The architecture of the multi-agent system is shown as Fig. 4. When a JFSS in an ASF system is operating under normal conditions, automatic control (AC) functions {such as feedback (FB), feed-forward (FF), and proportional-integral-differential (PID)} operate adequately to fix an adequate music notation. Each agent undertakes local solution of part of the problem and checks its constraint satisfaction based on the appropriately music notation and frequency knowledge database. It is also assumed that no single agent can accomplish global problem solving by itself. When an agent has produced its solution, it is not immediately made available outside the agent. The solutions are made available only after all the agents have finished their local problem solving. The local problem solving is thus carried out independently, because it is not influenced by the solutions constructed by other agents in the same cycle of problem solving. The process of cooperative problem solving can be summarized as follows:

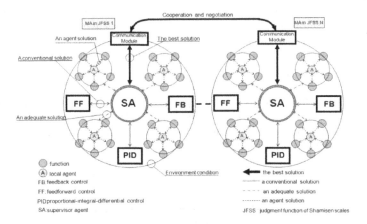

Fig. 4. Architecture of multi-agent system

(1) Each agent carries out some of its functions and constructs a solution by solving its own local problem independently.

(2) A supervisor agent (SA) selects an adequate solution by cooperating with other agents. Each SA shares the solution with other relevant SAs in order to get feedback.

(3) The first two steps are cyclically repeated until a better global solution can no longer be obtained.

(4) The JFSS uses the best solution obtained by the SA to estimate the right music notation.

(5) The JFSS then negotiates, via a communication module and cooperates with related the other JFSS to decide the further precise music notation according to the music notation & frequency knowledge database.

(6) During this negotiation, there will be some conflicts between the JFSSs. The details of how these conflicts are resolved are given in the next section.

4.3.2   Shamisen Scale Decipherment Process

Fig. 5 shows the scale decipherment software algorithm of Judgment Function of Shamisen Scales (JFSS) by the multi agent system using the scale or Tsubo identification device, which is the core. Firstly, the scale data sent by the acoustic source frequency analysis process is evaluated whether relevant sound exists by the scale decipherment process of each string, via entry module. If it is a single sound, its discrimination process activates. In the case of multi-layered sound, multiple discrimination processes activate simultaneously. Then the prescribed scale is identified according to the selective priority process and moves on to the scoring process via communication module in the order of entry. At this time, the scale assessment agent of each scale determines whether acoustic frequency entered corresponds the scale or Tsubo. This assessment process solves the issue described in the section 4.2 (1), whether the sound variation falls within certain time, continues or transfers to the next sound by comparing the scale information among the identification devices by the scale or Tsubo unit. This is based on the prescribed table of finger manipulation priority parameter as shown in Table 3, which expresses characteristics unique to Shamisen. In the case when the entered scale information is not clearly a single sound, the scale array conversion process particular to finger manipulation takes place according to the pre-registered patters such as the "Su or Sukui -bachi" or the "Ha or Hajiki-yubi."

The agents determine each scale and type ("Su" or "Ha") among themselves referring to the prescribed table of priority. When the agents compete or contradict among themselves in the process of comparison, they re-determine the scale and type by further comparing the priority invariables (parameters.) Also, if the scale is flat or sharp by 1/4 or 1/8 of a note, it is determined as dissonance and the scale adjustment function is activated to adjust to the normal scale.

4.3.3   Scale Identification Device Performance Flowchart

The performance flowchart of the scale identification is shown in Fig.6.

First, the step A identifies the vibrating string and its scale by comparing the acoustic source frequency entry and the table of all the scale frequency. If the entered acoustic frequency conflicts with any scale (N-scale of M-string) in Step B, it is determined as an un-corresponding sound or a noise and deleted then the next scale frequency is entered in Step D. If it does correspond to the said scale frequency (N-scale of M-string) in Step B, then Step E determines the technique whether it is particular to Shamisen such as finger manipulation or not. If the technique is standard, that is the technique not particular to Shamisen, it moves to Step H where the scale of the acoustic frequency entered is registered and enters the next acoustic frequency. If Step F determines it to be a technique particular to Shamisen, it transfers to Step G where comparison of priority order such as "Su or Sukui bachi" or "Ha or Hajiki-yubi" tales place and adjusts notes in accordance with the scale table. Then the adjusted scale is registered in Step H.

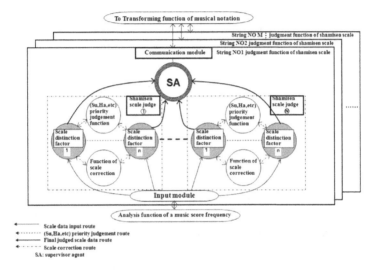

Fig. 5. Judgment function of Shamisen scales

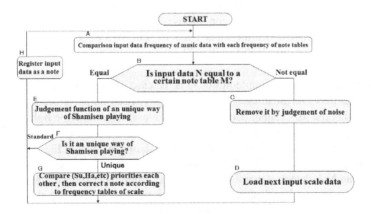

Fig. 6. Flowchart of scale distinction factor

4.3.4   Resolving Conflicts on Multiple Frequency Entry

During the function of step G in Fig.6, it has meant intermittent competition and had practical problems of miscounting notations by conflicts on multiple frequency entry due to the highly skillful and speed-up plays for Shamisen in this scoring system. Therefore this multi-agent method employing algorithms provides a realistic, viable and robust solution. Under such circumstances, the solution for comprehensive competition requires comparison of the geometric mean of the priority number of the parameter shown in Tables 3. The details are shown as follows using figures 4 to 7.

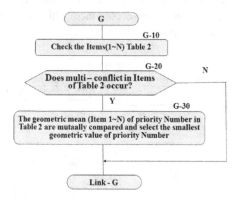

Fig. 7. Judgment flow of multi – conflict

In the beginning, as it has shown in Fig.5, the communication modules exchange data among the Strings number N judgment function of shamisen scale in order to prevent inconsistent assessment of priority number in Table 3. Conflicting judgments are identified by such as the (Su, Ha, etc) priority judgment function of the MAs, and a process to ensure one order of priority level is run by the MAs in accordance with the numbers a_1 to a_n listed in Table 3. However, the MA in JFSSs in Fig. 4 use an asynchronous communication method, and each MA is independent, so it is impossible for one MA to supervise all the conditions of the system on a real-time basis. Each MA therefore negotiates with the other MAs to achieve an adequate music notation level, but conflict among JFSSs will still occur, resulting in inconsistent overall solutions.

According to Fig.6, an example in step G is the case when conflicts occur among the highly skillful plays for Shamisen, especially highly speed-up plays occurring simultaneously, so that the MA cannot determine which music notation to be right. This deadlock between MAs is called an "infinite loop". To resolve such deadlocks in step G, decisions are made based on the control flow chart shown in detail in Fig. 7.

According to Fig. 7, each MA detects whether there are any Items (1-N) of Table 3. If an MA detects several conflicts there simultaneously, each MA compares the geometric mean (Item1 through N) of priority parameters (a_1 through a_n) in Table 3 among JFSSs and selects the largest geometric mean value of an MA. Normally, MA sorts out the JFSS by comparing geometric mean (or arithmetic mean) of the parameter numbers.

5   Experiment and Validation by the Test Model

We used an electronic Shamisen shown in Photo 1 and experimented in the composition shown in Fig. 1. The acoustic source for the experiment was a title called "Sansa Shigure." It is relatively slow music. However, it includes variety of pitch (such as "Ka or Kaeshi"), tempo, pattern refrains, all of which enable us to confirm organize linkage of the multi-agent. Therefore this particular title has become the acoustic source.

5.1   Shamisen Acoustic Source Frequency Analysis

Fig. 8 shows the characteristics of acoustic source frequency entered by the electronic Shamisen acoustic source without this method. Acoustic frequency is detected in the band of 100 Hz to several KHz as overtone peak and cluster of notes, which increase or decrease in response to the frequency. Scoring without the method of this research, as shown in Fig. 9, fundamental tones have been scored. However, noise and unnecessary notes, which are close to noise, were added and made it impossible to identify the original score of the acoustic source. The reasons for this are as follows:

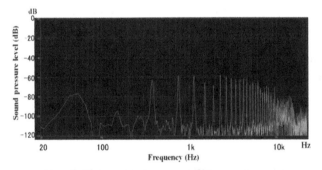

Fig. 8. Frequency peculiarity of an electronic Shamisen scale (Unedited method)

Fig. 9. Transformed score (Unedited method)

(1) Technique particular to Tsugaru Shamisen:

[1] Plectrum or bachi touches the string before the sound in question vibrates and the vibration sound of the plectrum hitting the string is recorded before and after the subject sound.

[2] As Shmaise does not have frets, in the process of reaching the desired scale by pressing Tsubo, slight off-sound or Suberi-on inaudible to human ears occur and recorded.

(2) Impact of electronic noise and resonance:

[1] Noise unique to electronic musical instrument such as 53 Hz hum sound or power supply noise is included in hardware.

[2] By improving the sensitivity of pickup microphone naturally create vibrating sound inaudible to human ears without vibrating the string by plectrum.

Low bass filter or band bass filter can solve the problems in the (2) of the above somewhat. However, the problem of (1) it is extremely difficult to solve, mainly because of fretless nature of Shamisen or any other similar instrument, regardless of the player's intention of deleting unnecessary sound to automatically create a score close to the original.

5.2   Application of This Research

Scoring took place under the same condition as above by applying this research method and electronic noise control filter. As a result, as shown in figures 10, 11 and 12, recognition rate of approximately 85 to 95 % of the original was scored. This result leads us to determine that "Multi-agent system of Shamisen scale Tsubo decipherment method" achieved certain degree of effectiveness.

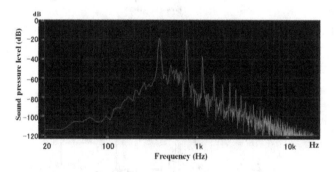

Fig. 10. Frequency peculiarity of an Electronic Shamisen scale (Proposed method)

Fig. 11. Western score (Proposed method, Nambu Zenifukiuta)

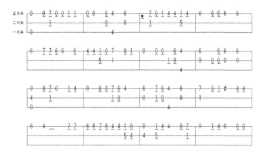

Fig. 12. Shamisen score (Proposed method, Nambu Zenifukiuta)

5.3   Review of the Experiment Results

Prior to the experiment, the same procedure using fret type string instrument other than Shamisen such as electronic guitar took place for measurement. As a result, resolution of rise and fall of the audio source frequency, which is the base point of the scale recognition improved intensely. We have also found that the recognition rate of the time lag reaching the steady vibration of particular to the string is far apart compared to that of keyboards such as piano. As expected, fretless string instrument is extremely difficult to score and especially the manipulation of fingers and plectrum or bachi unique to Tsugaru Shamisen requires more detailed design for commercialization of the product. These priorities will be worked on by improving the agent function and enhancing the scale recognition algorithm.

6   Conclusion

This research studied on the automatic scoring equipment in preservation of the traditional music of Tsugaru Shamisen by employing the multi-agent system to decipherment procedure of the scale or Tsubo corresponding entry audio source frequency and the scale analysis procedure of the unique sound of Tsugaru Shamisen. These procedures enable many of the music of the Japanese traditional musical instrument without frets to be scored, but not limited to musical instrument. In the future, the basic technique could be applicable to vocal scores. Furthermore, this research enables to create supportive environments and develop personal skills for elder people to realize successful ageing by using above method. Also, by using this instrument, it reminds dementia patients of good old scoreless folk music which they used to sing in their younger days. This will be definitely effective to cure these dementia patients by remaking folk music scores.

References

1. Sakauchi, H., Natsui, M., Tadokoro, Y.: Realization of Automatic Musical Transcription System Based on Comb Filters. IPSJ, 2007-MUS 71, 13–18 (2007)
2. Kinoshita, S.: Tsugaru Shamisen style book (2005)
3. Kosakaya, J., Tadokoro, H., Inazumi, Y.: Cooperative Control Technology with ITP Method for SCADA Systems. IEICE E91-D(8) (August 2008)

Visualization of Life Patterns through Deformation of Maps Based on Users' Movement Data

Hayato Yokoi, Kohei Matsumura, and Yasuyuki Sumi

Future University Hakodate, Hakodate, Hokkaido 041–8655, Japan
h-yokoi@sumilab.org

Abstract. This paper proposes a system for visualizing individual and collective movement within dense geographical contexts, such as cities and urban neighborhoods. Specifically, we describe a method for creating "spatiotemporal maps" deformed according to personal movement and stasis. We implement and apply a prototype of our system to demonstrate its effectiveness in revealing patterns of spatiotemporal behavior, and in composing maps that more closely correspond to the node-oriented "mental maps" traditionally used by individuals in the act of navigation.

1 Introduction

In this paper, we propose a system that models users' localized movement patterns to generate "spatiotemporal maps" of their pedestrian and vehicular movement within a city or urban neighborhood. These spatiotemporal maps are deformed according to the amount of time spent at locations and en route. The project aims to reveal awareness and dialogues not easily discerned from traditional geographic maps.

At present, most of the maps we use are formed by topographical projections of "geographical distance". Though these maps provide a baseline of valuable information, we must recognize that, when large numbers of people move between similar points on a map, travel time will depend more on local traffic conditions than on physical distance, and that, for the most part, our current topographical maps fail to reveal this dependency.

If you are asked for directions by travelers, you will tend to assemble a schematic map from your current position to the destination inawareness your head, and attempt to explain direction and orientation with respect to relevant nodes, such as the buildings and intersections en route. Generally, we regard these schematics as cognitive maps, assembled from various geographical or spatial memories and generally inaccurate in comparison to the topographical map. Nevertheless, they support consistently successful navigation to desired destinations. This is because a cognitive map, though geographically inaccurate, contains the minimum information required by human navigators, and as such, can be regarded as more streamlined and easier to use than the corresponding, geographically accurate, map.

T. Yoshida et al. (Eds.): AMT 2013, LNCS 8210, pp. 37–45, 2013.
© Springer International Publishing Switzerland 2013

We can assume that cognitive maps depend greatly on an individual's life patterns and his/her perception of time. As a result, we expect to be able to visualize the spatial structure and nodes of a city based on the cognitive maps of individuals living in that city.

2   Related Work

Ashbrook et al. demonstrated a system that automatically clusters GPS data, taken over an extended period, into meaningful locations at multiple scales [1]. In our approach, we consider that we can use the portion to extract landmarks for map deformation.

Agrawala et al. described algorithmic implementations of these generalization techniques within LineDrive, a real-time system for automatically designing and rendering route maps [2]. In this paper, we regard route maps as similar to our spatiotemporal maps, except that spatiotemporal maps are based on time, by way of movement data.

Patterson et al. demonstrated that by adding more external knowledge about bus routes and bus stops, accuracy is improved [3]. We want to determine whether there are new awareness by generating the spatiotemporal maps from human behavior.

Schoning et al. presented a study that discusses the suitability of various public maps to this task and evaluated whether geographically referenced photos can be used for navigation on GPS-enabled devices [4]. In the future we would like to create a system that generates spatiotemporal maps based on the built-in GPS of smartphones and similar mobile devices. For example, if a mobile user takes pictures for inclusion in PhotoMap, these pictures could be deformed into spatiotemporal maps based on the user's movement leading up to the photograph.

Shen et al. created a visual analysis tool called "MoviVis" that presents spatial and social information as a heterogeneous network [5]. The distance between any two points of the resulting maps is similar to the distance represented between landmarks of our spatiotemporal maps. Also, if the actual distance is far, it is close as distance sense.

3   Spatiotemporal Maps Generation System

3.1   System Configuration

Our spatiotemporal maps generation system visualizes users' movement patterns based on their GPS movement data, which must be captured using a GPS logger (using the model of user action proposed by Ashbrook et al. [1]). Based on this data, we create a time scale and spatial scale focused around the region and time span of activity. We then use threshold processing to determine the dwelling time of users, based on whether a mesh separated according to spatial location exceeds a certain time scale. The result is a set of points called nodes, joined by lines called links. The locations of these nodes are adjusted to reflect travel times along links and, finally, a general map is deformed to fit the adjusted nodes. The result is what we call a spatiotemporal map.

3.2   Features of the Spatiotemporal Maps

Our spatiotemporal maps are based on time-tracked movement data, and deformed by time-based depending on the life time of the individual.

Fig. 1. Example of a spatiotemporal map

Fig. 1 is an example of an actual spatiotemporal map. Note the deformation of the general map to reflect movement (i.e. motion and stasis) data.

– Spatiotemporal maps for individual users
  Because our maps reflect movement data, their deformations will vary according to individual users—i.e. where they live and work, and their means of transportation.
– Spatiotemporal maps for multiple users
  When mapping multiple users, places common to several users will become nodes. Note that some of these nodes can only emerge in the visualization itself, as aggregations of movement in space and time.

4   Prototype System

When mapping multiple users, places common to several users will become nodes. Note that some of these nodes can only emerge in the visualization itself, as aggregations of movement in space and time.

4.1   Collection and Storage of Movement Data

The data we want to collect must include the latitude, longitude, and time coordinates for individual users. We use the GPS data as movement data.

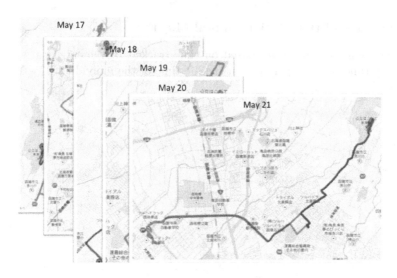

Fig. 2. Movement data everyday we collect

As shown in Fig. 2, GPS data is collected everyday. We are collecting GPS data now. We use GPS data collected in this way, visualize location, season, time, and more, and compare them.

All GPS data was initially stored in CSV format; however, this led to inefficient programming. Therefore, we extracted only the required CSV data, stored it in a relational database, and queried it as needed. We use the SQLite database and management system.

In our SQLite database, we created three simple tables: User, File and Data. The User table manages all user data, the File table includes userid and filepath columns to associate users with file-based data, and the Data table includes fileid, Latitude, Longitude, and Date columns for specifying space-time coordinates.

4.2 Creating Sparial Scales and Time Scales and Generating Nodes

We created the spatial mesh and time mesh. The spatial mesh is the mesh has a scale of two dimensional in the spatial direction. The time mesh is the mesh has a scale of one dimensional in the time direction. We tuned the spatial mesh and time mesh. As a result, the spatial mesh was in the 10m 10m, and the time mesh was in the 1 minute. Fig. 3 shows the stay time of the user in the spatial mesh. If the stay time is over the 1minute, we generate a point called a node in the spatial mesh from Fig. 3. The size of the node depends on the stay time of the user. Further, we display a link that connects between nodes. Links are calculated the sequence of connection between the nodes. It connects to the starting point and the end point as a link along the time axis in the movement. Fig. 4 is view that nodes and links are generated based on a spatial mesh.

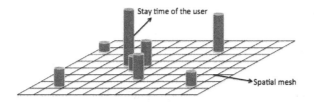

Fig. 3. Generation of nodes from a spatial mesh

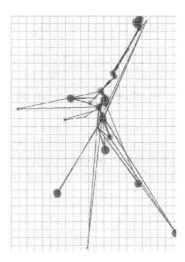

Fig. 4. Generation of nodes and links from a spatial mesh

4.3   Map Deformation

Next, nodes and links are repositioned, and a general map covering their geographical region is deformed accordingly.

- Restructuring of nodes and links

  The spatiotemporal maps are based on the amount of time spent at locations and en route. We consider that the links should to be based on the average of travel times. Therefore, we need to restructure the positions of nodes and links. We shift the position of the node that is mapping on the map in according to the travel time.

  We use a "a generalized solution of time-distance mapping" [6] as a reference. The problem of generating this time-distance map is how best to place a set of points given the time-distance between them on the plane. Shimizu et al. modeled this as a nonlinear least-square problem and proposed a generic solution on that basis. Their proposed solution is as follows.

$$\min \sum_{ij \in L} \left[(t_{ij} \sin \Theta'_{ij} - (x_j - x_i))^2 + (t_{ij} \cos \Theta'_{ij} - (y_j - y_i))^2 \right] \qquad (1)$$

- Map deformation

  To maintain useful context, it is important to deform the geographical map without critical loss of geographical information. To accomplish this, we perform image deformation using the moving least squares method proposed by Schaefer et al. [7].

  We try the map deformation by separating and distorting a map in the virtual mesh. In this deformation, we reconfigure the mesh by using the information the nodes displaced (x, y) (x', y'). We use the moving least squares method to reconfigure of the mesh. The moving least squares method corresponds to find the solution to the following.

$$\sum_i w_i |l_v(p_i) - q_i|^2 \qquad (2)$$

We can distort the map to affine transform the image by depending on the mesh reconfigured.

Fig. 5 shows an example deformation. The blue dots in the figure indicate control points, and the green dots indicate the adjusted location of those control points. Image deformation is accomplished by simply clicking a part of the image on the left and dragging to produce the image on the right.

4.4   Demonstration

To demonstrate the effectiveness of our prototype system, we used it to create several spatiotemporal maps. For this purpose, we found three users willing to have their GPS data collected and visualized for a one-month period.

Fig. 6 is a side-by-side comparison of the three users' spatiotemporal maps. Note that the maps differed significantly, even though the users lived and worked in the same city. All three users come and go different places. But it can be read from the Fig. 6 that left user has moved to the range of the left from the point A, middle user has moved to various locations around the point A, and right user has moved to the range of the right from the point A. This is because living space of person is different. This is a direct indication of the differences among the users' lifestyles—i.e. their dwelling places and transportation behaviors—resulting in different spatiotemporal deformations.

Fig. 7 is a comparison of spatiotemporal maps of the same user displaying different places. The left figure is the spatiotemporal map of user's hometown, the right figure is the spatiotemporal map of business trip destination. The user in the hometown walks around the neighborhood and goes quickly by a car to the long way as the airport. When the user moves by a car, the deformation of the spatiotemporal map is constant. But the user in the business trip destination moves by the train and walks to the destination from the station. When the user moves by the train, the deformation of the spatiotemporal map is also constant. When the user walks to the destination from the station, the spatiotemporal

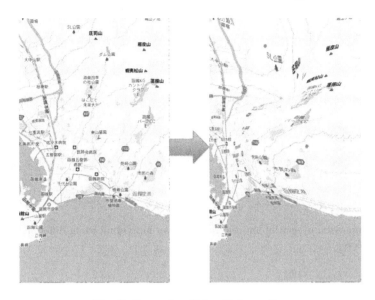

Fig. 5. Example of Map Deformation

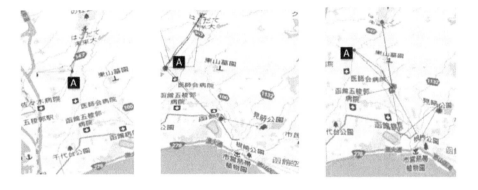

Fig. 6. Comparison of three users' spatiotemporal maps

map is distorted as fish-eye lens at the center of the station. We surmise that the different deformations are due to use of different means of transportation. In addition, we can guess also transport used by the region is different.

Fig. 8 is a comparison of spatiotemporal maps with different hours. The left figure is the spatiotemporal map of the day, the right figure is the spatiotemporal map of the evening. The road is quiet during the day, but the road is crowded in the evening. Since users tended to favor the same routes of travel, we can assume that the change in deformation is due to variations in traffic conditions, which may be hourly, daily or seasonal.

Fig. 7. Comparison of spatiotemporal maps for an individual using different means of transportation

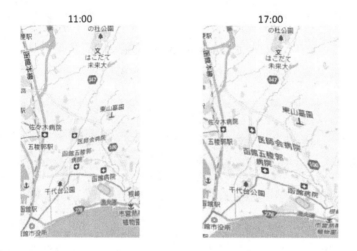

Fig. 8. Comparison of spatiotemporal maps with different hours

5   Conclusions and Future Work

In this paper, our purpose was to provide an insightful visualization of individual life patterns as they apply to movement in time and space. Specifically, we proposed a new kind of map deformation based on the movement and stasis of users within a city environment. By collecting data that reflects individual variations in destination, schedule, and transportation mode, and applying our prototype system to this data, we were able to generate a series of individual and multiple-user spatiotemporal maps. These maps made clear certain differences in individual and collective behavior over time, and helped to distinguish key features of the mental and transportation-related landscape.

In future work, we intend to focus on two important features.

– Automation of map deformation program
  Map deformation program that incorporates the method of Shimizu et al.
  did not work well. Therefore, we innovate the elements of the spring model
  in this program and distort to let a natural length of the spring be travel
  time. And another thing, we use Google Maps as images now and image
  deformation makes also characters distort. We want to devise such as divided
  into separate layers of map image and characters.
– Experiments on the "awareness due to show each other the maps"
  It is possible to be aware the part was not awaken by oneself by others
  by showing each other the generated spatiotemporal maps. For example,
  there are problems in the city and special local information. Otherwise, we
  can aware the differences between profile from showing each other. So, we
  should do an experiments on the "awareness due to show each other the
  spatiotemporal maps".

References

1. Ashbrook, D., Starner, T.: Learning Significant Locations and Predicting User Movement with GPS. In: Proceedings of the 6th IEEE International Symposium on Wearable Computers, pp. 101–108 (2002)
2. Agrawala, M., Stolte, C.: Rendering Effective Route Maps: Improving Usability
   Through Generalization. In: Proceedings of ACM SIGGRAPH 2001, pp. 241–250
   (2001)
3. Patterson, D.J., Liao, L., Fox, D., Kautz, H.: Inferring High-Level Behavior from
   Low-Level Sensors. In: Dey, A.K., Schmidt, A., McCarthy, J.F. (eds.) UbiComp
   2003. LNCS, vol. 2864, pp. 73–89. Springer, Heidelberg (2003)
4. Schoning, J., Cheverst, K., Lochtefeld, M., Kruger, A., Rohs, M., Taher, F.: PhotoMap: Using Spontaneously taken Images of Public Maps for Pedestrian Navigation
   Tasks on Mobile Devices. In: Proceedings of MobileHCI 2009 (2009)
5. Shen, Z., Ma, K.-L.: MobiVis: A Visualization System for Exploring Mobile Data.
   In: Proceedings of Pacific Visualisation Symposium 2008 (2008)
6. Shimizu, E., Inoue, R.: A Generalized Solution of Time-Distance Mapping. In: Proceedings of the 8th International Conference on Computers in Urban Planning and
   Urban Management (2003)
7. Schaefer, S., McPhail, T., Warren, J.: Image Deformation Using Moving Least
   Squares. In: Proceedings of ACM SIGGRAPH 2006, pp. 533–540 (2006)

Wi-Fi RSS Based Indoor Positioning Using a Probabilistic Reduced Estimator

Gang Shen and Zegang Xie

School of Software Engineering, Huazhong University of Science and Technology
Wuhan 430074, China
gang_shen@mail.hust.edu.cn, wexzg@163.com

Abstract. In this paper, we present an investigation of indoor objects positioning using the received Wi-Fi signal strength in the realistic environment with the presence of obstacles. Wi-Fi RSS based positioning is a promising alternative to other techniques for locating indoor objects. Two factors may lead to the low Wi-Fi RSS positioning accuracy: the existence of moving obstacles, and the limited number of available anchor nodes. We propose a novel approach to locating a target object in a given area by introducing a hidden factor for a reduced form of probabilistic estimator. This estimator is unbiased with the scalability in field size. With the selection of a Gaussian prior on this hidden factor characterizing the effects of RSS drop introduced by obstacles, we convert the positioning prediction into a maximum a posteriori problem, then apply expectation-maximization algorithm and conjugate gradient optimization to find the solution. Simulations in various settings show that the proposed approach presents better performance compared to other state-of-the-art RSS range-based positioning algorithms.

Keywords: Indoor positioning, Received Wi-Fi signal strength, Maximum a posteriori, Conjugate gradients.

1 Introduction

Due to the successful deployment of global positioning systems (GPS), location-based services (LBS) have been extensively developed for military, industrial, medical, household and personal uses. However, because the satellite signal cannot penetrate roofs and walls, alternatives to GPS, such as Wireless sensor networks (WSN) based positioning techniques, are a promising option for locating indoor objects. Without the loss of generality, the WSN based techniques can be classified by the measurements used for position calculation, for example, time of arrival (TOA), angle of arrival (AOA), and received signal strength (RSS). Compared to TOA and AOA, RSS based positioning methods have the advantages of low cost and low power, and usually there is no need for extra hardware (TOA requires high precision timer) [1][2]. But the RSS is affected not only by the distance between a pair of sender and receiver, but also the path along which the signal travels, therefore the RSS positioning accuracy are impacted by the high environmental noise.

T. Yoshida et al. (Eds.): AMT 2013, LNCS 8210, pp. 46–55, 2013.
© Springer International Publishing Switzerland 2013

The survey of earlier research in WSN positioning can be found in [1][3]. Roughly, the RSS based indoor positioning methods can be implemented either on the customer side or on the server [4]. A variety of wireless signals can be applied to form the WSN in positioning, e.g. laser, sonar, infrared, RFID, and more recently ZigBee, UWB and Wi-Fi. Among the afore mentioned signals, Wi-Fi is sensitive to the environment setup, thus easy to suffer from the environment variations [5][6]. Unlike the range based position algorithms that need to estimate the distances between objects, other proposals instead use indirect ways to locate the target, including centroid, DV-HOP, and APIT [7][8][9]. Usually range-less algorithms result in a lower location precision [10].

In this paper, we investigate the indoor positioning problems with changing obstacles on the signal propagation paths: given a collection of anchor nodes with known positions which periodically beacon their existence using Wi-Fi signals, a target needs to estimate its two dimensional position within a constrained area using only the information of anchor coordinates and the received signal strength indicators. The motivation of this research comes from the following facts: first, many hand-held devices such as smart phones and tablet computers have built-in Wi-Fi interfaces; and second, a large number of Wi-Fi access points have been installed in public facilities, shopping malls and offices, making Wi-Fi based indoor positioning free of additional infrastructure investment. But the major challenges in realistic environments also contain adversary factors: large fluctuation of RSS readings coming from the unstable signal propagation paths and varying environmental noise; low attenuation coefficient of the Wi-Fi signal (intended for data communication rather than location determination) and large measurement noise. In this paper, we apply the shadowing model of the Wi-Fi signal for the purpose of range estimate. To improve the performance of positioning, we introduce a hidden variable expressing the effect of obstacles on range estimation, and also convert the probabilistic positioning problem to a reduced objective function for optimization. To demonstrate the effectiveness of the proposed approach, we designed simulations and experiments and evaluated the performance of our algorithms in typical indoor positioning settings, as well as compared with other methods documented in recent research.

2   Assumptions and Models

Consider an indoor space deployed with a fixed number of anchor nodes (Wi-Fi routers), we note that these anchors may either be evenly distributed or randomly set up. The positions of these anchors are assumed known and will not change in the target positioning process. Suppose a target object is constrained to this area and is able to receive the beacon signals sent by the anchors, and the signals may be blocked by the moving obstacles on the line of sight paths between the target and some of the anchors, the objective is to estimate the position of the target using only the one-time RSS readings by the target as well as a map of the anchor layout.

In theory, the power of received signal, p at distance d can be characterized by the formula $p(d) = \frac{c}{d^n}$, where n is the loss factor in the open space and is a constant. The measured logarithm value of the received power, i.e., RSSI, can be represented as follows,

$$r(d) = r(d_0) - 10n \log_{10}(\frac{d}{d_0}) + x_\sigma \qquad (1)$$

In (1), d_0 is a reference distance (usually d_0 is taken as 1 meter), $r(d_0)$ is the RSSI value at d_0, and X_σ denotes a measurement noise. In the typical setting, it is assumed that X_σ follows the normal distribution $\mathcal{N}(0, \sigma^2)$. We assume the loss factor n is known to the target because the parameter can be pre-calibrated to satisfy a given precision criterion. In a realistic environment, the existence of moving obstacles will attenuate the RSSI readings by the target, rendering the model in (1) no longer applicable.

In [11], Ali-Rantala et al studied the impacts of different walls and concluded that the received signal strength varied from the wall thicknesses and materials. In [12] the authors proposed modification to the generic shadowing model by adding a term representing the impacts of walls in between the sender and receiver. To capture the impacts of obstacles, fingerprint is applied to train the environment-aware model [13][14]. Nevertheless, a disadvantage of this approach is that whenever the environment changes, the model has to be re-trained by collecting a new data set, in particular for the scenario with the presence of moving obstacles.

Similar to [11], we repeatedly measured the RSSI values of a wireless router at various distances in a computer lab. It can be seen that the RSSI values at a fixed distance follows normal distribution approximately. But if the range estimate error introduced by the obstacles is very big, it is no longer valid to model this error as a normally distributed measurement noise. At the working frequency of 2.4G, the decrease of RSSI reading is usually 7dBm to 12dBm [11].

Suppose the parameter n and $r(d_0)$ are also sent to the target, we may add a new random term Q to the shadowing model to reflect the impact of the existing obstacles.

$$r(d) = r(d_0) - 10n \log_{10}(\frac{d}{d_0}) + x_\sigma + Q \qquad (2)$$

Instead of directly model the variable Q in (2), we consider the its impact on the range estimate. Denote the estimated distance $\hat{d}(r)$ from a noisy RSSI reading $r(d)$, satisfying $10n \log_{10}(\frac{\hat{d}}{d_0}) = r(d_0) - r(d)$, while the real distance d is corresponding to the noise free RSSI $10n \log_{10}(\frac{d}{d_0}) = r(d_0) - \bar{r}(d)$, where $\bar{r}(d)$ can be thought of as the long term average of the measured received

$$\hat{d}(r) = d_0 10^{\frac{r(d_0)-r(d)}{10n}} = d_0 10^{\frac{r(d_0)-\bar{r}(d)-z_\sigma-Q}{10n}} = \gamma d e^{-\frac{\ln 10 z_\sigma}{10n}},$$

where $\gamma = 10^{-\frac{Q}{10n}} \geq 1$. It is straightforward to see, $\ln \frac{\gamma d}{d} \sim \mathcal{N}(0, (\frac{\ln 10}{10n}\sigma)^2)$ is a normal distributed random variable.

Let us first consider the case $\gamma = 1$, i.e., $Q = 0$ for all anchor nodes. Because $\ln \frac{d}{\hat{d}} \sim \mathcal{N}(0, (\frac{\ln 10}{10n}\sigma)^2)$, given a collection of RSSIs $\{r_i | 1 \leq i \leq N, N \geq 3\}$ from the anchors at (x_i, y_i) (not all anchors are collinear), the maximum likelihood estimation of the target position (x, y) is

$$(x^*, y^*) = \arg\max_{(x,y)} \prod_{i=1}^{N} e^{-[(\ln \frac{\sqrt{((x-x_i)^2+(y-y_i)^2)}}{\hat{d}_i(r_i)})^2 / (\frac{\ln 10}{10n}\sigma)^2]}$$

$$= \arg\min_{(x,y)} \sum_{i=1}^{N} (\ln \frac{\sqrt{((x-x_i)^2 + (y-y_i)^2)}}{\hat{d}_i})^2.$$

We simplify the natural logarithm function to its first order Taylor expansion at 1, let $\tilde{d}_i(x, y) = \sqrt{(x - x_i)^2 + (y - y_i)^2}$, then

$$(x^*, y^*) = \arg\min_{(x,y)} \sum_{i=1}^{N} (1 - \frac{\tilde{d}_i}{\hat{d}_i})^2. \tag{3}$$

We note that the sum in (3) is not convex because of the role played by square root in calculating distance. One reason to take a simplified form for the logarithm function is that the logarithm function has an unsmooth surface when (x, y) is close to (x_i, y_i), making the calculated distance approach zero, and thus cause the optimization algorithms fall into local minimum, while the function in (3) will not lead to such situation.

Theorem 1 (optimality). If $Q = 0$ in (2), there exists a constant k such that the real target position (x_t, y_t) minimizes the expectation

$$\phi(x, y | x_t, y_t) = E[\sum_{i=1}^{N} (k - \frac{\tilde{d}_i(x, y)}{\hat{d}_i})^2],$$

where $\hat{d}_i$ is the distance estimate to anchor node i for the readings of RSSIs from anchor nodes when a target is located at (x_t, y_t).

Now let us move to the cases that not all $\gamma = 1$. Suppose we know the exact impact of obstacles for each of the anchors, a corollary can be drawn.

Corollary. If $Q \neq 0$ in (2) for some anchor nodes, there exists a constant k such that (x_t, y_t) is the optimal solution that minimizes the expectation

$$\phi(x, y | x_t, y_t) = E(\sum_{i=1}^{N} (k - \gamma_i \frac{\tilde{d}_i(x, y)}{\hat{d}_i})^2)$$

where $\hat{d}_i$ is the distance estimate to anchor node i for the readings of RSSIs from anchor nodes when a target is located at (x_t, y_t).

Unfortunately, it is hard to measure γ_i without collecting a large number of data and γ_i may change from time to time in the presence of moving obstacles. A reasonable trade-off is to assume $\gamma = 1$, and select a good k using certain heuristics, to approximate the target position prediction. Given a collection of RSSIs from anchors for a target at unknown position (x_t, y_t), let function $f(x, y|k) = \sum_{i=1}^{N} (k - \frac{\tilde{d}_i(x,y)}{\hat{d}_i})^2$, we use the solution that minimizes $f(x, y|k)$ to estimate the target position, i.e.,

$$(x^*, y^*|k) = \arg\min_{x,y} \sum_{i=1}^{N} (k - \frac{\tilde{d}_i(x,y)}{\hat{d}_i})^2. \tag{4}$$

Let the prediction error be $\epsilon(k) = \sqrt{(x^* - x_t)^2 + (y^* - y_t)^2}$, we have the following scalable property. The proof relies on the fact that $\hat{d}_i \propto d_i$.

Theorem 2 (scalability). If the distances between N anchor nodes are scaled by a factor f, the expected prediction error of the solution to (4) is also scaled by f.

3   Probabilistic Reduced Estimation Algorithms

Since the parameters needed to calculate the exact value of k are not available, we have to find good heuristics to determine k. In this paper, we first choose a prior on k, and thus convert the positioning problem to a MAP problem, finally use the solution to the MAP problem as the estimated target position. As we have seen in the last section, the approximate estimation of the target position can be transformed to solving (4). Equivalently, we assume that all anchors send beacon signals independently, i.e., the RSSI measurements r_i from different anchor nodes are mutually independent, and the joint probability of location (x, y) in the reduced form is

$$P(x, y|r_i^N, k) = \prod_{i=1}^{N} P(x, y|r_i, k) \propto \prod_{i=1}^{N} e^{\frac{(k - \tilde{d}_i(x,y)/\hat{d}_i(r_i))^2}{s_1^2}},$$

for certain variance s_1^2. By speculating on the prior of k, $P(k)$, we obtain the following MAP problem

$$P(x, y|r_i^N) \propto (\prod_{i=1}^{N} \exp \frac{(k - \tilde{d}_i(x,y)/\hat{d}_i(r_i))^2}{s_1^2}) P(k),$$

and subsequently solve

$$(x^*, y^*) = \arg\max_{x,y} P(x, y|r_i^N).$$

We consider the following candidates of $P(k)$ and will use them to test the validity of the proposed algorithm in the simulations in Section 4: uniform prior, negative exponential prior and Gaussian prior with unknown mean.

1, *uniform prior*

The target position estimate $(x^*, y^* | r_{i=1}^N) = \arg\min_{x,y,k} \left(\sum_{i=1}^N (k - \frac{\tilde{d}_i(x,y)}{\tilde{d}_i(r_i)})^2 \right)$, subject to the area constraint and $0 < k \leq 1$.

2, *Gaussian prior with fixed mean*

Let k be independent of the RSSI data and follow a Gaussian distribution with mean of 1, $P(k) = \frac{1}{\sqrt{2\pi} s_2} \exp -(\frac{k-1}{s_2})^2$. The target position estimate

$(x^*, y^* | r_{i=1}^N) = \arg\min_{x,y,k} \left(\sum_{i=1}^N (k - \frac{\tilde{d}_i(x,y)}{\tilde{d}_i(r_i)})^2 + \lambda(1-k)^2 \right)$,

subject to the area constraint and $0 < k \leq 1$.

3, *Gaussian prior with unknown mean*

Let k follow $P(k) = \frac{1}{\sqrt{2\pi} s_2} \exp -(\frac{k-\mu}{s_2})^2$. The target position estimate

$(x^*, y^* | r_{i=1}^N) = \arg\min_{x,y,k} \left(\sum_{i=1}^N (k - \frac{\tilde{d}_i(x,y)}{\tilde{d}_i(r_i)})^2 + \lambda(k-\mu)^2 \right)$,

subject to the area constraint and $0 < k \leq 1$. Because μ is unknown, we apply EM algorithm and in the expectation step, we set $\mu = \frac{1}{N} \sum_{i=1}^N \frac{\tilde{d}_i(x,y)}{\tilde{d}_i(r_i)}$. The algorithm listed in Table 1 needs three preset control parameters: maximal number of iterations MAX, precision tolerance e, and prior weight λ.

We use conjugate gradient to solve the above nonlinear optimization problem in Step 3. In this paper, we take an iterative style implementation for the Gaussian-Newton conjugate gradient optimization. The fundamental idea of this technique is to approximate the nonlinear function with its Taylor expansion, using regression to minimize the sum square of residual errors. To be specific, the optimization problem is formulated as follows

$$\min_{x,y,k} f(x,y,k) = \min_{x,y,k} \sum_{i=1}^N \left(k - \frac{\tilde{d}_i(x,y)}{\hat{d}_i(r_i)} \right)^2 + \lambda(k - \mu_{n-1})^2$$

where $\tilde{d}_i(x,y) = \sqrt{(x-x_i)^2 + (y-y_i)^2}$, and subject to the conditions $|x| \leq X, |y| \leq Y, 0 < k \leq 1$.

Since the algorithms are linear in the number of anchor nodes, and Hessian has a fixed dimensionality, they apparently have scalable complexities.

4 Simulations and Evaluation

To evaluate the performances of the proposed approach, we tested the algorithms in a set of situations through Monte Carlo simulations. The factors considered in the experiment setup included the following:

1) size of the field, we tested the typical room with different sizes from 10m by 10m to 100m by 100m respectively

2) layouts of anchor nodes and the number of anchor nodes, we tested both the scenarios of systematically deployed anchor nodes and randomly placed anchor nodes in a few layouts with different number of anchor nodes

3) parameters of free space shadowing model and noise level, the loss factor was chosen as 2 and 3, the RSSI measurement variance was set from 1mdB to 10mdB

Table 1. EM based probabilistic reduced positioning algorithm

Step 1	Range estimation: set the distances to all anchor nodes $\hat{d}_i(r_i) = d_0 10^{\frac{r(d_0)-r(d)}{10n}}$ using $\{r_i | 1 \le i \le N\}$
Step 2	Initialization: $(x,y) = (x_0, y_0)$, set $\mu_0 = 1$, let counter $n = 1$
Step 3	Maximization: solve $(x_n, y_n | r_{i=1}^N) = \arg\min_{x,y,k} \left(\sum_{i=1}^N (k - \frac{\hat{d}_i(x,y)}{\hat{d}_i(r_i)})^2 + \lambda(k - \mu_{n-1})^2 \right)$
Step 4	Mean estimation: set $\mu_n = \dfrac{\sum_{i=1}^N (\frac{\hat{d}_i(x_n, y_n)}{\hat{d}_i(r_i)})^2}{\sum_{i=1}^N \frac{\hat{d}_i(x_n, y_n)}{\hat{d}_i(r_i)}}$
Step 5	Stop criteria: if $\sqrt{(x_n - x_{n-1})^2 + (y_n - y_{n-1})^2} \le e$ or $n \ge MAX$ return $(x^*, y^*) = (x_n, y_n)$; otherwise set $n = n + 1$, go to Step 3

4) ways of placing obstacles, we selected 3 modes of blocking effects: no block, weak block and strong block, with $Q = 0, -5, -10$ mdB respectively, and tested various combinations of the nodes in different modes
5) impacts of obstacles on received signal, the power drop Q was set in the range of -1 to -10 mdB.

We also compared the proposed approach with some other known techniques in literature. All simulations were implemented in Matlab and C programming language on a laptop computer with Intel Core CPU at 2G Hz and 2G memory. The RSSI readings were drawn randomly from a pool of simulated results using Monte Carlo method. Each trial was repeated for a number of times. For each studied scenario, the positioning error (defined as $\epsilon = \sqrt{(x^* - x_t)^2 + (y^* - y_t)^2}$, where (x^*, y^*) is the estimated position and (x_t, y_t) is the real position of the target) and the cumulative distribution functions of the error were used to evaluate the positioning performance. Despite that the function $f(x, y|k)$ is non-convex, it is smooth and often locally convex at the global optimum. In our experiments, the proposed conjugate gradient-based algorithm gave solutions very close to the real optimum.

One control parameter in the proposed approach that needs to be determined is the prior weight λ. We observed that compared to the uniform prior, Gaussian priors had smaller average estimate errors. And Gaussian EM presented robustness to the variation of λ. For a 20m by 20m with 9 anchors placed on a 3×3 grid, we tested with 5 anchors unblocked, 2 weakly blocked and 2 strongly blocked, and nodes in each mode were picked randomly, while a target was randomly located within the field. As demonstrated in Figure 1, Gaussian EM remained stable under different λ, and Gaussian prior with fixed mean was affected by λ. Therefore, we used Gaussian EM with $\lambda = 0.1$ for the evaluation and comparison with other methods, unless specified otherwise.

We compared the proposal with several established methods using the same parameters, including namely least square (LS) [15], maximum likelihood estimation (MLE) [16] and weighted MLE. To compare with these algorithms, we simulated a field of 20m by 20m, each test consisted of 500 randomly located

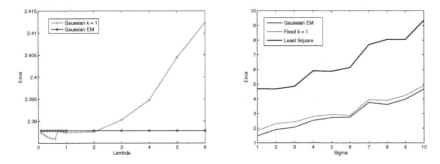

Fig. 1. Estimation errors under different parameters, left: λ, right: σ

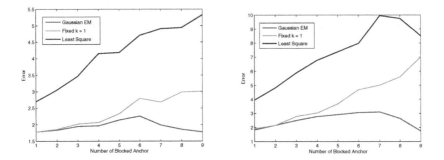

Fig. 2. Comparison with other algorithms, left: cumulative probability of error, right: estimate errors under different power drops caused by obstacles

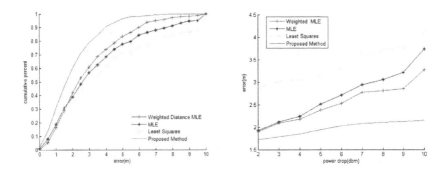

Fig. 3. Estimate errors under different number of blocked anchor nodes, left: $Q = -5$, right $Q = -10$

targets in the field. RSSI data were drawn using Monte Carlo method. The cumulative probability of position error is given in Figure 2, in which we set $\sigma = 4$, and each AN has 20% probability to be blocked by an obstacle that results in a 10dBm drop in RSSI. It is seen that the proposed algorithm outperformed all others in comparison (15% higher than the second best, weighted MLE). When the power drop was changed from 2 to 10dBm, the error grew slowly after the drop reached 8.

Tests were done with varying probability for a node to be blocked, ranging from 1 to 9 nodes, we see in Figure 3 that the proposed algorithm gives the best outcomes in all cases. At higher blockage probability, the advantage of the proposed approach is more remarkable.

5   Conclusions and Discussions

The precision of two-dimensional indoor positioning using RSS readings from Wi-Fi routers compromises by the obstacles along the propagation paths which cause the further attenuation of the transmitted power. By introducing a hidden factor, the approach proposed in this paper is able to adjust the solution automatically and generate an improved residual estimate error. A few assumptions on the priors of this factor are investigated. The introduction of the prior makes the distance estimating equivalent to a two-step random process: first a hidden variable is selected, and subsequently the normal distributed distances are estimated by each and every of anchor nodes. Solving the optimization problem of the maximum a posteriori with regard to both of the position and the hidden variable, a conjugate gradient algorithm gives a good approximate to the smooth non-convex function. In contrast to traditional fingerprint-ing way, the proposed approach does not require the collection of samples for model training, and it allows the environment to vary from time to time. This ensures the effectiveness and generality of the proposed approach. In addition, we compared this approach with a few established methods, showing that the proposed approach is stable generate better results, as well as stable to parameter variations and can be scalable in the number of anchor nodes. There are a few problems to be further investigated. In this paper, we assumed that all anchor nodes are identical and the RSS distributions have the same set of parameters. It is often more practical to consider different types of anchor nodes in applications. Moreover, as shown in the experiments, the proposed approach works better for more uniform anchor nodes, ideally in a single mode. However, if side information or history data are available, it can allow more accurate detection of the mode in which an anchor node is, and thus brings out higher positioning precision.

Acknowledgments. The work in this paper was supported by NSFC fund 61073095. The authors would like to thank the anonymous reviewers for their suggestions.

References

1. Liu, H., Darabi, H., Banerjee, P., Liu, J.: Survey of wireless indoor positioning techniques and systems. IEEE Transactions on Systems, Man, and Cybernetics, Part C: Applications and Reviews 37, 1067–1080 (2007)

2. Weng, Y., Xiao, W., Xie, L.: Total least squares method for robust source localization in sensor networks using TDOA measurements. International Journal of Distributed Sensor Networks (2011)
3. Mao, G., Fidan, B., Anderson, B.D.O.: Wireless sensor network localization techniques. Computer Networks. The International Journal of Computer and Telecommunications Networking, 2529–2553 (2007)
4. Swangmuang, N., Krishnamurthy, P.: Location Fingerprint Analyses Toward Efficient Indoor Positioning. In: Sixth Annual IEEE International Conference on Pervasive Computing and Communications, pp. 101–109 (2008)
5. Ladd, A.M., Bekris, K.E., Rudys, A.P., Wallach, D.S., Kavraki, L.E.: On the feasibility of using wireless Ethernet for indoor localization. Name IEEE Trans. Robotics and Automation, 555–559 (2004)
6. Xiang, Z., Song, S., Chen, J., Wang, H., Huang, J., Gao, X.: A wireless LAN-based in-door positioning technology. IBM J. Res. & Dev. (2004)
7. Wang, J., Urriza, P., Han, Y., Cabric, D.: Weighted centroid algorithm for estimating primary user location: Theoretical analysis and distributed implementation. IEEE Transactions on Wireless Communications (June 2011)
8. Chen, H., Sezaki, K., Deng, P., So, H.C.: An improved DV-hop localization algorithm for wireless sensor networks. In: Proc. IEEE Conference on Industrial Electronics and Applications (ICIEA 2008), Singapore, pp. 1557–1561 (2008)
9. Kumar, P., Reddy, L., Varma, S.: Distance measurement and error estimation scheme for RSSI based localization in Wireless Sensor Networks. In: Fifth IEEE Conference on Wireless Communication and Sensor Networks (WCSN), pp. 1–4 (2009)
10. Ahn, H.-S., Yu, W.: Indoor localization techniques based on wireless sensor networks. Mobile Robots State of the Art in Land, Sea, Air, and Collaborative Missions, 277–302 (2009)
11. Ali-Rantala, P., Ukkonen, L., Sydanheimo, L., Keskilammi, M., Kivikoski, M.: Different kinds of walls and their effect on the attenuation of radiowaves indoors. In: IEEE APS International Symposium, Columbus, OH, USA, pp. 1020–1023 (2003)
12. Chen, R.C., Lin, Y.C.: An Indoor Location Identification System Based on Neural Network and Genetic Algorithm. In: 3rd International Conferece on Awareness Science and Technology (iCAST), pp. 27–30 (2011)
13. Yang, Z., Wu, C., Liu, Y.: Locating in fingerprint space: wireless indoor localization with little human intervention. In: Proceedings of the 18th Annual International Conference on Mobile Computing and Networking (MobiCom 2012), pp. 269–280 (2012)
14. Arya, A.: An analysis of radio fingerprints behavior in the context of RSS-based boction fingerprinting systems. In: 2011 IEEE 22nd International Symposium on Date of Conference on Personal Indoor and Mobile Radio Communications (PIMRC), pp. 536–540 (2011)
15. Laaraiedh, M., Avrillon, S., Uguen, B.: Enhancing positioning accuracy through RSS based ranging and weighted least square approximation. In: Proceedings of POCA Conference, Antwerp, Belgium (2009)
16. Costa, J., Patwari, N., Hero III, A.O.: Distributed multidimensional scaling with adaptive weighting for node localization in sensor networks. ACM Trans. Sensor Netw. (2005)

Identifying Individuals' Footsteps Walking on a Floor Sensor Device

Kiryu Ibara[1], Kenta Kanetsuna[2,*], and Masahito Hirakawa[1]

[1] Interdisciplinary Graduate School of Science and Engineering, Shimane University, Japan
{s139501,hirakawa}@cis.shimane-u.ac.jp
[2] Interdisciplinary Faculty of Science and Engineering, Shimane University, Japan

Abstract. Studies of human-computer interaction have been broadened and dee- pened and powerful, novel gestural interfaces are of special interest in this field. This paper concerns the motion of feet on a floor sensor device. We have been investigating footstep tracking of individuals walking on the device and have pre- sented a particle filter-based framework. In this paper, we present a trial of extended system facilities, allowing the system to identify pairs of footsteps of in- dividuals walking on the floor sensor device. Three gait characteristics, stride length, footstep direction, and relative pressure values in foot regions, are consi- dered. In addition, an implementation of the improved particle filter framework with Walker's alias method is also described for speeding execution.

Keywords: Foot tracking, floor sensor, footstep pair, particle filter.

1   Introduction

As the field of human-computer interaction has grown tremendously, powerful and novel gestural interfaces have attracted considerable interest, and various systems and tools have been developed. Microsoft's Kinect is a good example, but it requires the user to stay within a limited area in his/her gesture demonstration. If we consider that the ability to walk upright is an evolutionary step and is one of the hallmarks of being human [1], it is worth considering gestures with movements on the floor, or human gait, as a means of interaction with a computer.

Studies of gait analysis are performed in health diagnostics, rehabilitation, and sports [2]. For example, the stride rate of elderly people tends to increase while their stride length decreases, which indicates a higher risk of falling [3]. In sports, gait analysis is applied to help athletes run faster and more efficiently, have better physio- logical treatment, and prevent serious injury. In addition, gait patterns convey com- plex states of mind. Happy people have an energetic walk and a faster walking speed. Those who feel anxious and fidgety might pace back and forth with shorter steps.

The authors have been conducting research in gait detection and analysis to recog- nize the physical and mental conditions of a walker [4, 5]. We adopt a floor mat

---

* Kenta Kanetsuna now works for Fujitsu Systems West Limited.

T. Yoshida et al. (Eds.): AMT 2013, LNCS 8210, pp. 56–63, 2013.
© Springer International Publishing Switzerland 2013

sensor as a motion capturing device instead of the computer vision techniques and wearable wireless sensors, which are widely used in the domain of gait analysis [6]. The advantages of using floor sensor devices include 1) the walker is not required to wear any special devices or be aware of their installation; 2) no additional space is needed in the system setup, whereas in the case of using vision technologies, cameras must be positioned some distance from the walker; and 3) captured data are not explicitly associated with privacy-sensitive characteristics (e.g., face and body).

However, detection and tracking of target foot regions while walking on a floor sensor device are difficult. One reason is that input data captured by the device carry only grayscale pressure values. Second, we walk bipedally. The feet move alternately and a gait cycle is divided into two phases: stance (the interval in which the foot is on the ground) and swing (the interval in which the foot is airborne). The foot contact images change depending on the phase. The system has to identify and track each of the feet properly, even though each foot appears again after its release from the device.

We first applied two-dimensional continuous dynamic programming to recognition of footsteps on the floor sensor device [4] and then proposed replacing it with a particle filter-based technique, which allows more responsive and reliable footstep tracking [5]. The system we developed can estimate the lower body motion of a person walking on the floor, where inverse kinematics is applied to compute joint angles in a body model, helping to recognize a walker's physical and mental status.

In this paper, we describe ideas to improve and extend footstep tracking facilities. This was motivated by the fact that the floor sensor device was expected to be of a reasonably large size and, therefore, it was irrational to restrict access to only one person on the device at a time. The system identifies pairs of shoes (i.e., individuals) on the device. It is not necessary for the individuals to walk in the same direction. They may turn in any direction and pass each other. In addition to this facility, an implementation of the improved particle-filter framework with Walker's alias method is described.

2   Related Work

Human gait detection and analysis using a pressure-sensitive floor have been carried out with [7] presenting a trial using floor sensor devices for retrieval and summarization of continuous archived multimedia data in a ubiquitous home environment. However, sensor density was coarse, and the system detected only the position and trajectory of walking humans. Other trials in [8-10] showed similar detection capability.

Leusmann et al. proposed an intelligent floor composed of 64 tiles, where each tile's corner is equipped with a piezoelectric sensor [11]. The system identifies human steps on the floor, specifically, the initial contact (heel strike and foot flat) phase and pre-swing (heel-off) phase, but detection of gait motion during actual walking was not discussed. None of the trials mentioned above consider situations where multiple people walk on the device at the same time and pass each other.

Srinivasan et al. [12] presented a pressure sensing floor with finer resolution (10 mm spatial resolution), similar to the one used in our study. The main focus lies on its hardware structure, and gait information analysis was not investigated.

GravitySpace [13] proposed by Bränzel et al. is capable of identifying users and their poses by sensing pressure imprints caused by gravity as it pulls people and objects against the floor. The system senses pressure on the basis of the frustrated total internal reflection (FTIR) technique using a camera located below the floor, making the system size quite large.

In summary, the existing trials using a pressure-sensitive floor have problems and limitations with the sensing resolution, abilities to extract gait features, the number of walkers, and the size in system setup. We aim to develop a smart floor system, which overcomes these challenges.

3 Foot Motion Tracking Using Particle Filter Framework

3.1 Application Scenarios

The National Institution of Population and Social Security Research in Japan reported that the proportion of elderly (65 years of age and over) out of the entire population will increase rapidly. The share was 23.0% as of 2010, and will move to 25.1% during 2013. By 2035, 33.4% of the total population will be elderly, increasing to 39.9% by 2060 [14]. Social security will become much more important and have a greater impact on our daily lives. To ensure stable growth, information and communication technology (ICT) could be of great use.

Home security systems based on gait pattern analysis would be helpful with an aging population, finding the state of health or aging of individuals by observing at a time series of walking pattern data and then taking preventative measures. Moreover, emergent adverse events such as falling to the floor in a private space (e.g., a restroom) can be detected.

Sensing technologies, such as cameras, may not be welcome at home. In fact, no one wants to be captured by cameras in private spaces, even though sophisticated services can be expected. Floor sensor devices do not cause the same anxiety. Moreover, unlike wearable sensor-based approaches, people are not required to wear any special devices or clothing at home.

In the medical domain, gait analysis can also work for rehabilitation by helping the patient to compensate for deficits. The patient knows how well a rehabilitation program is going. Another promising application of gait analysis is digital signage in which interactivity will serve as a key feature [15]. Implicit as well as explicit interaction is helpful. It would be useful to choose the content presented on a digital signage system, depending on the viewer's current emotion.

3.2 Foot Tracking with Two-Phase Particle Filters

The floor sensor device we used in our study is the Xiroku LL sensor, which adopts an electromagnetic induction mechanism. The sensor outputs a map of 8-bit pressure data whose spatial resolution is 10 mm. It is possible to connect up to 24 devices when a larger sensing area is needed. Figure 1 shows a snapshot of a working environment where three sensor mats are connected. People may wear their own shoes while walking on the device.

Fig. 1. Working environment with three floor sensor devices

Footstep tracking is performed using a two-phase particle filter-based technique in which two types of particle filter trackers are provided: an initialization tracker and an individual object tracker [16].

First, the initialization tracker starts an object search by spreading particles across the entire target space. When a possible footstep image appears, particles cluster around it. When a cluster corresponding to a foot in contact with the floor is found, the particle cluster is passed to an individual object tracker that tracks the foot until it disappears. A new initialization tracker is activated and starts searching for another footstep, realizing multiple footstep tracking. More details with the discussion of the recognition performance are available in [5].

4   Extension for Application of the System to Interactive Application Environments

4.1   Application of Walker's Alias Method to Particle Filter Processing

Since we are going to apply the system in interactive application environments, as explained in the previous section, it is important to speed the execution of the particle filters. Knowing that the re-sampling phase in the particle filter technique is computationally intensive, we apply Walker's alias method [17] in this study.

Figures 2 and 3 show results of exactly how the execution speed can be improved. Performance improvement in the re-sampling phase is given in Fig. 2, and that in the total particle filter process in Fig. 3. The ratio of improvement of the particle filter execution ranges from 1.75 to 1.99, depending on the number of feet on the device.

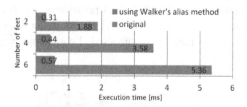

Fig. 2. Results of execution time improvement in the re-sampling phase

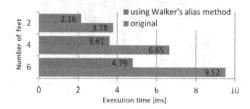

Fig. 3. Results of execution time improvement in the total particle filter process

4.2   Identifying Pairs of Footsteps

Methodology. When the gait analysis system is placed in a public space, several people can walk on the floor sensor device simultaneously. It is thus necessary to identify foot (shoe) pairs of individuals walking on the device before gait analysis, to make the system practical. The task cannot be simply performed by choosing a pair of adjacent footstep regions because people may pass each other. In our trial, we overcome this identification challenge by considering three gait characteristics. These are stride length, footstep direction, and relative pressure values.

- Stride length

If a footstep region is detected, the following footstep should appear within a certain limited range. We set the range to 80 cm on the basis of reference data in [18].

- Foot step direction

The person may change his/her walking direction. Range of motion is a combination of abduction of the foot and lateral rotation of the hip. The Japanese Orthopaedic Association and the Japanese Association of Rehabilitation Medicine suggest that normal values for abduction of the foot and lateral rotation of the hip are 10° and 45°, respectively.

Considering the stride length, we contemplate that the foot opposite to the one on the floor should appear in a fan-shaped area, as given in Fig. 4. Since each of the feet can turn out 55° (=10 + 45), the possible stepping range covers 110° in total. Here the direction of the base foot, indicated by an arrow in the figure, is determined as a straight line connecting weight centers of the initial foot touch region (i.e., the heel) and the foot region in the flat through midstance.

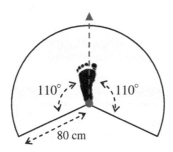

Fig. 4. Overall range of motion

- Relative pressure values at foot regions

During walking, the weight is transferred from one foot to the other. Specifically, the weight in one foot region increases in the heel strike, while the weight in the other foot region decreases accordingly in the heel off. We consider two foot regions as one pair if the above situation continues across three time frames.

This foot pair evaluation is activated when the weight center of a foot region shifts more than 1 cm after it touches the floor sensor device. The weight of the foot region is calculated by the average of the weights of particles, which are associated with the region in three consecutive frames, to cancel out fluctuations in captured data from the device.

Figure 5 shows the result of executing identification of pairs of footsteps under the conditions that initialization trackers and individual object trackers are ran with 3,600 and 1,000 particles, respectively. The floor sensor size is 60 cm × 180 cm. In the figure, footstep regions, which are identified as a pair, are indicated by a line.

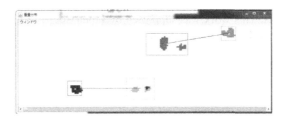

Fig. 5. Result of executing detection of pairs of footsteps

Experiments. We conducted a preliminary experiment to evaluate the performance of the scheme explained above. Nine university students participated in the experiment. The subjects were asked to walk on a floor sensor device with and without their own shoes. We also asked each pair of subjects to simultaneously walk on the device in two ways: in single file and passing each other. The walking area size was 60 cm × 180 cm.

Results of the experiment are summarized in Table 1. In the table, "missing" and "wrong" indicate cases that, for a foot located on the floor, a trial failed to find the right foot to be paired and assigned a foot not part of the pair, respectively. The percentage of correct answers is 96.2% for bare feet and 89.7% for feet with shoes. One reason for the failures is that the system missed a footstep at the very moment it came in contact with the device, resulting in missing directional information.

Table 1. Results of foot pair identification experiment (Number of footsteps)

	success	failure	
		missing	wrong
bare feet	25	1	0
with shoes	26	2	1
total	51	3	1

5   Conclusions

In this paper, we have presented a system of identifying pairs of walking footsteps as an extension of our floor sensor-based foot tracking framework toward achieving clinical practice and interactive human computer interaction research. Three gait characteristics, stride length, footstep direction, and relative pressure values at foot regions, were considered for this task. In addition, we described an implementation of the improved particle filter framework for efficient footstep tracking by applying Walker's alias method. Through experiments, we showed that the performance is good to run the system in practical interactive environments.

Further studies still remain, which include exploration of meaningful gait features originating in floor sensor devices, gait analysis based on those features, and development of smart applications.

Acknowledgements. This work was supported by Japan Society for the Promotion of Science (JSPS), Grant-in-Aid for Scientific Research (C) (23500149).

References

1. Hirst, K.K.: Bipedal Locomotion,
   http://archaeology.about.com/od/bterms/g/bipedal.htm
2. Tao, W., Liu, T., Zheng, R., Feng, H.: Gait Analysis using Wearable Sensors. Journal of Sensors 12, 2255–2283 (2012)
3. Tien, I., Glaser, S.D.: Structural Health Monitoring and Evaluation of Human Gait to Assist in the Diagnosis of Parkinson's Disease. In: 7th International Workshop on Structural Health Monitoring. DEStech Publications (2009)
4. Takai, H., Oikawa, M., Hirakawa, M.: Recognition of Human Gait using 2D Continuous Dynamic Programming towards Realization of Invisible User Interfaces. In: 2010 IEEE Region 10 Conference, pp. 269–274 (2010)
5. Oikawa, M., Tanijiri, N., Hirakawa, M.: Estimation of Lower Body Motion during Walking on a Floor Sensor Device. ICIC Express Letters 6(12), 3027–3031 (2012)

6. Hodgins, D.: The Importance of Measuring Human Gait. European Medical Device Technology 19(5), 42, 44-7 (2008)
7. De Silva, G.C., Yamasaki, T., Aizawa, K.: An Interactive Multimedia Diary for the Home. IEEE Computer 40(5), 52–59 (2007)
8. Chang, S., Ham, S., Kim, S., Suh, D., Kim, H.: Ubi-floor: Design and Pilot Implementation of an Interactive Floor System. In: 2010 Second International Conference on Intelligent Human-Machine Systems and Cybernetics, pp. 290–293. IEEE Press, New York (2010)
9. Visell, Y., Smith, S., Law, A., Rajalingham, R., Cooperstock, J.R.: Contact Sensing and Interaction Techniques for a Distributed, Multimodal Floor Display. In: 2010 IEEE Symposium on 3D User Interfaces, pp. 75–78. IEEE Press, New York (2010)
10. Valtonen, M., Mäentausta, J., Vanhala, J.: TileTrack: Capacitive Human Tracking Using Floor Tiles. In: IEEE International Conference on Pervasive Computing and Communications, pp. 1–10. IEEE Press, New York (2009)
11. Leusmann, P., Mollering, C., Klack, L., Kasugai, K., Ziefle, M., Rumpe, B.: Your Floor Knows Where You Are: Sensing and Acquisition of Movement Data. In: 12th IEEE International Conference on Mobile Data Management, pp. 61–66. IEEE Press, New York (2011)
12. Srinivasan, P., Birchfield, D., Qian, G., Kidané, A.: A Pressure Sensing Floor for Interactive Media Applications. In: ACM SIGCHI International Conference on Advances in Computer Entertainment Technology, pp. 278–281. ACM, New York (2005)
13. Bränzel, A., Holz, C., Hoffmann, D., Schmidt, D., Knaust, M., Lühne, P., Meusel, R., Richter, S., Baudisch, P.: GravitySpace: Tracking Users and Their Poses in a Smart Room using a Pressure-Sensing Floor. In: ACM SIGCHI Conference on Human Factors in Computing Systems, pp. 725–734. ACM, New York (2013)
14. Population Projections for Japan,
    http://www.ipss.go.jp/site-ad/index_english/esuikei/gh2401e.asp
15. Want, R., Schilit, B.N.: Interactive Digital Signage. IEEE Computer 45(5), 21–24 (2012)
16. Matsumura, R., Okumura, K.: Object Detection and Tracking using Particle Filter. Bulletin of Oshima National College of Maritime Technology 41, 75–85 (2008) (in Japanese)
17. Walker, A.J.: An Efficient Method for Generating Discrete Random Variables with General Distributions. ACM Transactions on Mathematical Software 3(3), 253–256 (1977)
18. Yamasaki, M., Sato, H.: Human Walking: With Reference to Step Length, Cadence, Speed and Energy Expenditure. Journal of Anthropological Society of Nippon 98(4), 385–401 (1990) (in Japanese)

Detection and Presentation of Failure of Learning from Quiz Responses in Course Management Systems

Toshiyasu Kato[1] and Takashi Ishikawa[2]

[1] Graduate School of Engineering, Nippon Institute of Technology, Saitama, Japan
c3115001@cstu.nit.ac.jp
[2] Nippon Institute of Technology, Saitama, Japan
tisikawa@nit.ac.jp

Abstract. In this paper, we propose a method to detect failure of learning of students using quiz responses in a course management system Moodle. Failure of learning is defined as a situation in which the correct answer rate of a learning topic is significantly lower than the correct answer rate of other topics. In the research, the researchers identified the presence or absence of failure of learning in actual classes to evaluate the usefulness of the proposed method. The results revealed that students checked the quiz results significantly more in the experimental group than in the control group, and that more instruction was given to the experimental group.

Keywords: Course management system, Failure of learning detection, Learning analytics.

1 Introduction

Course management systems are able to accumulate a vast amount of information; this accumulated collection of education data is very valuable for analyzing students' behavior [9]. In order to properly utilize the learning and instruction activities, the large amount of these accumulated data is processed, analyzed and data mining, presented in an easy-to-understand way, are carried out. Data mining in course management systems is the automatic extraction of implicit and interesting patterns from the historical data on learning activities, for example, lessons, forums, and quizzes [4]. Data mining for quiz responses is able to detect failure of learning. Failure of learning is defined as a situation in which students attempt to solve problems but, in encountering unexpected problems, are unable to progress smoothly [7]. Issues in failure of learning of course management systems are the determination of failure of learning in real time, and the method of presentation to teachers and students.

The purpose of this paper is to propose a method to detect failure of learning of students in a course using quiz responses in a course management system Moodle [5, 6]. Failure of learning detection to apply the proposed method is to evaluate the usefulness by experiments using actual classes. The failure of learning in this paper means an item significantly lower compared to the overall correct answer rate for

T. Yoshida et al. (Eds.): AMT 2013, LNCS 8210, pp. 64–73, 2013.
© Springer International Publishing Switzerland 2013

each learning topics of quiz responses. The significance of this research is that it can grasp the students in lack of understanding by detecting the failure of the learning.

2   Data Mining of Quiz Responses

Moodle is an open-source course management learning system that helps educators create effective online learning communities [8]. It is an alternative to proprietary commercial online learning solutions and is distributed at no cost under open-source licensing [2]. Moodle quizzes are created using the following format:

- Multiple Choice: Select one or more correct answers.
- True / False: Indicate if true or false.
- Essay: Enter a word or phrase.
- Calculated: Enter a numeric value.

  The procedure for creating and implementing a quiz is as follows:

1. Creation of questions: In this case, specify the format for each question, and enter the correct response based on question content and format.
2. Registration and implementation of data and questions: In this case, enter the score of each problem in the time limit period.
3. Conduction of a quiz.
4. Checking of responses.

  Moodle allows the teacher to access certain types of statistical information, for example, statistical quiz reports, which permit item analysis. It presents processed quiz data in a way that is suitable for analyzing and judging the performance of students on each question. The data mining of quiz responses can detect failure of learning for the entire class as well as for each student, and allows for comparisons between the two.

3   Determination of Failure of Learning

The purpose of determination of failure of learning is to determine failure of learning of both entire class and individuals. To determine unsuccessful learning, questions are assigned to each learning unit and a learning topic. A question is defined as one that appears on the Moodle quizzes. It is necessary to enter the names of learning units and learning topics and map them for each question in advance; data mining techniques are then applied to quiz responses. The learning contents are equal to the questions of quizzes, and the question types are limited to single-answer, multiple-choice questions, and true/false questions.

3.1   Determination for the Class

The failure of learning of the class is derived from a comparison of the correct answer rate in the classifications of items, such as learning, and a significantly lower unit.

A chi-square test of independence of statistical methods is employed to determine the failure of learning. It assesses if the values of row and column items are independent [1]. In (1), χ^2 is a statistic that shows the difference between the expected frequency and the actual observations. O_i and E_i are respectively the observed frequency and the expected frequency. In this case, O_i is the correct answer rate of the target items. A target item is a learning unit, a learning topic, or a quiz question. E_i is the correct answer rate of the parent classification of target items.

$$\chi^2 = \sum_{i=1}^{k} \frac{(O_i - E_i)^2}{E_i} \tag{1}$$

The algorithm of determination for the class is as follows:

1. Enter the learning unit, learning topic, or question as an item, as well as the significance level.
2. The responses of participants to the quiz questions, which correspond to the entered items and their parent classification items, are obtained from Moodle.
3. The correct answer rate is calculated for each item.
4. In the case where the correct answer rate of an item is lower than the correct answer rate of its parent classification, a chi-squared value is calculated from that item's correct/incorrect answer rate and its parent classification correct/incorrect answer rate. In addition, the degree of freedom in the test is 1.
5. In case the null hypothesis is rejected, the item is outputted from the chi-squared value and the significant level.

3.2   Determination for Individual Students

The failure of learning of the individual student is a situation where the number of correct responses is significantly lower than the average number of correct responses of the entire class. A chi-square test is used to determine the failure of learning.

The algorithm of determination for an individual student is as follows:

1. Enter the learning unit or learning topic, as well as the significance level.
2. The responses of experiment participants to the quiz questions, which correspond to the entered items, are obtained from Moodle.
3. The number of correct responses of individual students is calculated.
4. The average numbers of correct responses of all the participants are calculated.
5. In the case where the number of correct responses of an individual student is lower than the average number of correct responses of the entire class, a chi-square value is calculated from that student's number of correct/incorrect responses and the entire class' average number of correct/incorrect responses. In addition, the degree of freedom in the test is 1.
6. In case the null hypothesis is rejected, the data for the student are determined using the chi-square value and the significance level.

The determination of failure of learning employs hypergeometric distribution in cases there is only one classification. The hypergeometric distribution is a discrete probability distribution, which arises in relation to random sampling (without re-placement) from a finite population [3]. The probability mass function P of the hypergeometric distribution is the probability that X takes a random variable x in re-peated attempts. This distribution consists of a population with N elements, as well as M elements of those that have attribute A. When taking a sample of size k from this population, if X is the number of those that have attribute A, then the probability dis-tribution of X is given in (2). Thus, the probability mass function P refers to the prob-ability of k number of students with failure of learning.

$$P(X = x) = \frac{\binom{M}{x}\binom{N-M}{k-x}}{\binom{N}{k}} \tag{2}$$

The algorithm of determination for an individual student in quiz questions is as follows:

1. Enter the question and the significance level.
2. The responses of experiment participants, which correspond to the entered ques-tion, are obtained from Moodle.
3. The hypergeometric distribution value is calculated from the number of partici-pants and the number of their correct responses. In addition, N = the number of candidates and M = the number of correct responses; the value is applied to (2) is $k = 1, x = 0$.
4. In case the hypergeometric distribution value is lower than the significance level, the data of the students who have incorrect responses are outputted.

4   Presentation of Failure of Learning

The purpose of presentation of failure of learning is to provide items of failure of learning on teachers and students. Here, item is either a learning units, learning topics, and questions. The teacher's user interface is shown in Fig. 1-4 and that of students in Fig. 5 and 6. An exercise support system [10], which is discussed in Section 5, is used during class exercise time. The failure of learning data gathered through the algo-rithms mentioned in Section 3 can subsequently be displayed in visual form for the teacher and for the students. As an example, Question 3.3-1 in Fig. 6 displays a failure of learning in red because the students answered incorrectly while 87.0% of the class answered in Fig. 3.

The teacher's method of presentation is to display the following message in red on the screen of the system: "The number of students who encountered failure of learn-ing in today's quiz" (Fig. 1). By clicking the link in the panel with the displayed number of students, a list of student IDs is shown. The IDs of students with failure of learning are displayed in red (Fig. 2). The "Quiz Responses" link shown in Fig. 1 leads to a page where the quiz responses of the entire class are shown. In this page, the correct answer rates for learning topics, for which students demonstrate a failure of learning, are displayed in red (Fig. 3). From the correct answer rate in the quiz

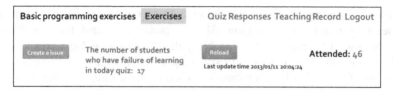

Fig. 1. Show that there is a failure of learning of students to teachers

Appropriate students Quiz result of 1/11		17 / examinees 46	Teaching record
When you click the student ID, we will show the quiz results for individual students. The color red of student IDs represents the students who have failure of learning for any item.			
Student ID	Full name	Seat Number	Correct answer
1115xx1	Tarou Nikkou	-	0.0 %
1115xx2	Jirou Nikkou	-	0.0 %
1115xx3	Saburou Nikkou	-	50.0 %

Fig. 2. A list of students with failure of learning

Learning items	Question number	Correct answer	
	Question 3.3-1	87.0 %	
Repetition (while)	Question 3.3-2	71.7 %	69.6 %
	Question 3.3-3	50.0 %	

Fig. 3. Quiz responses of the entire class

Basic programming exercises Exercises Quiz responses **Students with inc. ans.** Logout			Teaching record
Students with incorrect ans. Learning Unit: Method 28 / examinees 46			
When you click the student ID, we will show the quiz results for individual students. The color red of student IDs represents the students who have failure of learning for any item.			
Student ID	Full name	Seat Number	Correct answer
1115xx1	Tarou Nikkou	-	0.0 %
1115xx2	Jirou Nikkou	-	0.0 %
1115xx3	Saburou Nikkou	-	27.8 %

Fig. 4. A list of students who have the incorrect answer

responses of the entire class, a list of IDs of those who have incorrect responses is shown. In this list, the student ID of those who have failure of learning is exhibited in red (Fig. 4).

For students the following message is displayed in red on the exercise selection screen of the exercise support system: "Failure of learning in the responses to this quiz" (Fig. 5). A "Quiz Responses" link is also exhibited in red on the top right corner of the screen. By clicking the red link, personal quiz responses are shown with the correct answer rate of the learning topic; the failure of learning is displayed in red (Fig. 6).

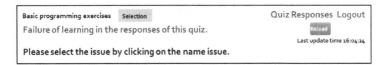

Fig. 5. Show that there is a failure of learning to students

Learning items	Question number	Correct answer	
	Question 3.3-1	×	
Repetition (while)	Question 3.3-2	×	33.3 %
	Question 3.3-3	○	

Fig. 6. Quiz responses to individual student

5   Evaluation

5.1   Objective

The objective of the evaluation is to assess the usefulness of failure of learning detection in order to support teachers' instruction and the review of student. For teachers, the hypotheses about the usefulness of this approach are (a), (b), and (c); for students, they are (a) and (d):

(a) There is a difference in the number of times students check the quiz responses.
(b) Students who have failure of learning receive extra instruction from teachers.
(c) The teachers believe that such instruction is effective.
(d) There is a difference in the review action taken toward quiz responses.

Hypothesis (a) can be verified using the chi-square test of the view recording quiz results. Hypothesis (b) can be confirmed from the instruction log. Hypothesis (c) can be verified using teachers' responses to the survey results. Hypothesis (d) can be substantiated using the difference in review action of students.

5.2   Method

Target Course. The course was a six-week, two-hour course, entitled "Basic Programming Exercises," which consisted of a lecture, quiz, and exercise phases. It was held in the fall semester for first-year students of the university. A five-minute quiz was given after each lecture.

Learning Contents. For the experiment, two sections of the course, with two different teachers, were used (Table 1). Both sections covered the same learning content (Table 2). At the end of the experiment, we conducted a questionnaire survey with both the students and the teachers.

Table 1. Tested classes

Teacher	Number of Students	Number of Teaching Assistants
A	47	1
B	35	1

Table 2. Learning contents of the experiment

Lesson items	Learning unit	Learning topic
1. Logical operator		Logical operator
2. Repetition (for)	Repetition and	Repetition process
3. Repetition (while)	conditional branching	Repetition process
4. Array		Array
5. Overview of class	Base of class	Class, Instance, Reference
6. Case study of class		Examples of use of the class

Method of Hypothesis Verification. The method of verification compare the difference with and without the presentation of the failure of learning, divided into the student experimental and control group. The student ID numbers of the experimental group are shown in red and that of the control group in blue (Fig. 2 and 4). In addition, the message is not presented in the exercise support system screen (Fig. 5). Experimental and control groups are not allowed to make changes to the computer selected at random at the start of the experiments. Hypothesis of difference are compared to the corresponding percentage of the number that is different from the number of experimental and control groups. Teachers and students do not know who is belonging to either group.

Prototype System. The prototype system was implemented as a web application. The database server and the web server both ran on Linux; the languages used for development were MySQL and PHP.

5.3   Results

Number of Views of Quiz Responses. In the experimental group, the quiz responses were viewed by 90 times (39.5%), whereas in the control group, teachers were viewed by 23 times (9.4%) (Table 3). The experimental group is statistically significant at 5% level of significance.

With regard to view, the experimental group, the quiz responses were viewed by 164 times, whereas in the control group, students were viewed by 122 times (49.8%) (Table 4). The experimental group is statistically significant at 5% level of significance. The experimental group is statistically significant at 5% level of significance.

Number of Instructions Given to Students Who Experienced Failure of Learning.
The instructions of the experimental group are 57 times (25.0%) and that of the control group and 34 times (13.9%) that of the number of teachers (Table 5). The experimental group is the most statistically significant at 5%.

Table 3. Number of views of quiz responses of individual students in the teachers

Experiments	Experimental group			Control group		
	Examinees	Views	Views rate	Examinees	Views	Views rate
1	39	8	20.5	42	8	19.0
2	40	16	40.0	42	3	7.1
3	37	10	27.0	38	10	26.3
4	38	16	42.1	41	0	0.0
5	36	13	36.1	42	2	4.8
6	38	27	71.1	40	0	0.0
Total	228	90	39.5	245	23	9.4

Table 4. Number of views of quiz responses of students

Experiments	Experimental group			Control group		
	Number of Views	Views	Views rate	Number of Views	Views	Views rate
1	16	25	64.1	13	25	59.5
2	12	17	42.5	12	20	47.6
3	12	19	51.4	6	8	21.1
4	16	40	105.3	11	22	53.7
5	14	34	94.4	13	25	59.5
6	19	29	76.3	9	22	55.0
Total	89	164	71.9	64	122	49.8

Table 5. Number of instructions for students with failure of learning

Type of instruction	Experimental group			Control group		
	Class	Individual	No need	Class	Individual	No need
	11	46	17	0	34	17
Total		57			34	

User Survey. The survey results indicate the percentage by the number of people who have true positive responses in the control group and the experimental group (Table 6). Table 7 shows the results of the survey for teachers. The presentation of failure of learning is changing instruction.

User Benefits. The teacher understands the learning topics of a lack of understanding as an entire class. Students understand the lack of understanding of their own learning.

Table 6. Results of the survey for students

Group	Q.1 Check re-sponses	Q.2 Is diffi-culty	Q.3 Adequacy of presen-tation	Q4. Check of items	Q5. Re-view	Q6. Received instruc-tion
Experimental	83.3	86.1	88.9	72.2	66.7	11.7
Control	76.9	76.9	82.1	71.8	66.7	7.7

Table 7. Results of the survey for teachers

Teacher	Q.1 Check responses	Q.2 Is difficulty	Q.3 Adequacy of presentation	Q4. Instruc-tion change	Q.5 Effective instruction
A	Check all	Always check	Little reasonable	Little change	Little effective
B	Check all	Little check	Little reasonable	Changed	Little effective

Summary of Results. In the experimental group, which was presented with failure of learning, the number of teachers and students confirm the failure of learning is significantly high. Also, there was also a difference in the number of instructions and the confirmation action of the learning topics.

5.4   Discussion

Usefulness of Failure of Learning Detection. The presentation of the failure of learning, teachers are instructed check the students who stumbled statistically significantly in the experimental group. Additionally, more instruction was provided to the experimental group and the experimental group checked the quiz results more than the control group. The amount of instruction of teachers is improved as an opportunity on failure of learning, has the effect of communication of teachers and students is promoted. From these results, it can be concluded that the failure of learning detection using the proposed algorithm is useful for instruction and for students to check the learning topics.

Factor Analysis of Failure of Learning. It is necessary to provide information on the factor analysis of the lack of understanding, which was taken from the confirmation of learning of teachers and students. The model of level of understanding supports the discovery of the failure of learning factors. From these, the presentation of factors related view based on a model of comprehension to support the discovery of factors failure of learning is issues.

6   Conclusion

In this paper, we proposed a method to detect failure of learning of students in a course using quiz responses in a course management system Moodle. This method identifies failure of learning when the correct answer rate of a learning topic in quiz responses is significantly lower than the correct answer rate of other learning topics. The proposed method evaluated the usefulness in actual classes. The results of the evaluation, a significantly large number of teachers and students have checked the failure of learning. Additionally, there was also a difference in the number of instruction given to students.

Future work will involve the development of a function to analyze the failure of learning factors, as well as one in the course management system that generalizes failure of learning detection. A method of effectively displaying failure of learning should be devised.

Acknowledgment. The authors of this paper are grateful to Professor Hiroshi Takase and Lecturer Masashi Katsumata of the Nippon Institute of Technology for their cooperation in the evaluation experiment.

References

1. Bull, C.R., Bull, R.M., Rastin, B.C.: On the Sensitivity of the Chi-square Test and its Consequences. Meas. Sci. Technol. 3, 789–795 (1992)
2. Graf, S., Kinshuk, K.: Providing Adaptive Courses in Learning Management Systems with Respect to Learning Styles. In: Bastiaens, T., Carliner, S. (eds.) Proceedings of World Conference on E-Learning in Corporate, Government, Healthcare, and Higher Education 2007, pp. 2576–2583 (2007)
3. Huang, X.Z.: The HypergeometricDistribution Model for Predicting the Reliability of Softwares. Microelectron Reliab. 24, 11–20 (1984)
4. Klosgen, W., Zytkow, J.: Handbook of Data Mining and Knowledge Discovery. Oxford University Press, New York (2002)
5. Moodle, http://moodle.org/
6. Open University of Japan:H21-22 Survey on the Promotion of ICT Use Education. Center of ICT and Distance Education (2011)
7. McCartney, R., Eckerdal, A., Mostrom, J.E., Sanders, K., Zander, C.: Successful Students' Strategies for Getting Unstuck. SIGCSE Bull. 39, 156–160 (2007)
8. Roessling, G., Kothe, A.: Extending Moodle to Better Support Computing Education. SIGCSE Bull. 41, 146–150 (2009)
9. Romero, C., Ventura, S., Garcia, E.: Data Mining in Course Management Systems: Moodle Case Study and Tutorial. Computers and Education 51, 368–384 (2007)
10. Kato, T., Ishikawa, T.: Design and Evaluation of Support Functions of Course Management Systems for Assessing Learning Conditions in Programming Practicums. In: International Conference on Advanced Learning Technologies 2012, pp. 205–207 (2012)

Gamification of Community Policing: SpamCombat

Alton Y.K. Chua and Snehasish Banerjee

Wee Kim Wee School of Communication and Information
Nanyang Technological University
altonchua@ntu.edu.sg, snehasis002@e.ntu.edu.sg

Abstract. The purpose of this paper is two-fold. First, it seeks to introduce the conceptual prototype of SpamCombat, a Web application that helps combat spam through gamification of community policing. Second, it attempts to evaluate SpamCombat by identifying factors that can potentially drive users' behavioral intention to adopt. A questionnaire seeking quantitative and qualitative responses was administered to 120 participants. The results indicate that behavioral intention to adopt SpamCombat is generally promising. Most participants appreciated the novelty of SpamCombat in supporting community policing to promote a spam-free cyber space. However, participants felt that using SpamCombat could be time-consuming.

Keywords: gamification, community policing, spam, behavioral intention.

1   Introduction

The upsurge of Web 2.0 in recent years has revolutionized the role of online users by allowing them to freely engage in knowledge sharing through exchange of ideas and opinions [1]. One way for users to share knowledge comes in the form of community policing, whereby they voluntarily bear the responsibility to police the cyber space for benefits of their peers [2, 3]. For example, users of MySpace, a popular social networking site, can participate in community policing by flagging off inappropriate content posted by others [4]. For the purpose of this paper, community policing refers to the voluntary practice of marking inappropriate content as spam by users to promote a spam-free cyber space.

Advancements in Web technologies in recent past have led to the emergence of a variety of online games [5]. These can range from ordinary Flash-based animation games to sophisticated Massively Multiplayer Online Games (MMOG) that allow users to play, interact, co-operate with or compete against thousands of peers in the virtual world [6]. One of the most recent forms of online games includes games with a purpose [7]. These involve the use of games in non-gaming contexts to promote user engagement in performing tasks that are trivial yet mundane for humans but challenging for computers through what is known as gamification [8]. For example, gamified applications have been deployed to accomplish image labeling tasks that humans perform more easily and accurately than computers [9].

T. Yoshida et al. (Eds.): AMT 2013, LNCS 8210, pp. 74–83, 2013.
© Springer International Publishing Switzerland 2013

However, gamified Web applications are usually not designed for community policing. Conversely, Web applications that support community policing may not offer entertainment. Hence, this paper proposes a Web application that gamifies community policing to combat spam. This is necessary because Web provides ample opportunities for individuals and businesses with malicious intentions to create spam indiscreetly as well as directing traffic to irrelevant websites [10]. Thus far, detection of such spam has been mostly studied using text-based approaches or link-based approaches [11, 12]. However, the gradual sophistication of spamming techniques often renders these approaches ineffective [13, 14]. Such shortcomings of the current state-of-the-art techniques may be mitigated by human intervention, framed through the twin lenses of community policing and gamification. This approach not only taps into users' altruism and commitment towards online communities for spam detection [15], but also enlivens the task through entertainment.

Given the dearth of Web applications that gamify community policing coupled with the potential of detecting spam through human intervention, the purpose of this paper is two-fold. First, it seeks to introduce the conceptual prototype of SpamCombat, a Web application that helps combat spam through gamification of community policing. Second, it attempts to evaluate SpamCombat by identifying factors that can potentially drive users' behavioral intention to adopt.

The rest of the paper proceeds as follows. Section 2 reviews the literature on community policing, gamification and behavioral intention to adopt. Section 3 offers an overview of SpamCombat. Section 4 explains the methods while section 5 presents the results. This is followed by the discussion and the conclusion.

2   Literature Review

2.1   Community Policing

Community policing in the context of this paper refers to users' proclivity to mark inappropriate content and combat spam in the cyber space for the benefits of other users. Given the ease with which users can create content frivolously without any quality control coupled with the malicious practices of individuals and businesses to contribute inappropriate content indiscreetly [10], the state-of-the-art spam detection techniques are gradually becoming less effective [13, 14]. Hence, it could be a timely endeavour to leverage on users' community policing efforts to combat spam.

As users play a very active role online, they should also be seen to play a part in policing the Internet [16]. Although such a form of online vigilante is not very widespread to date, there have been a few instances of community policing intended to weed out inappropriate content. As early as 1995, a group called CyberAngels was formed to actively police the cyber space and promote netiquette [17]. In the modern Internet-blitzed society, such community policing is gradually being opened to all users. For example, review platforms such as TripAdvisor.com present the question *"Problem with this review?"* alongside each submitted review. If users doubt the authenticity of reviews, they can report their concerns. Such a collaborative approach of community policing could be beneficial to combat spam on the Web in the long run.

2.2   Gamification

Gamified applications represent fictitious or artificial situations governed by rules that structure users' actions with respect to an objective, which is to win or to overcome an obstacle [18]. They can be designed for various purposes. They may be treated as tools for individual entertainment, means for social interaction, strategies for publicity campaigns, platforms for experimenting new design concepts or pedagogical media for learning and knowledge sharing [19]. In any case, gamified applications must necessarily offer entertainment to users [20].

There exist five characteristics of gamified applications that contribute to users' entertainment. These include goals and rules, feedback, challenge, social interaction, and rewards [21]. First, goals define the objective to be achieved while rules organize the virtual environment to specify how the goals can be accomplished [22]. Second, feedback measures users' current performance and map their progress against the goals [23]. Third, challenge issued progressively within the game encourages users to surpass themselves by improving their previous performance [24]. Fourth, social interaction motivates users to socialize and combine their efforts in achieving specific goals [25]. Finally, provision of rewards is essential to restore users' positive gaming experiences so that they would again return to the application [26].

2.3   Behavioral Intention to Adopt

Behavioral intention to adopt can be defined as a measure of users' inclination to embrace a new technology, service, application or system [6]. Being highly related to the cognitive psychology of users, behavioral intention to adopt is increasingly regarded as a necessary condition before the large scale implementation of any information technology application [26]. To analyze behavioral intention, there exist several models such as the Theory of Reasoned Action [27], the Theory of Planned Behavior [28], and the Technology Acceptance Model [29].

Drawing collaboratively from such models, this paper evaluates behavioral intention to adopt SpamCombat based on five factors, namely, perceived utility, perceived enjoyment, perceived ease of use, social influences and user preferences. Perceived utility refers to users' perceptions on the extent to which the application can effectively serve its intended purpose of task completion [30]. Perceived enjoyment is the extent to which users derive fun and perceive an enjoyable experience from the application [31]. Perceived ease of use denotes the extent to which users conceive an application as user-friendly with a smooth learning curve, and that it can be used with minimal effort [30]. Social influences measure the degree to which users' behavioural intentions are shaped by their attitudes and subjective norms [28, 29]. Finally, user preferences refer to the set of pre-dispositions shared by users based on their subjective opinions that one system is better than another [32].

3   SpamCombat: Design Overview

SpamCombat is a social Web-based prototype that gamifies community policing to combat spam. As shown in Figure 1, it appears as a simple floating toolbar that runs on top of the Web browser and can be accessed unobtrusively by users surfing the Internet. The application is designed to support four prominent features, namely, (1) flagging spam, (2) creating assignments, (3) building profiles, and (4) leader boards.

Fig. 1. Floating toolbar in SpamCombat

First, users can flag off a part or whole of the screen and designate the content as spam if they suspect its authenticity and credibility. Figure 2 shows the screen shot of the user interface for flagging spam. Once users confirm that the content is spam, the selected area of the website is marked grey. If numerous users flag the same content as spam, the specific portion is eventually hidden. Thus, spam is eliminated collaboratively by tapping into users' commitment for community policing. To induce a feeling of personal rewards, users earn a brownie each time they flag out spam. On the other hand, users who frivolously flag arbitrary areas as spam are penalized.

Fig. 2. Flagging spam through SpamCombat

Second, users with sufficient number of brownies can create assignments. An assignment refers to a series of thematically similar Web pages that users wish to recommend to others. Users can view assignments created by other users and optionally accept one to play. Those who chose to accept an assignment are brought through the series of Web pages specified by the respective creator. After completing the assignment, other users can review it through ratings, votes and comments. Users earn brownies for creating, accepting and completing assignments.

Third, users can build their own profiles citing their accomplishments in weeding out spam, brownies earned, as well as creating and accepting assignments. In addition, users can view the ratings, votes and comments of all assignments completed by other users through their profiles. Moreover, users can also create a friend list. When any friend on the list is online, a synchronous chat feature is enabled to support social interaction.

Fourth, the prototype also has a leader board listing users with the most number of spam flagged, brownies earned, assignments created, and assignments completed. Friends of users who appear on the leader board are notified. The leader board serves as a feedback and challenge for others to perform better. The desire of leading the leader board to leverage one's social reputation entices users to play SpamCombat.

Moreover, SpamCombat supports three additional features meant to proliferate fun. These include bomb-planting, treasure-laying and shield-buying. As users surf the Web, they can plant a bomb or place a treasure chest on any URL. To minimize the damage of any potential bomb, users can also buy shields to protect themselves. With a shield, users who stumble on a URL planted with a bomb do not lose any brownies. On the other hand, users who encounter a treasure are rewarded with brownies. To use each of these features however, users must have sufficient number of brownies.

4   Methodology

For the purpose of this study, a questionnaire seeking both quantitative and qualitative responses was developed. The purpose was to investigate the extent to which perceived utility, perceived enjoyment, perceived ease of use, social influences and user preferences could influence users' behavioral intention to adopt SpamCombat. Questionnaires were selected given that they constitute a popular data collection instrument in technology acceptance research, particularly if statistical strengths of factors influencing intention to adopt need to be established [33].

The questionnaire included three segments. The first contained questions related to demographics, while the second covered 25 items. Items 1 to 4 measured perceived utility, items 5 to 9 measured perceived enjoyment, items 10 to 12 measured perceived ease of use, items 13 to 18 measured social influences, and items 19 to 22 measured users' preferences for SpamCombat. Finally, items 23 to 25 were meant to measure the dependent variable, behavioral intention to adopt SpamCombat. The third segment of the questionnaire was intended to complement the numerical responses with open-ended comments. Specifically, participants were asked to comment on features that they liked and disliked about SpamCombat.

A total of 120 graduate students (66 males and 54 females) with an average age of 28 years participated in the study. Participation was voluntary and anonymous. Of all the participants, 97 were working professionals who mostly hailed from the IT industry. The rest were enrolled in full-time graduate studies. All participants were active Internet users. Among them, 48 reported to play online games regularly while 112 actively shared and browsed content in social media platforms such as Facebook, MySpace and YouTube.

The data collection was conducted in two stages. In the first, stage, the prototype of SpamCombat was introduced to the participants. To help them get acquainted with the way SpamCombat is used, the four prominent features, namely, (1) flagging spam, (2) creating assignments, (3) building profiles, and (4) leader boards (as described in

Section 3), were extensively demonstrated. In the second stage, the questionnaire was administered to the participants to seek their responses.

5   Results

Table 1 shows the descriptive statistics of the five independent variables, namely, perceived utility (PU), perceived enjoyment (PE), perceived ease of use (PEU), social influence (SI), user preference (UP) and the dependent variable, namely, behavioral intention to adopt (BI). The Cronbach's Alpha for all variables were greater than 0.7, suggesting acceptable levels of internal consistency reliability.

Table 1. Descriptive statistics and Cronbach's alpha

Variables (N = 120)	Mean ± SD	Cronbach's alpha
Perceived Utility (PU)	3.35 ± 0.66	0.77
Perceived Enjoyment (PE)	3.37 ± 0.71	0.79
Perceived Ease of Use (PEU)	3.18 ± 0.72	0.85
Social Influence (SI)	3.25 ± 0.62	0.76
User Preference (UP)	3.32 ± 0.59	0.75
Behavioral Intention to Adopt (BI)	3.11 ± 0.68	0.82

Multiple regression analysis was used to investigate the extent to which the five factors were associated with behavioral intention to adopt SpamCombat. Overall, there exists statistical evidence to support the relationship between the five independent variables, namely, PU, PE, PEU, SI and UP, and the dependent variable, BI [$F(5, 114) = 20.57$, $p < 0.001$]. The combined effects of the independent variables accounted for 57.20 % of the variability in the dependent variable. All the five independent variables were found to significantly influence BI ($p < 0.05$). In particular, the order of strength of association betweeen the independent variables and the dependent variable is as follows: UP, SI, PEU, PU and PE. The results of the regression analysis are presented in Table 2.

Table 2. Results of the regression analysis (N = 120)

Variables	Coefficients	Std. Error	t-value	Significance
PU	0.23	0.13	1.77	0.03*
PE	0.16	0.26	1.135	0.04*
PEU	0.42	0.22	1.92	0.02*
SI	0.39	0.11	3.63	< 0.001*
UP	0.55	0.12	4.69	< 0.001*

Based on the qualitative responses, most participants appreciated the novelty of SpamCombat in supporting community policing to promote a spam-free cyber space. For instance, participant 35 expressed that the prototype "...*could be highly effective to fight spam*" while participant 30 liked the concept of "...*an active community that collaboratively shares web page links to deal with fake content*". Participant 85 noted that "*SpamCombat offers substantial incentives to continue playing and highlighting spam*". Furthermore, participant 71 remarked that "*Planting bombs and laying treasure are really fun and promote intention to play*".

On the other hand, the most unanimous dislike of SpamCombat was the concern about the time spent in playing. Most participants expressed that SpamCombat cannot be used during busy hours as it hinders smooth surfing of the Internet. For instance, participant 26 related that "... *flagging spam and creating assignments, though interesting, are time crunching and I may not be inclined to use it when I am busy surfing for important information.*"

6   Discussion

Three findings could be gleaned from this paper. First, the behavioral intention to adopt SpamCombat appears generally promising. At a conceptual level, there appears to be support for the proof-of-concept to gamify community policing to combat spam. Consistent with prior research (eg. [34, 35]), performance expectancy and effort expectancy emerged as crucial determinants for behavioral intention to adopt a new technology. Performance expectancy is the extent to which users find an application useful to achieve a desired performance while effort expectancy is the degree to which users find an application engaging to use [36]. While the former is pre-dominantly derived from perceived usefulness, the latter is significantly driven by perceived ease of use [37]. The statistically significant effects of perceived usefulness and perceived ease of use testifies the role of performance expectancy and effort expectancy as significant antecedents for behavioral intention to adopt.

Second, in terms of community policing, most participants saw their role in using SpamCombat to promote a spam-free cyber space for the online communities. Such a social collectivistic attitude could be attributed to users' altruism and commitment to online groups [1, 15]. Altruistic behavior was evident among participants who exhibited the desire to help one another through recommending sites of similar interests. For example, participants 7 and 10 related that SpamCombat encourages users to filter appropriate information and "*instills a sense of community building*". Their commitment to group also surfaced as they appreciated the concept of flagging inappropriate content from the cyber space over a period of time. In particular, participant 105 remarked "*Never expected ordinary users like us to play a role in combating spam...that too collaboratively*".

Third, in terms of gamification, three game characteristics of SpamCombat that were mostly appreciated include rewards, feedback, and challenge. Specifically, participants liked the feature of brownie rewards and felt that it not only encouraged community policing activities but also enticed them to compete with others. Tangible rewards are known to reinforce challenge and excitement within users by seeking adjustments in their gaming behavior [38]. Feedback in the form of profile views and leader boards offered participants a summary of their past accomplishments, thereby allowing them to reflect on their performance. Feedback thus helped users' reduce the discrepancy between their goals and performance [23]. The optimal challenge to compete with others on flagging the most amount of spam, creating and completing the most number of assignments, and earning the most number of brownies seemed to thrill participants. Such a feature fostered users' efforts to improve their performance in the application [39].

7   Conclusion

This paper introduced SpamCombat, a Web application that helps combat spam through gamification of community policing, and investigated users' behavioral intention to adopt such applications. A questionnaire seeking quantitative and qualitative responses was administered to 120 participants. The results indicate that behavioral intention to adopt SpamCombat is generally promising. Most participants appreciated its novelty in supporting community policing to promote a spam-free cyber space. However, participants felt that using SpamCombat could be too time-consuming.

The paper is significant on two counts. First, it illustrates the feasibility of designing Web applications that could gamify community policing, thus providing new venues for Web designers to explore. Second, it redefines online collaboration by introducing pervasive Web applications that blend with any online community and offer a new perspective in community policing to combat spam.

However, a limitation of the paper should be acknowledged. Since SpamCombat was a conceptual prototype, participants did not have the opportunity to use a functional system for evaluation. Future research could replicate the experimental study by allowing a larger cohort of participants to interact with a fully functional system. Future research could also look into the extent to which gamification of community policing is effective in combating spam. By combining gamification and community policing, this paper highlights the potential of such applications in harnessing the synergy of online communities to perform tasks that could even be challenging for computers.

Acknowledgements. This work is supported by Nanyang Technological University Research Grant Number M58060012.

References

1. van den Hooff, B., de Ridder, J.A.: Knowledge sharing in context: The influence of organizational commitment, communication climate and CMC use on knowledge sharing. Journal of Knowledge Management 8(6), 117–130 (2004)
2. Gill, M.: Crime at work: Studies in security and crime prevention. Perpetuity Press, Leicester (1994)

3. Williams, M.: Policing and cybersociety: The maturation of regulation within an online community. Policing and Society: An International Journal of Research and Policy 17(1), 59–82 (2007)
4. Wanas, N., Magdy, A., Ashour, H.: Using automatic keyword extraction to detect off-topic posts in online discussion boards. Paper presented at Proceedings of Content Analysis for Web 2.0, Madrid, Spain (2009)
5. Hsu, C.L., Lu, H.P.: Why do people play on-line game? An extended TAM with social influence and flow experience. Information & Management 41(7), 853–868 (2004)
6. Guo, Y., Barnes, S.: Why people buy virtual items in virtual worlds with real money. The Data Base for Advances in Information Systems 38(4), 70–76 (2007)
7. Von Ahn, L., Dabbish, L.: Designing games with a purpose. Communications of the ACM 51(8), 58–67 (2008)
8. Deterding, S., Sicart, M., Nacke, L., O'Hara, K., Dixon, D.: Gamification: Using game-design elements in non-gaming contexts. In: Proceedings of the 2011 Annual Conference on Human Factors in Computing Systems, pp. 2425–2428. ACM (2011)
9. Prestopnik, N., Crowston, K.: Exploring collective intelligence games with design science: A citizen science design case. Paper presented at the ACM Group Conference (2012)
10. Ntoulas, A., Najork, M., Manasse, M., Fetterly, D.: Detecting spam web pages through content analysis. In: Proceedings of the 15th International Conference on World Wide Web, Edinburgh, Scotland, pp. 83–92 (2006)
11. Ortega, F.J., Macdonald, C., Troyano, J.A., Cruz, F.: Spam detection with a content-based random-walk algorithm. Paper presented at the Proceedings of the 2nd International Workshop on Search and Mining User-generated Contents, Toronto, Canada, pp. 45–52 (2010)
12. Zhou, B., Pei, J.: Link spam target detection using page farms. ACM. ACM Transactions on Knowledge Discovery from Data 3(3), 13–50 (2009)
13. Cheng, Z., Gao, B., Sun, C., Jiang, Y., Liu, T.Y.: Let web spammers expose themselves. In: Proceedings of the 4th ACM International Conference on Web Search and Data Mining, Hong Kong, pp. 525–534 (2011)
14. Jindal, N., Liu, B.: Review spam detection. Paper presented at the Proceedings of the 16th International Conference on World Wide Web, pp. 1189–1190 (2007)
15. Hew, K.F., Hara, N.: Knowledge sharing in online environments: A qualitative case study. Journal of the American Society for Information Science and Technology 58(14), 2310–2324 (2007)
16. Wall, D.S.: Catching cyber criminals: Policing the Internet. International Review of Law, Computers & Technology 12(2), 201–218 (1998)
17. Scheffler, S.: Angels of cyber space: The CyberAngels cyber crime unit is watching over victims of internet crimes. Law Enforcement Technology 33(4), 32–35 (2006)
18. Sauvé, L., Renaud, L., Kaufman, D., Marquis, J.S.: Distinguishing between games and simulations: A systematic review. Educational Technology & Society 10(3), 247–256 (2007)
19. Laine, T.H., Islas, S.C.A., Joy, M., Sutinen, E.: Critical factors for technology integration in game-based pervasive learning spaces. IEEE Transactions on Learning Technologies 3(4), 294–306 (2010)
20. Perry, D., DeMaria, R.: David Perry on game design. Charles River Media, Hingham Massachusetts (1999)
21. Prensky, M.: Digital Game-Based Learning. McGraw-Hill, New York (2001)
22. Garris, R., Ahlers, R., Driskell, J.E.: Games, motivation, and learning: A research and practice model. Simulation & Gaming 33(4), 441–467 (2002)

23. Wagner, E.D.: In support of a functional definition of interaction. American Journal of Distance Education 8(2), 6–29 (1994)
24. Squire, K.: Video games in education. International Journal of Intelligent Simulations and Gaming 2(1), 49–62 (2003)
25. Hinske, S., Lampe, M., Magerkurth, C., Röcker, C.: Classifying pervasive games: on pervasive computing and mixed reality. In: Concepts and Technologies for Pervasive Games - A Reader for Pervasive Gaming Research, vol. 1. Shaker Verlag, Aachen (2007)
26. Lopez-Nicolas, C., Molina-Castillo, F.J., Bouwman, H.: An assessment of advanced mobile services acceptance: Contributions from TAM and diffusion theory models. Information & Management 45(6), 359–364 (2008)
27. Sheppard, B.H., Hartwick, J., Warshaw, P.R.: The theory of reasoned action: A meta-analysis of past research with recommendations for modifications and future research. Journal of Consumer Research 15(3), 325–343 (1988)
28. Ajzen, I.: The theory of planned behavior. Organizational Behavior and Human Decision Processes 50(2), 179–211 (1991)
29. Venkatesh, V., Davis, F.D.: A theoretical extension of the technology acceptance model: Four longitudinal field studies. Management Science 46(2), 186–204 (2000)
30. Davis, F.D.: Perceived usefulness, perceived ease of use, and user acceptance of information technology. MIS Quarterly 13(3), 319–339 (1989)
31. Kim, Y., Oh, S., Lee, H.: What makes people experience flow? Social characteristics of online games. International Journal of Advanced Media & Communication 1(1), 76–92 (2005)
32. Lee, S., Koubeka, R.J.: The Impact of Cognitive Style on User Preference Based on Usability and Aesthetics for Computer-Based Systems. International Journal of Human-Computer Interaction 27(11), 1083–1114 (2011)
33. Wu, J.H., Wang, S.C.: What drives mobile commerce? An empirical evaluation of the revised technology acceptance model. Information & Management 42(5), 719–729 (2005)
34. Chiu, C.M., Wang, E.T.G.: Understanding web-based learning continuance intention: The role of subjective task value. Information & Management 45(3), 194–201 (2008)
35. Nov, O., Ye, C.: Resistance to change and the adoption of digital libraries: An integrative model. Journal of the American Society for Information Science and Technology 60(8), 1702–1708 (2009)
36. Marchewka, J.T., Liu, C., Kostiwa, K.: An application of the UTAUT model for understanding student perceptions using course management software. Communications of the IIMA 7(2), 93–104 (2007)
37. Venkatesh, V., Morris, M.G., Davis, G.B., Davis, F.D.: User acceptance of information technology: Toward a unified view. MIS Quarterly 27(3), 425–478 (2003)
38. Charles, D., Charles, T., McNeill, M., Bustard, D., Black, M.: Game-based feedback for educational multi-user virtual environments. British Journal of Educational Technology 42(4), 638–654 (2011)
39. Squire, K.: Changing the game: What happens when video games enter the classroom. Innovate: Journal of Online Education 1(6) (2005)

Tackling the Correspondence Problem

Closed-Form Solution for Gesture Imitation by a Humanoid's Upper Body

Yasser Mohammad[1,2] and Toyoaki Nishida[1]

[1] Kyoto University
[2] Assiut University

Abstract. Learning from demonstrations (LfD) is receiving more attention recently as an important modality for teaching robots and other agents new skills by untrained users. A successful LfD system must tackle several problems including the decision about what and whom to imitate but, ultimately, it needs to reproduce the skill it learned solving the *how to imitate* problem. One promising approach to solving this problem is using Gaussian Mixture Modeling and Gaussian Mixture Regression for reproduction. Most available systems that utilize this approach rely on kinesthetic teaching or require the attachment of special markers to measure joint angles of the demonstrator. This bypasses the correspondence problem which is accounting for the difference in the kinematic model of the demonstrator and the learner. This paper presents a closed-form analytic solution to the correspondence problem for an upper-body of a humanoid robot that is general enough to be applicable to many available humanoid robots and reports the application of the method to a pose copying task executed by a NAO robot using Kinect recorded data of human demonstrations.

1 Introduction

Service robots are starting to get their place in our everyday lives. This means that these robots will have to interact more with untrained humans who will need to influence the behavior of these robots without knowledge of programming, kinematics or dynamics. Learning from demonstrations is a promising technology for service robots [1] [2] [3] because it allows the robot to acquire new skills without explicit programming.

For a robot to be able to learn from a demonstration, it must solve many problems. Most important of these problems are the following four challenges:

- *Action Segmentation.* How can the learner segment the continuous stream of actions (e.g. motion in the trajectory space) perceived from the demonstrator into discrete *behaviors* (e.g. a tennis serve, opening a door, ...)?
- *Behavior Significance for Imitation.* How to know the interesting behaviors that it should imitate? What of the actions and state components (e.g. pose) of the demonstrator is related to the behavior to be imitated? This encapsulates the *what* and *whom* problems in [4].
- *Correspondence Problem.* How can the sensory input and actions of the demonstrator be mapped to the corresponding spaces of the learner?

T. Yoshida et al. (Eds.): AMT 2013, LNCS 8210, pp. 84–95, 2013.
© Springer International Publishing Switzerland 2013

– *Behavior Generation.* How can the learned behavior be generated by the robot (after being mapped to the sensorimotor space of the learner)?

In previous research we focused on the first two of these challenges [5][6] through the fluid imitation engine [7] which allows the robot to learn from continuous unsegmented streams of human motions. Only navigation tasks were used in these systems because of the simplicity of the correspondence and behavior generation challenges in this domain which allowed us to focus on the segmentation and behavior significance evaluation challenges. In this paper we focus on solving the correspondence problem for the case of an upper-body of a humanoid robot. Humanoids were selected because they provide a high-dimensional complex mechanism to evaluate the scalability of our approach. Humanoids also elicit anthropomorphizing reactions that may help users understand their behavior more naturally. The reason that we focus on upper-body humanoids rather than full-body cases is that it allows us to convert the problem to a single frame pose correspondence problem which is possible to solve in closed form. For a full humanoid, it is possible to do the same but the resulting poses may not be stable and constraints related to the center-of-mass calculations and dynamics of motion must be taken into account which is outside the scope of this paper.

The main focus of this paper is on the solution to the correspondence problem and for this reason, we use a copycat task in which behavior generation is direct replay of the recorded trajectory of the demonstrator. Nevertheless, a more complex behavior generator like the one proposed in [8] can be utilized.

Nehaniv and Dautenhahn [9] provided an early work in the correspondence problem in which it was defined as a relational homomorphism between the transformational semigroups of the automata representing the demonstrator and the learner. One important aspect of this formalization for our purposes is the explicit role of the observer in the definition of correspondence which is a feature inherited in the proposed definition of the problem. A drawback of this formalization is that it requires the specification of equivalences between combined agent-environment states to induce the transformational semigroup. While finding these equivalence classes is easy for navigation tasks (the source of several examples in [9]) and in purely environment-directed imitation (e.g. filling a cup), it is hard (and may not even be possible) in more complex manipulation tasks and even purely kinematic tasks like learning sign language or reproduction of gestures using LfD.

A related but different problem that is known for long time in robotics is the inverse kinematics problem. In this case, the input to the system is the homogeneous transformation of the end effector and its output is one (or more) vectors of joint variables (all angles in the case of humanoids) that can put the end effector (extremity) in the required transformation relative to the base frame (which is usually the torso of the robot). Inverse Kinematics solvers do not in general constrain the location of any points of the robot body other than the end effector while searching for a solution and this makes them inadequate for gesture imitation.

Asfour and Dillmann [10] proposed a solution to the correspondence problem for the upper-body of a humanoid robot called ARMAR. Their solution exploits the fact that the robot has a redundant 7DoFs arm. The formulation of the problem was based on the decomposition of the workspace of the arm and on the analytical description of the redundancy of the arm. The solution obtained was characterized by its accuracy and low cost of computation. The algorithm targeted manipulation tasks and for this reason it allowed the elbow position to vary from the demonstrated position. While this is a useful use of redundancy in manipulation tasks, in gesture imitation tasks it is not possible because the elbow position is an important constraint over allowed solutions.

Pitt et al. [11] proposed a solution to the inverse kinematics problem based on conformal geometry and automatic code generation utilizing symbolic computation. The proposed approach is general and can be effective for solving the inverse kinematics problem for humanoids. The disadvantage of this approach for our purposes is that – again– it does not constraint the elbow position which is crucial in effective gesture imitation.

Recently, Nunez et al. [12] proposed a solution to the inverse kinematics problem for a Bioloid humanoid robot. The focus of this work was on the lower limbs of the robot not the upper body. The solution proposed for the upper-body uses the rotational parts of the transformation matrix for the extremities and does not constrain the elbow joint. For these reason, it is difficult to apply it to gesture imitation.

Ali et al. [13] provided a closed form solution for the inverse kinematics problem that applies to several humanoid robots. This solution requires the knowledge of the position and orientation of the robot extremity which is the normal problem of all inverse kinematics solvers for gesture imitation. It also assumes that the elbow is a 1-DoF joint while the shoulder is spherical. While this arrangement is common in several humanoids, it is not the case for many others (e.g. NAO which is increasingly used in humanoid research and Robovie R3, ...). The proposed solution can be used with both types of humanoids with minor modifications.

Tee et al. [14] proposed a modular approach for solving the correspondence problem numerically by decomposing the upper-body into a set of 2-DoF and 3-DoF components and solving them recursively. The main advantage of this approach is the inclusion of a term that is automatically activated near singular configurations to increase the robustness to these configurations. The proposed approach on the other hand uses a closed-form solution and deals with singular configurations as special cases within the same symbolic framework.

The main contributions of this paper are an observer-aware definition of the correspondence problem suitable for both goal-directed and trajectory-directed LfD and an exact solution of the problem for an upper-body humanoid robot for trajectory directed LfD. The proposed solution is – by the definition of the problem – robot specific but modifying it to other humanoids is a matter of changing few well-defined parameters rather than complete rethinking of the modeling process as in geometric solutions to the problem [15]. It is also a closed-form solution in contrast to numerical approaches as in [14] and Jacobian pseudo-inverse [16].

The rest of the paper is organized as follows: Section 2 defines the correspondence problem and section 3 details the proposed solution for upper-body humanoids. Section 4 provides a proof-of-concept evaluation experiment to show the applicability of the proposed approach and section 5 concludes the paper.

2 The Correspondence Problem

Argall et al. [17] define the correspondence problem as:

Identification of a mapping between the teacher and the learner that allows the transfer of information from one to the other.

This problem can be divided into three subproblems: observational mapping, external mapping and learner's mapping. Assuming that the demonstrator's state is a D dimensional vector, the observed signal is O dimensional while the learner's state is L dimensional we can define these mappings as follows: Observational mapping ($f_d^o : \mathbb{R}^D \to \mathbb{R}^O$) refers to the mapping between the demonstrator's state and observed/recorded data (record mapping in [17]). External mapping ($f_o^e : \mathbb{R}^O \to \mathbb{R}^E$) refers to the mapping between observed demonstrator's behavior and the E dimensional corresponding externally expected behavior of the learner. Learner's mapping ($f_l^e : \mathbb{R}^L \to \mathbb{R}^E$) refers to the mapping between the learner's state and corresponding externally observable learner's behavior. These three mappings are *forward* mappings in the sense that they are usually one-to-one mappings and relatively easy to find in closed form using kinematics analysis. We define f_x^{y-1} as the inverse mapping corresponding to f_x^y and it can be a one-to-many mapping and is usually relatively difficult to find in closed form.

The main difference between this decomposition of the problem and the decomposition proposed in [17] is that we assume that the correspondence happen between externally observable behaviors of the demonstrator and the learner rather than between their states which allows us to represent the agent states independent on the task and helps in simplifying goal-directed LfD. The difference between these two definitions of the problem may not be clear in many situations (e.g. when the observational and learner's mappings f_e^l and f_d^o are trivial) but consider the task of learning to use a spoon to put sugar into a cup using learning by demonstration. Using our terminology, we define the state of the demonstrator and learner by their joint angles which is the most natural for control and the observable behavior by the relative homogeneous transformation from the spoon to the cup which is the most natural representation of the task itself on the observational level. On the other hand, given the record and embodiment mapping decompositions of [17], the *state* of the demonstrator and the learner must contain a reference to the external transformation between the spoon and the cup which is task specific.

Another related definition of the correspondence problem can be found in [9]. One problem of this approach is that the state of the system combines both agent's and environmental state which is applicable and may be even desirable for environmental change directed imitation (called effect imitation in [9]) but it complicates the control problem for trajectory imitation which is the target of this paper.

Using the aforementioned mapping definitions, we can now define the correspondence problem more concretely as:

Definition 1 (Correspondence Problem). *Given an external mapping f_o^e, find the external correspondence mapping f_o^l which is defined as:*

$$f_o^l = f_e^l \bigodot f_o^e = f_l^{e-1} \bigodot f_o^e \qquad (1)$$

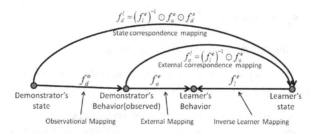

Fig. 1. Commutative Diagram of the mapping involved in the proposed definition of the correspondence problem

where $\odot$ is the mapping concatenation operator defined such that applying the mapping $x \odot y$ corresponds to applying the mapping y followed by the mapping x. In many cases this operator will correspond to a matrix multiplication operation but the definition is general enough to encapsulate more complex mapping concatenation operations including nonlinear ones.

Once this mapping is found and given the demonstrator's observational mapping (f_d^o) it is trivial to find the *state correspondence mapping* f_d^l which is defined as $f_d^l = f_o^l \odot f_d^o$. In the remaining of this paper we will use the terms *correspondence mapping* to refer to both external and state correspondence mappings. This state correspondence mapping is what was referred to as the embodiment mapping in [17]. Fig. 1 shows a commutative diagram of the proposed decomposition of the correspondence problem into the aforementioned four basic mappings.

As defined, the correspondence problem is robot dependent because it depends on the learner's mapping f_l^e which is necessarily robot dependent. For this reason, there can be no general solution to this problem without further constraints. In this paper we provide a closed-form solution for finding f_l^e that is general enough to be applicable with only parameter change to most upper-body humanoid robots.

Another feature of the correspondence problem is its dependence on the mapping f_o^e which is necessarily dependent on the sensors used for capturing the demonstrations (e.g. Mocap sensors, cameras, depth cameras, accelerometers, etc). The decomposition of the mapping into separated f_l^e and f_o^e helps alleviating this problem as f_o^e can be calculated for each of these types of sensors. By combining f_l^e of various robots and f_o^e of various sensors, it is possible to build a modular library that is the next best thing to a general solution to the problem.

From dentition 2, it is clear that the solution of the correspondence problem does not explicitly depend on the observational mapping (f_d^o) which means that the form factor of the demonstrator need not be modeled. This – slightly nonintuitive – feature of the proposed definition makes sense because the act of observing the action of the demonstrator (through the observational mapping f_d^o) abstracts away any aspects of the demonstrator's state that is not relevant to observed quantities but in the same time translates the relevant parts of this state into what the learner perceives as an observed behavior.

3 Proposed Solution

In section 1, the correspondence problem was decomposed into two subproblems: calculation of the external mapping and the learner's mapping and it was shown that this decomposition helps separating the aspects of the correspondence problem related to the sensing modality of the demonstrations from aspects related to the form factor of the learner. The following two subsections provide details of the proposed approach to solve these two subproblems. Because we use homogenous transformations for the rest of this paper, the mapping concatenation operation $\odot$ is simply a matrix multiplication for the forward mappings and a set of equations for the inverse mappings that are usually nonlinear as will be shown in section 3.2.

3.1 External Mapping

The external mapping (f_o^e) depends on the kind of sensors used to capture the demonstration – as this determines the observational space – and the observer's notion of equivalence between agent and environmental states. For this reason, no general solution to this problem can be given.

One possible external mapping is to map the effects on the environment of the demonstrator's actions to actions of the robot that achieve the same effects. This is best captured by the homomorphism formulation of the transformational groups proposed in [9]. This kind of mapping can be of use in effect imitation. In this paper we limit ourselves to low level pose/trajectory mappings in which the pose of the human demonstrator at every single time-step is mapped to a corresponding actuation command of the learner robot. This low level mapping is well-posed enough in the case of upper-humanoids that it can be calculated generally in closed form and is useful in cases of trajectory imitation (e.g. gesture imitation).

For the case of upper-humanoids, a very simple – and usually effective – approach is to consider a robot's pose equivalent to a human pose if –and only if– the relative positions and orientations of the head, shoulders, elbows, and extremities (hands/wrists) in some fixed coordinate system are the same as those for the human demonstrator up to a scaling factor of the upper and lower arm links. This pose equivalence is useful in cases of trajectory imitation (e.g. imitating gaze behavior, gesture learning, sign-language learning, teleoperation tasks , etc). One advantage of this mapping is that its input (positions of different joints) is currently easily calculated from a variety of sensor types (e.g. Mocap sensors like phaseSpace [18], Kinect sensors, D-cameras, Laser scanners, accelerometers through double integration etc).

In the most limited case, only relative positions are available or both positions and link orientations are available. Because of the kinematic constraints of the arm chains, a single orientation (e.g. of the extremity) is enough to define the mapping completely. If no such orientation is available, then it can easily be proven that at most 4 DoFs of each arm can be calculated fixing completely the orientation of the first limb (upper-arm) as well as the position of the extremity of each arm but without constraining the orientation of the extremity. In the pose matching framework employed in this paper, orientations relative to the robot's torso transform without any change implying that f_o^e is unity for them. We focus now on position data.

In the rest of this paper t, s, e, w,h stand for torso, shoulder, elbow, wrist and hand and adding a subscript l or r stands for left or right joint. A_i^r refers to the position of point i in frame r, T_i^j is the homogenous transformation which transforms any point from frame i to frame j, and R_i^j is the rotation matrix that transform any point from frame i to frame j. For the rest of this section, we use $\hat{x}$ to represent the frame on the learner's body corresponding to the frame x on the demonstrator's body. It is well known that the columns of R_i^j are the direction cosines of the three coordinate axes of frame i represented as vectors in frame j. It is also well known that:

$$T_i^j = \begin{bmatrix} R_i^j & O_i^j \\ 000 & 1 \end{bmatrix} \text{ and } T_i^{j-1} = \begin{bmatrix} R_i^{j-1} & -R_i^{j-1}O_i^j \\ 000 & 1 \end{bmatrix}$$

We assume that the positions of the shoulders (A_{ls}^g, A_{rs}^g), elbows (A_{le}^g, A_{re}^g), and hands/ wrists (A_{lh}^g, A_{rh}^g) are measured in some global frame of reference G. We also assume that the homogeneous transformation of the demonstrator's torso in the same reference frame is give (T_t^g). Because left and right arms are treated exactly the same, we will drop the l and r subscript letters for the rest of this section. Robot's upper and lower arm lengths (U and L) and the location of left and right shoulders in the robot's torso frame $A_s^{\hat{t}}$ are assumed fixed and known which is always the case for humanoids made of noncompliant materials and even for most mechanically achievable compliant designs.

The first step to find the required mapping is to project all points to the torso frame of the demonstrator using $A_i^t = T_t^{g-1} A_i^g$ applied to the eight input positions. We then calculate the required mapping using:

$$A_e^{\hat{i}} = A_s^{\hat{i}} + \frac{U}{\|A_e^t - A_s^t\|} \left(A_e^t - A_s^t\right) \text{ and } A_h^{\hat{i}} = A_e^{\hat{i}} + \frac{L}{\|A_h^t - A_e^t\|} \left(A_h^t - A_e^t\right)$$

If this is the only available information then we can only calculate 4 DoFs for each arm of the robot because $A_e^{\hat{s}}$ and $A_h^{\hat{e}}$ each has one length constraint.

If any rotation matrix is known in the global frame of reference g, we transform it to the torso frame of reference using: $R_i^t = T_t^{g-1} R_i^g$ and then it transforms without change to the learners frame of reference ($R_i^{\hat{t}} = R_i^t$). The mapping f_o^e and its inverse are now uniquely determined.

This simple linear transformation method takes into account the kinematic constraints of the learner except joint ranges and maximum speeds which will be dealt with in the learner's mapping and behavior generation respectively.

3.2 Learner's Mapping

The learner's mapping (f_l^e) is a forward mapping that takes as input the state of the learner (joint angles of the upper body in the case of an upper humanoid) and returns the external behavior that is to be corresponded to the observed behavior of the demonstrator. In order to solve the correspondence problem, the inverse learner's mapping (f_l^{e-1}) is what needs to be calculated (see definition 2).

Fig. 2(a) shows the kinematic chains of a general 23 DoFs humanoid robot and the torso frame orientation used throughout this paper. Solving the inverse kinematic for

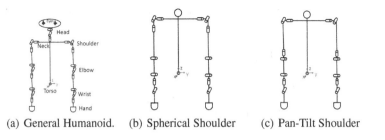

(a) General Humanoid.   (b) Spherical Shoulder   (c) Pan-Tilt Shoulder

Fig. 2. The Kinematic chain of the upper body of humanoid robots for which the learner's mapping is designed

the general case humanoid robot is complex and several simplifying assumptions must be made to arrive at a closed form solution. The head and eyes can each be modeled by a pure orientation matrix and so their inverse problem is an inverse orientation problem that can be solved in closed form using the equations in [19]. Due to lack of space this solution is not repeated here.

The arm has 9 DoFs in the general chain but most robots use one of the two arrangements presented in Fig. 2(b) or Fig.2(c). The final spherical joint at the wrist does not affect the position of the elbow or the wrist. This allows us to decompose the problem into two problems. One inverse orientation problem for the wrist that can be solved as in [19] if the final hand orientation in the wrist coordinate frame is known and another inverse kinematic problem for the shoulder and elbow joints. Assuming that we could find the first 6 DoFs, we can use forward kinematics to find the homogeneous transformation of the wrist assuming zero values for its 3 DoFs (T_w^s). Using this information and the hand transformation found using the external mapping (T_h^t), the rotation achieved by the 3DoFs of the wrist can be found as: $R_w^h = R_h^{t\,-1} R_s^t R_w^s$. With all of these details out of the way, we can concentrate on the shoulder and the elbow joints to complete the solution.

From this point on we will drop using $\hat{i}$ to refer to learner's joints and will use i instead and will drop the distinction between right and left arms because of their symmetry.

Fig. 2(b) and Fig. 2(c) give the two mechanisms used by most humanoid robots that we call the *spherical shoulder* and the *pan-tilt shoulder* configurations. The only difference between these two configurations is that the third joint is located differently in the two arrangements (i.e. in the shoulder in spherical shoulder configuration and in the elbow in pan-tilt configuration). For example, NAO and Robovie R3 robots use the pan-tilt configuration while ASIMO, HRP2 and HOAP2 use the spherical configuration. We exploit the similarity of these two designs in the proposed solution by focusing our attention on the pan-tilt configuration and encoding the difference in a transformation matrix that changes the location of the third joint to the shoulder position for the case of spherical shoulder.

The parameters of our system are: robot's upper and lower arm lengths (U and L) and the location of left and right shoulders in the robot's torso frame A_s^t, the transformation from the end of the upper-arm to the beginning of the lower arm T_u^l and the final

Table 1. The DH parameters of the full arm, upper arm and lower arm for a pan-tilt shoulder type humanoid (e.g. NAO)

Joint	1_F	2_F	3_F	4_F	1_U	2_U	1_L	2_L
θ_i	θ_1^F	θ_2^F	θ_3^F	θ_4^F	θ_1^U	θ_2^U	θ_1^L	θ_2^L
d_i	0	0	U	0	0	0	0	0
α_i	$\pi/2$	$\pi/2$	$-\pi/2$	0	$\pi/2$	0	$-\pi/2$	0
a_i	0	0	0	L	0	U	0	L

transformation of the robot's end effector in the wrist frame (T_{end}^w). We use the standard Denavit-Hartenberg convention for specifying kinematic parameters.

The first step is to find the DH parameters of the upper-arm, lower-arm and full-arm kinematic chain. Table 1 gives the DH parameters of the pan-tilt type robot. The second step is to find the forward transformation of the Upper-arm, Lower-arm and Full-arm of the robot called $^U T_e^s$, $^L T_w^e$, and $^F T_w^s$ using the DH parameters. The forward learner's mapping is then found as $f_l^e = {}^F T_w^s$

Because the number of DoFs encoded in the elbow position are the same as the number of DoFs of the upper arm robot, it is possible to find a closed form solution for the inverse kinematics of this robot. This leads to the following equations for the first two angles:

$$\theta_2^U = \sin^{-1}\left(\frac{X_e^t(y) - A_s^t(y)}{U}\right) \tag{2}$$

If θ_2 was zero then the robot is in a singular position and the value of θ_1 does not affect the position of the elbow. In this case we temporarily set θ_1 to zero and may revise it later based on the orientation of the wrist. Otherwise, the first angle can be calculated as:

$$\theta_1^U = \cos^{-1}\left(\frac{X_e^t(z) - A_s^t(z)}{U \sin \theta_1^U}\right) = \cos^{-1}\left(\frac{X_e^t(x) - A_s^t(x)}{U \cos \theta_1^U}\right) \tag{3}$$

Because there are two ways to find this angle, we select the most numerically robust of them by using the first equality if $\sin \theta_1^U$ was greater that $1/\sqrt{2}$ otherwise the second equality is used.

Once we have θ_1^U and θ_2^U, we use the forward transform of the upper arm to get the position of the wrist in the elbow's coordinate frame using :$^U A_w^e = {}^U T_e^{s-1} A_w^s$. This forward kinematic step simplifies the problem of finding the final two angles tremendously.

The next step is to use the transformation from the upper to the lower arm in order to find the position of the wrist in the elbow coordinate of the lower arm: $^L A_w^e = T_u^{lU} A_w^e$.

Finally the lower arm robot's forward kinematics transformation ($^L T_w^e$) is analyzed to find the relation between the position of the wrist and the final two angles which leads to two possible cases: If the final hand transformation was only rotational (i.e. $T_h^w(1:3,4) = 0$) the following equations can be used to recover the final two angles:

$$\theta_2^L = \tan^{-1}\left(-\frac{A_h^e(x)}{\sqrt{A_h^e(y)^2 + A_h^e(z)^2}}\right) \tag{4}$$

$$\theta_1^L = \cos^{-1}\left(\frac{A_h^e(z)}{L\sin\theta_1^L}\right) = \cos^{-1}\left(\frac{A_h^e(y)}{L\cos\theta_1^L}\right) \tag{5}$$

If the final hand transformation had a translational component, the situation becomes slightly more complicated as we will have two solutions as shown in equation 6. In this case, we must rely on the human constraint on the elbow joint to find a unique solution.

$$\theta_2^L = \pm\sin^{-1}\left(-\frac{A_h^e(x)}{L}\right) \tag{6}$$

The final angle θ_3^L is found by solving the two equations in two variables resulting from substituting θ_4^L into $LT_w^e(2:3,4)$ resulting on the following equation for the case when $A_h^e(z) \neq 0$ (similar equations can be used when other translations exist in T_h^w):

$$\theta_1^L = \tan^{-1}\left(\frac{\left(A_h^e(z) - Lc_1\sin\theta_2^L\right)/A_h^w(z)}{c_1}\right) \tag{7}$$

where $c_1 = (A_h^e(y)L\sin\theta_2^L - A_h^e(z)A_h^w(z))/(A_h^w(z)^2 + (L\sin\theta_2^L)^2)$.

The final step is to find angels of the full arm using the found angles for the upper and lower arm for which the following identities can be proven trivially: $\theta_1^F = \theta_1^U$, $\theta_2^F = \theta_2^U$, $\theta_3^F = \theta_1^L$, and $\theta_4^F = \theta_2^L$. Using equations 2 to 7 and these identities, the complete inverse learner's mapping $f_l^{e^{-1}}$ is now found. Notice that the process described here is to be carried out only once.

4 Evaluation

NAO is a humanoid robot designed by Aldebaran. We use a version 3.2 robot which has 15.9 offset in the Z direction of each hand. The parameters of our system were set from the robot's manual as:

$$\begin{bmatrix} U \\ L \end{bmatrix} = \begin{bmatrix} 100 \\ 108.55 \end{bmatrix}, A_{ls}^t = \begin{bmatrix} 0 \\ 98 \\ 100 \end{bmatrix}, A_{rs}^t = \begin{bmatrix} 0 \\ -98 \\ 100 \end{bmatrix}, T_{end}^h = \begin{bmatrix} 1 & 0 & 0 & 0 \\ 0 & 1 & 0 & 0 \\ 0 & 0 & 1 & -15.9 \\ 0 & 0 & 0 & 1 \end{bmatrix}, T_l^u = \begin{bmatrix} 0 & 0 & 1 & 0 \\ 1 & 0 & 0 & 0 \\ 0 & 1 & 0 & 0 \\ 0 & 0 & 0 & 1 \end{bmatrix}$$

Fig. 3. Example pose imitations from the copycat task experiment

To evaluate the proposed approach, we use a copycat task in which a real NAO robot tries to copy the pose of a human subject captured through a single Kinect sensor using the skeleton tracking capability of Microsoft Kinect SDK. We only pass position data to the system which applies the external mapping described in section 3.1 followed by the inverse learner's mapping as described in section 3.2. No calibration steps are needed for the system and the subject can just step in the sensing area of the Kinect. Fig. 3 shows three shots of the experiment. The first shot shows both arms in two different singular positions. Nevertheless, the system was able to accurately copy the pose.

To provide a quantitative measure of the quality of the proposed system, we compared it with the inverse kinematics solution provided in the NAOqi library (the official SDK of NAO) on 3456 frames collected by moving the hands randomly in front of a Kinect. The error was calculated as:

$$error = \frac{1}{4} \left(\frac{\left\| N_{lh}^t - f_o^e \left(H_{lh}^t \right) \right\|}{2(U+L)} + \frac{\left\| N_{le}^t - f_o^e \left(H_{le}^t \right) \right\|}{2L} + \frac{\left\| N_{rh}^t - f_o^e \left(H_{rh}^t \right) \right\|}{2(U+L)} + \frac{\left\| N_{re}^t - f_o^e \left(H_{re}^t \right) \right\|}{2L} \right)$$

This error term calculates the observable difference between the elbow and hand joint positions and the corresponding positions of the demonstration in the robot's space. To remove the dependence on upper and lower arm lengths, we normalize elbow error by elbow length and hand error by hand length, then average over left and right arms. The maximum possible value for this error measure is one. Because the IK solver can sometimes provide multiple solutions, we use the one with least error in the evaluation. When the IK solver fails to find a solution (which happens frequently due to errors in localization of the hand) we penalize it by an error equal to the robot's full arm length. The average error value for the IK solver was 0.412 (std. dev. 0.24) while it was only 0.01 (std. dev. 0.003) for the proposed system.

5 Conclusions

This paper presents a solution of the correspondence problem for trajectory imitation (e.g. gesture imitation) for upper-body humanoid robots. The proposed solution is closed-form and takes into account singularities of the kinematic chains. Because the system uses only position data, it can be used with most motion-capture devices. The proposed system can be applied to many available humanoid robots with small or no modifications after setting the kinematic parameters of the robot like upper and lower arm lengths and position of the torso frame to be considered the base of all kinematic chains. The proposed method was evaluated in a copycat task using Kinect records of human motion imitated by a real NAO robot and its performance was compared quantitatively to the IK solver of NAOqi and was found to result in lower errors in the locations of arm joints.

References

1. Aleotti, J., Caselli, S.: Grasp programming by demonstration: A task-based quality measure. In: Robot and Human Interactive Communication, RO-MAN 2008, pp. 383–388 (2008)
2. Argall, B.D., Chernova, S., Veloso, M., Browning, B.: A survey of robot learning from demonstration. Robotics and Autonomous Systems 57(5), 1–15 (2009)

3. Abbeel, P., Coates, A., Ng, A.Y.: Autonomous Helicopter Aerobatics through Apprenticeship Learning. The International Journal of Robotics Research 29(13), 1608–1639 (2010)
4. Nehaniv, C., Dautenhahn, K.: Mapping between dissimilar bodies: Affordances and the algebraic foundations of imitation. In: Demiris, J., Birk, A. (eds.) Proceedings European Workshop on Learning Robots 1998, EWLR-7 (1998)
5. Mohammad, Y., Nishida, T., Okada, S.: Unsupervised simultaneous learning of gestures, actions and their associations for human-robot interaction. In: Intelligent Robots and Systems, IROS, pp. 2537–2544. IEEE (2009)
6. Mohammad, Y., Nishida, T.: Learning interaction protocols using augmented baysian networks applied to guided navigation. In: Intelligent Robots and Systems, IROS, pp. 4119–4126. IEEE (2010)
7. Mohammad, Y., Nishida, T.: Fluid imitation: Discovering what to imitate. International Journal of Social Robotics 4(4), 369–382 (2012)
8. Calinon, S., Guenter, F., Billard, A.: On learning, representing, and generalizing a task in a humanoid robot. IEEE Transactions on Systems, Man, and Cybernetics. Part B, Cybernetics 37(2), 286–298 (2007)
9. Nehaniv, C., Dautenhahn, K.: Mapping between dissimilar bodies: A ordances and the algebraic foundations of imitation. In: EWLR 1998, pp. 64–72 (1998)
10. Asfour, T., Dillmann, R.: Human-like motion of a humanoid robot arm based on a closed-form solution of the inverse kinematics problem. In: Intelligent Robots and Systems, IROS, pp. 1407–1412. IEEE (2003)
11. Pitt, J., Hildenbrand, D., Stelzer, M., Koch, A.: Inverse kinematics of a humanoid robot based on conformal geometric algebra using optimized code generation. In: IEEE Humanoids 2008, pp. 681–686 (2008)
12. Nunez, J.V., Briseno, A., Rodriguez, D.A., Ibarra, J.M., Rodriguez, V.M.: Explicit analytic solution for inverse l.kinematics of bioloid humanoid robot. In: IEEE-Brazilian Robotics Symposium and Latin American Robotics Symposium, SBR-LARS, pp. 33–38 (2012)
13. Ali, M.A., Park, H.A., Lee, C.G.: Closed-form inverse kinematic joint solution for humanoid robots. In: Intelligent Robots and Systems, IROS, pp. 704–709. IEEE (2010)
14. Tee, K.P., Yan, R., Chua, Y., Huang, Z.: Singularity-robust modular inverse kinematics for robotic gesture imitation. In: 2010 IEEE International Conference on Robotics and Biomimetics, ROBIO, pp. 920–925. IEEE (2010)
15. Zannatha, J.I., Limón, R.C.: Forward and inverse kinematics for a small-sized humanoid robot. In: International Conference on Electrical, Communications, and Computers, CONIELECOMP 2009, pp. 111–118. IEEE (2009)
16. Wang, J., Li, Y.: Inverse kinematics analysis for the arm of a mobile humanoid robot based on the closed-loop algorithm. In: International Conference on Information and Automation, ICIA 2009, pp. 516–521. IEEE (2009)
17. Argall, B.D., Chernova, S., Veloso, M., Browning, B.: A survey of robot learning from demonstration. Robotics and Autonomous Systems 57(5), 469–483 (2009)
18. http://www.phasespace.com
19. Paul, R.P., Shimano, B., Mayer, G.E.: Kinematic control equations for simple manipulators. IEEE Transactions on Systems, Man, and Cybernetics 11, 449–455 (1981)

Learning and Utilizing a Pool of Features in Non-negative Matrix Factorization

Tetsuya Yoshida

Graduate School of Information Science and Technology,
Hokkaido University
N-14 W-9, Sapporo 060-0814, Japan
yoshida@meme.hokudai.ac.jp

Abstract. Learning and utilizing a pool of features for a given data is important to achieve better performance in data analysis. Since many real world data can be represented as a non-negative data matrix, Non-negative Matrix Factorization (NMF) has recently become popular to deal with data under the non-negativity constraint. However, when the number of features is increased, the constraint imposed on the features can hinder the effective utilization of the learned representation. We conduct extensive experiments to investigate the effectiveness of several state-of-the-art NMF algorithms for learning and utilizing a pool of features over document datasets. Experimental results revealed that coping with the non-orthogonality of features is crucial to achieve a stable performance for exploiting a large number of features in NMF.

1 Introduction

Since many real world data can be represented as a non-negative data matrix, various research efforts have been conducted on Non-negative Matrix Factorization (NMF) [7,8,12,2,1]. Different from eigenvalue analysis such as PCA (Principal Component Analysis), localized features, which are not necessarily orthogonal to each other, can be learned by NMF thanks to the non-negativity constraint. NMF has been successfully used in image analysis [7,8]. In addition, since documents are usually represented in terms of the non-negative frequencies of terms, NMF can also be used for document analysis [12].

Since another data representation of instances can be learned by NMF, it can be regarded as a representation learning method [6]. The objective of representation learning is to learn a pool of feature from data so that the performance in data analysis can be improved by exploiting the learned features [9]. However, when the number of features is increased, the constraint imposed on the features can hinder the effective utilization of the learned representation [9].

In order to exploit the representation learned by NMF, we investigate three kinds of approaches in this paper: preprocessing of data [12,6], incorporation of additional constraints [2,1], and post-processing of the learned representation [13]. Extensive experiments are conducted to evaluate their performance over document datasets. Experimental results revealed that preprocessing of

T. Yoshida et al. (Eds.): AMT 2013, LNCS 8210, pp. 96–105, 2013.
© Springer International Publishing Switzerland 2013

data can lead to better performance when the number of features is small. However, the performance may rapidly degrade as the number of features increases. On the other hand, even when a large number of features is learned and used, coping with the non-orthogonality of features through additional constraints or post-processing can lead to achieving a stable performance. Furthermore, it is revealed that exploiting the metric of the feature space with post-processing is effective for performance improvement.

Notation. The symbols used in this paper are summarized in Table 1. We assume that data instances are originally represented in a p-dimensional standard Euclidian space. A bold normal uppercase letter is used for a matrix, and a bold italic lowercase letter for a vector. For a matrix $\mathbf{X}$, $\mathbf{X}_{ij}$ stands for an element in $\mathbf{X}$, tr$(\mathbf{X})$ stands for the trace of $\mathbf{X}$, and $\mathbf{X}^T$ stands for its transposition. For a non-singular matrix $\mathbf{A}$, its inverse matrix is represented as $\mathbf{A}^{-1}$.

Table 1. Notation

p	number of attributes
q	number of features
n	number of instances
k	number of clusters

Oraganization. Section 2 explains the problem setting of NMF and several NMF algorithms. Section 3 reports the evaluation of the algorithms over document datasets. Section 4 discusses the results and related work. Section 5 summarizes our contributions and suggests future directions.

2   Non-negative Matrix Factorization

Non-negative Matrix Factorization (NMF) [8] factorizes a non-negative matrix $\mathbf{X} = [\boldsymbol{x}_1, \cdots, \boldsymbol{x}_n] \in \mathbb{R}_+^{p \times n}$ into a product of two non-negative matrices $\mathbf{U} = [\boldsymbol{u}_1, \cdots, \boldsymbol{u}_q] \in \mathbb{R}_+^{p \times q}$ and $\mathbf{V} = [\boldsymbol{v}_1, \cdots, \boldsymbol{v}_n] \in \mathbb{R}_+^{q \times n}$ as:

$$\mathbf{X} \simeq \mathbf{U}\mathbf{V} \tag{1}$$

Minimization of the following objective function is conducted to obtain the matrices $\mathbf{U}$ and $\mathbf{V}$:

$$J_0 = ||\mathbf{X} - \mathbf{U}\mathbf{V}||^2 \tag{2}$$

where $||\cdot||$ stands for a norm of a matrix. We focus on Frobenius norm $||\cdot||_F$ [8] in this paper.

2.1   Representation Learning with NMF

In addition to image analysis [7,4], NMF has also been used for document analysis, especially for document clustering [12,2]. In previous approaches, the number of features are set to the number of clusters, and the matrix $\mathbf{V}$ is interpreted as a cluster indicator matrix [12,2].

By approximiating a data matrix $\mathbf{X}$ as a product $\mathbf{U}\mathbf{V}$ in eq.(2), each data $\boldsymbol{x}_i$ (i-th column of $\mathbf{X}$) is approximated as a linear combination of columns of $\mathbf{U}$ as $\boldsymbol{x}_i \simeq \sum_{\ell=1}^{q} \mathbf{V}_{\ell i} \boldsymbol{u}_\ell$. This means that, each column $\boldsymbol{u}_\ell$ plays the role of a learned feature, and i-th column $\boldsymbol{v}_i$ of $\mathbf{V}$ playes the role of coefficients for the learned features to represent the instance $\boldsymbol{x}_i$. Thus, in addition to the above interpretation, the matrix $\mathbf{V}$ can also be regarded as a new data representation in a low-dimensional feature space[1], which is spanned by the learned features.

In order to learn an effective representation under a pool of features in NMF, three kinds of approaches can be considered: preprocessing of data [12,6], incorporation of additional constraints [2,1], and post-processing of the learned representation [13].

2.2 Preprocessing of Data

Preprocessing of data means that, before applying some data analysis algorithm, the given representation of the instances is converted with a rather simple transformation. This approach is employed in the framework of spectral learning [6]. In spectral learning, symmetric Graph Laplacian in Ncut [11] is used to construct a new representation in [6].

The weighting scheme in Ncut [11] is also employed in [8]. For a given data matrix $\mathbf{X}$, this method considers a graph whose weighted adjacency matrix $\mathbf{W}$ is represented as the Gram matrix of the data matrix. Then, by defining the degree vector $\boldsymbol{d}$ of the graph as $d_i = \sum_j (\mathbf{W})_{ij}$, the diagonal matrix $\mathbf{D}=diag(\boldsymbol{d})$ is constructed[2]. Finally, by denoting the inverse matrix of $\mathbf{D}$ as $\boldsymbol{\Gamma}$, the data matrix $\mathbf{X}$ is converted into $\mathbf{X}\boldsymbol{\Gamma}^{1/2}$. The standard NMF algorithm is applied to the converted data for the factorization.

The key idea in [8] is the preprocessing of the data matrix $\mathbf{X}$ into $\mathbf{X}\boldsymbol{\Gamma}^{1/2}$. This method is called WNMF in this paper. It is also reported that WNMF is effective for document clustering [12].

2.3 Incorporation of Additional Constraints

In order to learn orthogonal features within the framework of NMF, minimization of the following objective function was proposed [2]:

$$J_O = ||\mathbf{X} - \mathbf{U}\mathbf{V}||^2 \text{ such that } \mathbf{U}^T\mathbf{U} = \mathbf{I} \tag{3}$$

where $\mathbf{I}$ is a unit matrix. The difference from the original objective function J in eq.(2) is the additional constraint ($\mathbf{U}^T\mathbf{U} = \mathbf{I}$) imposed on the minimization: this constraint requires that the learned features should be orthogonal to each other. We call this algorithm as ONMF in this paper.

Besides the non-orthogonality of features, for better interpretability of the factorized matrices, another method was proposed in [1]. This method restricts each column of $\mathbf{U}$ to be a convex combination of columns in $\mathbf{X}$ as:

$$J_c = ||\mathbf{X} - \mathbf{U}\mathbf{V}||^2 \text{ such that } \mathbf{U} = \mathbf{X}\mathbf{A} \tag{4}$$

where $\mathbf{A}$ is also a non-negative matrix to restrict each feature as a convex combination of instances. With this constraint, when the number of features is set to

---

[1] Usually $q \ll p$.
[2] The matrix $\mathbf{D}$ is called the degree matrix of the graph.

Table 2. TREC datasets

dataset	# attr.	#clusters
hitech	126372	6
reviews	126372	5
sports	126372	7
la12	31372	6
k1b	21839	6
ohscal	11465	10

Table 3. TREC datasets (imbalanced)

dataset	# attr.	#clusters	#instances
tr11	6429	9	414
tr12	5804	8	313
tr23	5832	6	204
tr31	10128	7	927
tr41	7454	10	878
tr45	8261	10	690

the number of clusters, the learned features can be interpreted as cluster means. This method is called convexNMF in this paper.

2.4 Post-processing of Learned Representation

Instead of incorporating additional constraints and inventing learning rules under the additional constraints, we have proposed an approach for rectifying the representation learned by NMF [13]. The key idea is the utilization of the matrix $\mathbf{U}$, which is also learned by NMF. By normalizing each column $\boldsymbol{u}_\ell$ such that $\boldsymbol{u}_\ell^T \boldsymbol{u}_\ell = 1$ in the factorization, the metric $\mathbf{M}$ of the feature space is estimated in this method as:

$$\mathbf{M} = \mathbf{U}^T\mathbf{U}, \quad s.t. \quad \boldsymbol{u}_l^T\boldsymbol{u}_l = 1, \quad \forall l = 1, \dots, q \tag{5}$$

Since the metric $\mathbf{M}$ in eq.(5) is symmetric positive semi-definite, it can be uniquely decomposed by the Cholesky decomposition as $\mathbf{M} = \mathbf{T}^T\mathbf{T}$, where $\mathbf{T} \in \mathbb{R}^{q \times q}$ is an upper triangular matrix [3]. Finally, another data representation of instances is obtained by utilizing $\mathbf{T}$ as a linear mapping as:

$$\mathbf{V} \mapsto \mathbf{T}\mathbf{V} \tag{6}$$

Thus, this method conducts a post-processing of the representation $\mathbf{V}$ learned by NMF, and construct another representation $\mathbf{T}\mathbf{V}$.

3 Evaluation

3.1 Experimental Setting

Datasets. Based on previous work [12,9], the effectiveness of the approaches in Section 2 were evaluated for document clustering over TREC datasets[3]. In the following experiments, 12 datasets in Table 2 and Table 3 were used. Each document in these datasets is represented in the standard vector space model based on the frequencies of terms.

For the datasets in Table 2, 50 documents were sampled from each cluster to create a sample, and 10 samples were created for each dataset. The created samples are balanced in terms of the number of instances per cluster. On the other

---

[3] http://glaros.dtc.umn.edu/gkhome/cluto/cluto/download

hand, it is known that learning from imbalanced data is a difficult problem [5]. Thus, we also used the datasets in Table 3, which are very imbalanced in terms of the number of instances per cluster. To preserve the imbalance in the data, we did not conduct sampling for these datasets.

Evaluation Measure. For each dataset, the cluster assignment was evaluated with respect to the Normalized Mutual Information (*NMI*) [10]. Let C, $\hat{C}$ stand for the random variables over the true and assigned clusters. *NMI* is defined as:

$$NMI = \frac{I(\hat{T};T)}{(H(\hat{T}) + H(T))/2} \qquad (7)$$

where $H(\cdot)$ is Shannon Entropy, $I(;)$ is Mutual Information. *NMI* in eq.(7) corresponds to the accuracy of cluster assignment. The larger *NMI* is, the better the clustering assignment is.

NMF Algorithms. We used the NMF algorithms explained in Section 2, namely: NMF [8], WNMF [12], ONMF [2], and convexNMF [1]. WNMF corresponds to the preprocessing of data in Section 2.2. Both ONMF and convexNMF correspond to the algorithms with additional constraints in Section 2.3, In addition, the post-processing in Section 2.4 was applied to the representation, constructed by each algorithm.

Evaluation Procedure. When learning and utilizing a pool of features by NMF algorithms, the main parameter is the number of features in the factorization. Thus, the number of features was varied in the following experiments. As the standard clustering algorithms based on Euclidian space, kmeans and skmeans were applied to the learned representation matrix $\mathbf{V}$ from each NMF algorithm, and the post-processed representation $\mathbf{TV}$ in eq.(6). When applying these partitioning based clustering algorithms, we used the number of clusters in each dataset.

With the iterative update of matrices, the learned matrices depends on their initialization. Thus, we conducted 10 random initialization for the same data matrix. Furthermore, since both kmeans and skmeans are affected from the initial cluster assignment, clustering was repeated 10 times with a random initial assignment. For the iterative update of matrices $\mathbf{U}$ and $\mathbf{V}$ in eq.(1), the number of maximum iterations for matrix updates was set to 30.

3.2   Results for Balanced Data

Results for the balanced data in Table 2 are shown in Fig. 1 and Fig. 2. In these figures, the horizontal axis corresponds to the number of features, the vertical one to *NMI*. The reported results are the average of 10 samples in each dataset [4]. In the legend, solid lines with circles correspond to NMF, dotted lines with triangles to WNMF, dashed lines with squares to ONMF, and twodash lines

---

[4] Since 100 runs were conducted for each sample, the average of 1,000 runs is reported for each dataset.

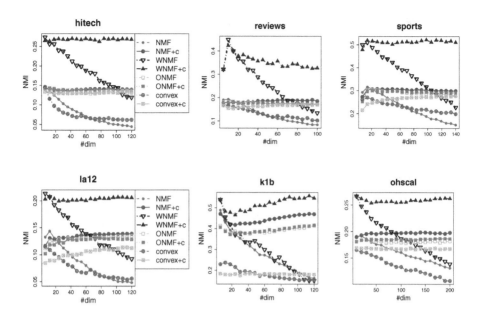

Fig. 1. Results for balanced data in Table 2 (*NMI*, with kmeans)

with crosses to convexNMF (shown as convex in the legend in Fig. 1 and Fig. 2.). For each NMF algorithm, +c stands for the results with the post-processing in Section 2.4.

When kmeans was used for clustering (Fig. 1), the performance of NMF, WNMF and convexNMF degraded as the number of features increased. Especially, although WNMF (with the preprocessing in Section 2.2) showed the best performance, its performance rapidly degraded. Thus, these algorithms were not capable of learning and utilizing a pool of features. On the other hand, with the post-processing in Section 2.4 (namely NMF+c, WNMF+c and convex+c), they showed a stable performance, even when a large number of features was learned and used. Similar stable performance was observed in ONMF with the orthogonal constraint in eq.(3). However, the performance of ONMF+c was almost the same with that of ONMF.

When skmeans was used (Fig. 2), the performance of NMF and WNMF also degraded as the number of features increased (but the performance degradation was not observed in convexNMF). However, their performance was improved by utilizing the post-processing in Section 2.4. In addition, compared with the results in Fig. 1 (with kmeans), better performance was observed in all the algorithms. This performance improvement would be because skmeans was originally proposed for clustering high-dimensional data such as documents. In addition, the performance of the algorithms with constraints (namely ONMF and convexNMF) was rather stable with respect to the change in the number of features.

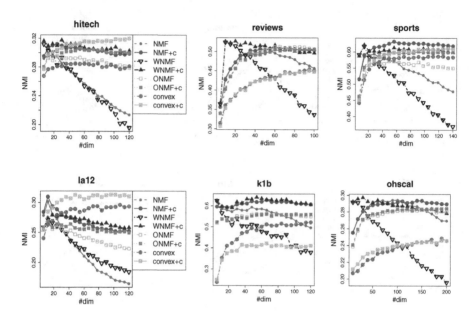

Fig. 2. Results for balanced data in Table 2 (*NMI*, with skmeans)

3.3   Results for Imbalanced Data

Results for the imbalanced data in Table 3 are shown Fig. 3 and Fig. 4.

Contrary to the results in Fig. 1, when kmeans was used for clustering, there were no significant changes in the clustering performance with respect to the number of features. Among the algorithms, WNMF (with the preprocessing in Section 2.2) showed the best performance. However, performance improvement with the post-processing (with +c) was not so remarkable for the imbalanced data.

On the other hand, when skmeans was used, better clustering performance was observed in all the NMF algorithms. Furthermore, contrary to the results in Fig. 3, the post-processing in Section 2.4 (with +c) showed performance improvement for NMF, WNMF and convexNMF, especially in large datasets (namely tr31, tr41 and tr45). Similarly, as in the results in Section 3.3, WNMF+c, which is based on both the pre-processing in Section 2.2 and post-processing in Section 2.4, showed better performance than those of ONMF and convexNMF (with additional constraints). Unfortunately, the post-processing in Section 2.4 was not effective to ONMF.

4   Discussion

In [9], generic features are learned based on NMF [9], but the representation based on the learned features was not effective for document analysis. In order to learn effective features and the representation under the learned features, three kinds of approaches in Section 2 were investigated.

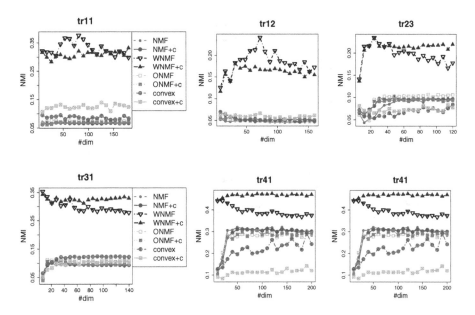

Fig. 3. Results for imbalanced datasets in Table 3 (*NMI*, with kmeans)

Preprocessing of data in Section 2.2 showed the best performance when the number of features was set to the number of clusters. However, its performance rapidly degraded as the number of features was increased. Thus, it was not so effective for exploiting a large number of features. On the other hand, when additional constraints are imposed (Section 2.3), the clustering performance was rather stable with respect to the number of features, especially for imbalanced data. However, they could not outperform WNMF (with the preprocessing). Finally, the post-processing in Section 2.4 was effective to most NMF algorithms to improve their performance (except for ONMF). Furthermore, their performance was stable, even when the number of features was increased.

Our current conjecture for the stable and improved performance with the post-processing in Section 2.4 is as follows. By apply NMF to a data matrix, each instance x_i is represented as v_i with respect to the learned features. Then, v_i is mapped to $\mathbf{T}v_i$ through the linear mapping in eq.(6). Thus, the square distance of any pair of instances is calculated as:

$$(\mathbf{T}v_i - \mathbf{T}v_j)^T(\mathbf{T}v_i - \mathbf{T}v_j) = (v_i - v_j)^T \mathbf{T}^T \mathbf{T}(v_i - v_j)$$
$$= (v_i - v_j)^T \mathbf{M}(v_i - v_j)$$

where $\mathbf{M}$ is the estimated metric of the learned feature space in Section 2.4. Thus, as in the Mahalanobis generalized distance (which has been widely used in pattern recognition), it is possible to exploit the square distance based on the metric of the learned feature space. This enables to coping with the non-orthogonality of features learned by NMF.

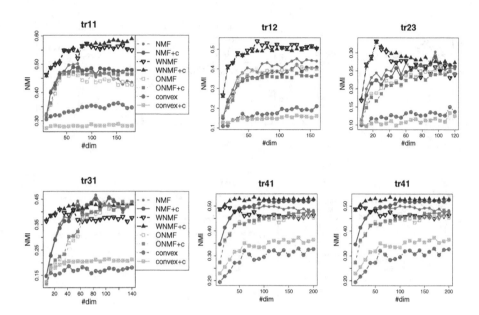

Fig. 4. Results for imbalanced datasets in Table 3 (*NMI*, with skmeans)

5   Concluding Remarks

Since localized features as well as a data representation with respect to the features can be learned by Non-negative Matrix Factorization (NMF), NMF can be used as a representation learning method. This paper investigated the effectiveness of NMF for learning and utilizing a pool of features over document datasets. Performance of NMF algorithms for different number of features was analyzed in terms of the preprocessing of date, incorporation of additional constraints, and post-processing of the representation learned by NMF. Experimental results revealed that coping with the non-orthogonality of features is crucial to achieve a stable performance for exploiting a large number of features. Furthermore, it was revealed that exploiting the metric of the feature space with post-processing is effective for performance improvement. We plan to conduct further investigation with other real world datasets such as image datasets.

Acknowledgments. This work is partially supported by the grant-in-aid for scientific research (No. 24300049) funded by MEXT in Japan and the Murata Science Foundation.

References

1. Ding, C., Li, T., Jordan, M.I.: Convex and semi-nonnegative matrix factorizations. IEEE Transactions on Pattern Analysis and Machine Intelligence 32(1), 45–55 (2010)

2. Ding, C., Li, T., Peng, W., Park, H.: Orthogonal nonnegative matrix tri-factorizations for clustering. In: Proc. KDD 2006, pp. 126–135 (2006)
3. Harville, D.A.: Matrix Algebra From a Statistican's Perspective. Springer (2008)
4. Hoyer, P.O.: Non-negative matrix factorization with sparseness constraints. Journal of Machine Learning Research 5, 1457–1469 (2004)
5. Japkowicz, N., Stephen, S.: The class imbalance problem: A systematic study. Intelligent Data Analysis 6(5), 429–450 (2002)
6. Kamvar, S.D., Klein, D., Manning, C.D.: Spectral learning. In: Proc. of IJCAI 2003, pp. 561–566 (2003)
7. Lee, D.D., Seung, H.S.: Learning the parts of objects by non-negative matrix factorization. Nature 401, 788–791 (1999)
8. Lee, D.D., Seung, H.S.: Algorithms for non-negative matrix factorization. In: Proc. NIPS 2001, pp. 556–562 (2001)
9. Raina, R., Battle, A., Lee, H., Packer, B., Ng, A.: Self-taught learning:transfer learning from unlabeled data. In: Proc. ICML 2007, pp. 759–766 (2007)
10. Strehl, A., Ghosh, J.: Cluster ensembles — a knowledge reuse framework for combining multiple partitions. J. Machine Learning Research 3(3), 583–617 (2002)
11. von Luxburg, U.: A tutorial on spectral clustering. Statistics and Computing 17(4), 395–416 (2007)
12. Xu, W., Liu, X., Gong, Y.: Document clustering based on non-negative matrix factorization. In: Proc. SIGIR 2003, pp. 267–273 (2003)
13. Yoshida, T.: Cholesky decomposition rectification for non-negative matrix factorization. In: Kryszkiewicz, M., Rybinski, H., Skowron, A., Raś, Z.W. (eds.) ISMIS 2011. LNCS, vol. 6804, pp. 214–219. Springer, Heidelberg (2011)

Theoretical Analysis and Evaluation
of Topic Graph Based Transfer Learning

Tetsuya Yoshida and Hiroki Ogino*

Graduate School of Information Science and Technology,
Hokkaido University
N-14 W-9, Sapporo 060-0814, Japan
{yoshida,hiroki}@meme.hokudai.ac.jp

Abstract. Various research efforts have been invested in machine learning and data mining for finding out patterns from data. However, even
when some knowledge may be learned in one domain, it is often difficult
to re-use it for another domain with different characteristics. Toward
effective knowledge transfer between domains, we proposed a transfer
learning method based on our transfer hypothesis that two domains have
similar feature spaces. A graph structure called a topic graph is constructed by using the learned features in one domain, and the graph is
used as a regularization term. In this paper we present a theoretical analysis of our approach and prove the convergence of the learning algorithm.
Furthermore, the performance evaluation of the method is reported over
document clustering problems. Extensive experiments are conducted to
compare with other transfer learning algorithms. The results are encouraging, and show that our method can improve the performance by
transferring the learned knowledge effectively.

1 Introduction

Transfer Learning is a machine learning framework for realizing the re-use of
data or knowledge [10]. Based on the notation in [10], the domain in which the
knowledge is learned is called the source domain, and the other domain is called
the target domain in this paper. Various methods have been proposed in the
literature [3,8,6,14]. We have also proposed a method based on Non-negative
Matrix Factorization (NMF) [7,13,5]. Since NMF enables to learn features from
a non-negative data matrix, our method tries to preserve the feature space which
is spanned by the learned features with NMF [9].

In this paper we present a theoretical analysis of our approach and prove the
convergence of the learning algorithm. Furthermore, the performance evaluation
of the method is reported over document clustering problems. Extensive experiments were conducted in the evaluation to compare with other transfer learning
algorithms, and the results show that our method can improve the performance
of document clustering by exploiting the knowledge learned in the source domain. Especially, performance improvement can be achieved not because the

---

* Currently in NTT DATA Corporation.

T. Yoshida et al. (Eds.): AMT 2013, LNCS 8210, pp. 106–115, 2013.
© Springer International Publishing Switzerland 2013

available amount of instances is simply increased, but because it can exploit the knowledge learned in the source domain.

Notation. We use bold normal uppercase letters for matrices, and bold italic lowercase letters for vectors. For a matrix $\mathbf{X}$: $\mathbf{X}_{ij}$ stands for ij-th coordinate of $\mathbf{X}$, $\operatorname{tr}(\mathbf{X})$ stands for the trace of $\mathbf{X}$, and $\mathbf{X}^T$ stands for its transposition. $\mathbf{I}$ stands for a unit matrix, and $\mathbf{1}$ stands for a vector where all the coordinates are equal to 1. For a non-singular matrix $\mathbf{A}$, its inverse matrix is represented as $\mathbf{A}^{-1}$. We assume that a data matrix, which represents instances in a domain, is non-negative. In this paper the matrices for the source domain are represented with subscript s, and the matrices for the target domain are represented with subscript t. For example, $\mathbf{X}_s$ and $\mathbf{X}_t$ represent a data matrix in the source domain and a data matrix in the target domain, respectively.

2   Theoretical Analysis of a Transfer Learning Method

2.1   Topic Graph Based NMF for Transfer Learning

We have proposed a transfer learning method based on Non-negative Matrix Factorization (NMF) [7,13,5]. In NMF, a non-negative data matrix $\mathbf{X}$ is approximately factorized as

$$\mathbf{X} \simeq \mathbf{U}\mathbf{V} \tag{1}$$

where $\mathbf{U}$ and $\mathbf{V}$ are non-negative matrices. The columns of $\mathbf{U}$ correspond to features, and those of $\mathbf{V}$ correspond to coefficients in the factorization. Our transfer hypothesis is the similarity (preservation) of feature spaces between the source domain and the target domain. This is formalized as the minimization of the following objective function [9]:

$$J_1 = ||\mathbf{X}_t - \mathbf{U}_t\mathbf{V}_t||^2 + \nu||\mathbf{U}_s - \mathbf{U}_t||^2 \tag{2}$$

where ν is a regularization parameter, and $||\cdot||$ is a matrix norm. The first term represents the approximate factorization of a data matrix in the target domain, and the second term represents our transfer hypothesis (preservation of feature spaces) in transfer learning. In this paper we focus on Frobenius norm $||\cdot||_F$ [7] as a matrix norm $||\cdot||$.

When each column in $\mathbf{U}$ is normalized, we may consider the relation of features (columns of $\mathbf{U}$) and represent the relation as a graph. The weighted adjacency matrix of the graph is defined as $\mathbf{W} = \mathbf{U}^T\mathbf{U}$ where $\boldsymbol{u}_\ell^T \boldsymbol{u}_\ell = 1$, $\forall \ell$. Here, each $\boldsymbol{u}_\ell$ is a column in $\mathbf{U}$. We call the graph as a topic graph, since a feature learned by NMF usually corresponds to a topic in document analysis. By using the topic graph, minimizing J_1 in eq.(2) is equivalent to minimizing the following objective function [9]:

$$J_2 = ||\mathbf{X}_t - \mathbf{U}_t\mathbf{V}_t||^2 + \nu \operatorname{tr}(\mathbf{U}_t\mathbf{L}_s\mathbf{U}_t^T) \tag{3}$$

where $\mathbf{D}_s$ is the degree matrix of topic graph $\mathbf{W}_s$ in the source domain, $\mathbf{L}_s = \mathbf{D}_s - \mathbf{W}_s$ is the graph Laplacian of $\mathbf{W}_s$ [12]. For minimizing J_2 in eq.(3), we proposed an algorithm TNT (Topic Graph based NMF for Transfer Learning), which repeatedly applies the following multiplicative update rules:

$$(\mathbf{U}_t)_{ij} \leftarrow (\mathbf{U}_t)_{ij} \frac{(\mathbf{X}_t \mathbf{V}_t^T + \nu \mathbf{U}_t \mathbf{W}_s)_{ij}}{(\mathbf{U}_t \mathbf{V}_t \mathbf{V}_t^T + \nu \mathbf{U}_t \mathbf{D}_s)_{ij}} \tag{4}$$

$$(\mathbf{V}_t)_{ij} \leftarrow (\mathbf{V}_t)_{ij} \frac{(\mathbf{U}_t^T \mathbf{X}_t)_{ij}}{(\mathbf{U}_t^T \mathbf{U}_t \mathbf{V}_t)_{ij}} \tag{5}$$

2.2   Theoretical Analysis of TNT

The following theorem holds for the algorithm:

Theorem 1. *The objective function in eq.(3) is non-increasing under the above update rules, and since it is non-negative, the algorithm converges.*

The objective function J_2 in eq.(3) is certainly bounded from below by zero. To prove Theorem 1, we need to show that J_2 is non-increasing under the update rules in eq.(4) and eq.(5). Since the second term in eq.(3) is only related to $\mathbf{U}_t$, we have exactly the same update formula for $\mathbf{V}_t$ as in the original NMF update rules [7]. Please refer to [7] for the details.

In order to prove Theorem 1, we use the notion of auxiliary function [7].

Definition 1. *A function $G(u, u^{(t)})$ is an auxiliary function for a function $F(u)$ if the conditions:*

1) $G(u, u^{(t)}) \geq F(u)$
2) $G(u, u) = F(u)$

are satisfied. For any auxiliary function, the following lemma holds:

Lemma 1. *if G is an auxiliary function of F, then F is non-increasing under the following update:* [7]

$$u^{(t+1)} = \underset{u}{\mathrm{argmin}}\, G(u, u^{(t)}) \tag{6}$$

Proof. $F(u^{(t+1)}) \leq G(u^{(t+1)}, u^{(t)}) \leq G(u^{(t)}, u^{(t)}) = F(u^{(t)})$ □

In the following, each element $(\mathbf{U}_t)_{ab}$ in $\mathbf{U}_t$ is also represented as u_{ab}. In addition, F_{ab} stands for the terms in eq.(3) which are only related to $(\mathbf{U}_t)_{ab}$. Now, we consider the partial derivative of eq.(3) with respect to each $(\mathbf{U}_t)_{ab}$, and define the following symbols:

$$F'_{ab} \overset{\triangle}{=} (\frac{\partial J_2}{\partial \mathbf{U}_t})_{ab} = (-2\mathbf{X}_t \mathbf{V}_t^T + 2\mathbf{U}_t \mathbf{V}_t \mathbf{V}_t^T + 2\nu \mathbf{U}_t \mathbf{L}_s)_{ab} \tag{7}$$

$$(\frac{\partial F'_{ab}}{\partial \mathbf{U}_t})_{ab} = 2(\mathbf{V}_t \mathbf{V}_t^T)_{bb} + 2\nu (\mathbf{L}_s)_{bb} \tag{8}$$

Lemma 2. *The following function is an auxiliary function of F_{ab}:*

$$G(u, u_{ab}^{(t)}) = F_{ab}(u_{ab}^{(t)}) + F_{ab}'(u_{ab}^{(t)})(u - u_{ab}^{(t)})$$
$$+ \frac{(\mathbf{U}_t\mathbf{V}_t\mathbf{V}_t^T)_{ab} + \nu(\mathbf{U}_t\mathbf{D}_s)_{ab}}{u_{ab}^{(t)}}(u - u_{ab}^{(t)})^2 \tag{9}$$

Proof. We show that G is an auxiliary function of F_{ab}. Apparently, $G(u, u) = F_{ab}(u)$ holds. Thus, we show that $G(u, u') \geq F_{ab}(u)$ holds. Based on the Taylor expansion of $F_{ab}(u)$, the following holds:

$$F_{ab}(u) = F_{ab}(u_{ab}^{(t)}) + F_{ab}'(u_{ab}^{(t)})(u - u_{ab}^{(t)})$$
$$+ [(\mathbf{V}_t\mathbf{V}_t^T)_{bb} + \nu(\mathbf{L}_s)_{bb}](u - u_{ab}^{(t)})^2 \tag{10}$$

By comparing eq.(9) and eq.(10), we show that the following holds:

$$\frac{(\mathbf{U}_t\mathbf{V}_t\mathbf{V}_t^T)_{ab} + \nu(\mathbf{U}_t\mathbf{D}_s)_{ab}}{u_{ab}^{(t)}} \geq (\mathbf{V}_t\mathbf{V}_t^T)_{bb} + \nu(\mathbf{L}_s)_{bb} \tag{11}$$

Now, the following inequalities hold:

$$(\mathbf{U}_t\mathbf{V}_t\mathbf{V}_t^T)_{ab} = \sum_{k=1}^{q}(\mathbf{U}_t)_{ak}^{(t)}(\mathbf{V}_t\mathbf{V}_t^T)_{kb} \geq (\mathbf{U}_t)_{ab}^{(t)}(\mathbf{V}_t\mathbf{V}_t^T)_{bb} \tag{12}$$

$$\nu(\mathbf{U}_t\mathbf{D}_s)_{ab} = \nu\sum_{k=1}^{q}(\mathbf{U}_t)_{ak}^{(t)}(\mathbf{D}_s)_{kb} \geq \nu(\mathbf{U}_t)_{ab}^{(t)}(\mathbf{D}_s)_{bb} \tag{13}$$

$$\geq \nu(\mathbf{U}_t)_{ab}^{(t)}(\mathbf{D}_s - \mathbf{W}_s)_{bb} = \nu(\mathbf{U}_t)_{ab}^{(t)}(\mathbf{L}_s)_{bb} \tag{14}$$

Thus, we can show the following inequality:

$$\frac{(\mathbf{U}_t\mathbf{V}_t\mathbf{V}_t^T)_{ab} + \nu(\mathbf{U}_t\mathbf{D}_s)_{ab}}{u_{ab}^{(t)}} \geq \frac{(\mathbf{U}_t)_{ab}^{(t)}(\mathbf{V}_t\mathbf{V}_t^T)_{bb} + \nu(\mathbf{U}_t)_{ab}^{(t)}(\mathbf{L}_t)_{bb}}{u_{ab}^{(t)}} \tag{15}$$

$$\geq (\mathbf{V}_t\mathbf{V}_t^T)_{bb} + \nu(\mathbf{L}_t)_{bb} \tag{16}$$

From the above, since $G(u, u^{(t)}) \geq F_{ab}(u)$ holds, Lemma 2 holds.   □

Since eq.(9) is an auxiliary function, from Lemma 1, the following holds:

$$u_{ab}^{(t+1)} = u_{ab}^{(t)} - u_{ab}^{(t)}\frac{F_{ab}'(u_{ab}^{(t)})}{2(\mathbf{U}_t\mathbf{V}_t\mathbf{V}_t^T)_{ab} + 2\nu(\mathbf{U}_t\mathbf{D}_s)_{ab}} \tag{17}$$

$$= u_{ab}^{(t)}\frac{(\mathbf{X}_t\mathbf{V}_t^T + \nu\mathbf{U}_t\mathbf{W}_s)_{ab}}{(\mathbf{U}_t\mathbf{V}_t\mathbf{V}_t^T + \nu\mathbf{U}_t\mathbf{D}_s)_{ab}} \tag{18}$$

Therefore, the update rule in eq.(18) is non-increasing with respect to eq.(3). Thus, Theorem 1 is proved.

Table 1. The hierarchy in 20 Newsgroup data

top category	sub category	symbol	top category	sub category	symbol
comp	graphics	comp-1	sci	crypt	sci-1
	os.ms-windows.misc	comp-2		electronics	sci-2
	sys.ibm.pc.hardware	comp-3		med	sci-3
	sys.mac.hardware	comp-4		space	sci-4
	windows.x	comp-5			
rec	autos	rec-1	talk	politics.guns	talk-1
	motorcycles	rec-2		politics.mideast	talk-2
	sport.baseball	rec-3		politics.misc	talk-3
	sport.hockey	rec-4		religion.misc	talk-4

Table 2. Datasets for two top categories (four sub-categories)

id	Dataset	Domain	Clusters			
a	comp vs. rec	Source	comp-1	comp-4	rec-2	rec-4
		Target	comp-2	comp-5	rec-1	rec-3
b	comp vs. sci	Source	comp-1	comp-2	sci-1	sci-2
		Target	comp-4	comp-5	sci-3	space
c	comp vs. talk	Source	comp-1	comp-5	talk-2	talk-4
		Target	comp-2	comp-3	talk-1	talk-3
d	rec vs. talk	Source	rec-1	rec-2	talk-1	talk-3
		Target	rec-3	rec-4	talk-2	talk-4
e	sci vs.talk	Source	sci-2	sci-3	talk-3	talk-4
		Target	sci-1	sci-4	talk-1	talk-2

Regularization via Pairwise Relation. The objective function in eq.(3) was extended by adding another regularization term in eq.(3):

$$J_3 = ||\mathbf{X}_t - \mathbf{U}_t\mathbf{V}_t||^2 + \nu\operatorname{tr}(\mathbf{U}_t\mathbf{L}_s\mathbf{U}_t^T) + \lambda\operatorname{tr}(\mathbf{V}_t^T\mathbf{L}_t\mathbf{V}_t) \qquad (19)$$

where λ is another regularization parameter. Even when eq.(19) is used, we can also derive the corresponding multiplicative update rules as in eq.(4) and eq.(5). Since the partial derivative of $\mathbf{U}_t$ and that of $\mathbf{V}_t$ are independent, the same update rule in eq.(4) can be used for $\mathbf{U}_t$; the update rule in [1] can be used for $\mathbf{V}_t$ to minimize (19), which is represented as

$$(\mathbf{V}_t)_{ij} \leftarrow (\mathbf{V}_t)_{ij} \frac{(\mathbf{U}_t^T\mathbf{X}_t + \lambda\mathbf{W}_t\mathbf{V}_t^T)_{ij}}{(\mathbf{U}_t^T\mathbf{U}_t\mathbf{V}_t + \lambda\mathbf{D}_t\mathbf{V}_t^T)_{ij}} \qquad (20)$$

Thus, we can also show the similar theorem for the update rules even for the objective function in eq.(19) as in Theorem 1.

3 Evaluations

3.1 Experimental Settings

Datasets. Based on previous work [8,6], we conducted experiments on the 20 Newsgroup dataset (20NG)[1], which has been widely used in document analysis.

---

[1] http://people.csail.mit.edu/~jrennie/20Newsgroups/
(20news-18828 was used).

Table 3. Datasets for three top categories (six sub-categories)

id	Dataset	Domain	Clusters					
f	comp vs. rec vs. sci	Source	comp-1	comp-2	rec-1	rec-2	sci-1	sci-2
		Target	comp-3	comp-4	rec-3	rec-4	sci-3	sci-4
g	comp vs. rec vs. talk	Source	comp-1	comp-2	rec-1	rec-2	talk-1	talk-2
		Target	comp-3	comp-4	rec-3	rec-4	talk-3	talk-4
h	comp vs. sci vs. talk	Source	comp-1	comp-2	sci-1	sci-2	talk-1	talk-2
		Target	comp-3	comp-4	sci-3	sci-4	talk-3	talk-4
i	rec vs. sci vs. talk	Source	rec-1	rec-2	sci-1	sci-2	talk-1	talk-2
		Target	rec-3	rec-4	sci-3	sci-4	talk-3	talk-4

This contains seven top categories, and 20 sub-categories are included under the top categories. However, since three top categories include only one sub-category respectively, we used the remaining 4 top categories ({comp,rec,sci,talk}) and the sub-categories in these top categories in the subsequent experiments. Sub-categories and their symbols in this paper are summarized in Table 1. Following the procedure in [8,6], we divided the data so that no sub-category is included in both the source domain and the target domain. The used datasets are shown in Table 2 and Table 3.

When a sufficiently large amount of data is available in the target domain, there is no need to use transfer learning. Thus, transfer learning is useful for the situation where only a limited number of instances is available in the target domain. In order to simulate the above situation, the amount of data in the source domain was set to four times larger than that in the target domain, and 25 documents were sampled from each sub-category in the target domain. By repeating this process, 10 samples were created for each dataset in Table 2 and in Table 3. For each sample, we conducted stemming using the porter stemmer[2] and MontyTagger[3], removed stop words, and selected 2,000 words with large mutual information [2].

The following two tasks were conducted in the subsequent experiments: i) top-category clustering problem, and ii) sub-category clustering problem. For example, for the dataset "comp vs. rec vs. sci" (id "f") in Table 3, the task i) is to assign the documents in the target domain into three clusters (since the number of top-categories is three) by using the available documents from the source domain. On the other hand, in task ii), the number of clusters is set to six when clustering the documents in the target domain.

Evaluation Measure. The cluster assignment was evaluated with respect to the Normalized Mutual Information (NMI) [11]. Let C, $\hat{C}$ be random variables over the true and assigned clusters, respectively. NMI is defined as $NMI = \frac{I(\hat{C};C)}{(H(\hat{C})+H(C))/2}$ where $H(\cdot)$ is the Shannon entropy, $I(;)$ is the mutual information. $NMI \in [0,1]$ corresponds to the accuracy of the cluster assignment in document clustering. Larger NMI values indicate better clustering results.

Methods. We evaluated the performance of the following seven methods. In order to evaluate the effectiveness of **TNT** based on NMF, the following three NMF

---

[2] http://www.tartarus.org/~martin/PorterStemmer
[3] http://web.media.mit.edu/~hugo/montytagger

algorithms were used: 1) NMF [7], 2) WNMF [13], 3) GNMF [1], and evaluated its effectiveness. As other transfer leaning methods, we compared with 4) SDT [8] 5) MTrick [14]. Furthermore, 6) skmeans [4] and 7) Ncut [12] were evaluated as baseline for document clustering. Since the above methods are partitioning based clustering methods, we assumed that the number of clusters (denoted as k) is specified based on the ground truth label in the dataset.

Parameters. Cosine similarity, which has been widely used in text processing, was used as the pairwise similarity measure. Although SDT andMTrick assume that the cluster label is given to each data instance in the source domain, no label information was used in the subsequent experiments. Based on [8], the coefficient for the target domain was set to 0.025 in SDT, and kmeans was used for conducting clustering.

Following the description in [14], the number of word clusters was set to 50 in MTrick, and the coefficient for the target domain was set to 1.5. Since logistic regression based on the cluster label in the source domain may not be used, one of the matrices was initialized using skmeans, but other matrices were initialized as described in [14]. The parameters in GNMF were set based on [1] [4]. Based on our preliminary experiments. ν was set to 0.15 and λ to 1.5 in eq.(19).

Evaluation Procedure. NMF may be considered as a dimensionality reduction method for constructing a new data representation $\mathbf{V}_t$. In the subsequent experiments, document clustering was conducted by applying skmeans to the constructed $\mathbf{V}_t$. In order to remove the influence of the matrix initialization in NMF, we conducted 10 random initialization for the same data matrix. This process was repeated in 10 samples for each dataset [5]. The number of maximum iterations for conducting multiplicative updates in NMF was set to 30 in the following experiments.

In addition, we compared with the previous approach for document clustering with NMF, which is based on the indicator matrix interpretation of $\mathbf{V}$ in eq.(1) [13,5]. In the conventional NMF and WNMF, an instance is assigned to the cluster with the maximal coefficient in the corresponding column in $\mathbf{V}$. However, since the results of GNMF was very poor, skmeans was applied to the constructed representation $\mathbf{V}_t$ by GNMF.

3.2   Performance Evaluations

Comparisons with other methods are summarized in Table 4 and Table 5. Table 4 corresponds to task i) (top-category clustering problem where the number of clusters k was set to the number of top categories); Table 5 corresponds to task ii) (sub-category clustering problem). The number of features q (#features in these tables) was set to the number of clusters k in the compared NMF based methods (MTrick, NMF, WNMF, GNMF, NMF-All, WNMF-All, GNMF-All). On the other hand, although the performance depends on the number of features, the results for different number of features were shown in these tables.

---

[4]   The number of neighbors was set to 10, and the regularization parameter to 100.

[5]   The average of 100 runs is reported in each dataset.

Table 4. Comparison (k(#clusters) was set to#top-categories, *NMI* (with skmeans))

#features	Method	a	b	c	d	e	f	g	h	i
k	NMF+T	0.540	0.335	0.747	0.366	0.141	0.334	0.404	0.332	0.210
(ours)	WNMF+T	0.352	0.232	0.544	0.290	0.138	0.292	0.371	0.303	0.183
	GNMF+T	**0.581**	**0.355**	0.755	0.442	0.166	0.376	0.477	0.386	0.250
$k \times 10$	NMF+T	0.169	0.105	0.312	0.211	0.083	0.197	0.235	0.184	0.156
(ours)	WNMF+T	0.216	0.099	0.379	0.179	0.084	0.168	0.253	0.221	0.116
	GNMF+T	0.500	0.298	**0.761**	**0.579**	**0.200**	**0.436**	**0.532**	0.407	0.358
$k \times 30$	NMF+T	0.109	0.077	0.178	0.142	0.051	0.165	0.180	0.138	0.116
(ours)	WNMF+T	0.210	0.110	0.167	0.120	0.158	0.105	0.158	0.186	0.134
	GNMF+T	0.487	0.286	0.717	0.562	0.164	0.429	0.522	**0.419**	**0.363**
other	MTrick	0.425	0.255	0.718	0.437	0.164	0.374	0.433	0.348	0.248
transfer	SDT	0.233	0.140	0.409	0.159	0.047	0.096	0.172	0.198	0.083
w/o	NMF	0.416	0.263	0.648	0.373	0.144	0.334	0.377	0.322	0.244
transfer	WNMF	0.162	0.135	0.423	0.228	0.111	0.266	0.288	0.250	0.194
	GNMF	0.261	0.114	0.568	0.252	0.078	0.265	0.307	0.275	0.181
(union)	NMF-All	0.332	0.166	0.617	0.090	0.072	0.126	0.232	0.233	0.070
	WNMF-All	0.203	0.071	0.398	0.084	0.092	0.112	0.210	0.160	0.072
	GNMF-All	0.349	0.162	0.635	0.235	0.086	0.159	0.347	0.341	0.117
baseline	skmeans	0.152	0.066	0.277	0.139	0.056	0.156	0.189	0.157	0.100
	Ncut	0.341	0.219	0.475	0.345	0.190	0.365	0.420	0.361	0.292

Note that the rows for "k (ours)" corresponds to the situation where the number of features was set to k as in the compared methods[6]. For calculating *NMI*, the role of the source domain and the target domains in Table 2 and Table 3 were also exchanged, and their average performance with skmeans is shown in Table 4 and Table 5[7].

In Table 4, by comparing our method (the rows for "k (ours)") and the conventional NMF based clustering algorithms (the rows for "w/o transfer"), the results show that our method improves performance via transfer learning. Furthermore, GNMF+T outperformed all the other compared methods. Especially, it showed better performance with large number of features (the rows for "$k \times 10$" and "$k \times 30$"). On the other hand, the performance of NMF+T and WNMF+T decreased when large number of features was used. However, NMF+T showed equivalent performance with MTrick when the number of features was set to the number of clusters (the row for "k"). Although WNMF+T could not outperform MTrick, but it outperformed SDT.

Similar results were obtained in Table 3 as well. As in Table 4, GNMF+T outperformed all the other compared methods. On the other hand, in task ii), the performance of NMF+T and WNMF+T were almost equivalent to that of the conventional NMF and WNMF, and could not show the improvement. However, they still outperformed SDT (but not MTrick).

Performance improvement might be obtained not by the effect of transfer learning, but sorely due to the increase in the number of instances. Performance improvement in the latter situation may occur if the data instances in the source and the target domain are sufficiently similar. In order to investigate this, we

---

[6] The rows for "$k \times 10$" corresponds to the situation where the number of features was set to 10 times larger than the number of clusters.

[7] When q is increased, the representation of $\mathbf{V}_t$ becomes high-dimensional. Thus, skmeans showed slightly better performance than kmeans.

Table 5. Comparison (k(#clusters) was set to #sub-categories, *NMI* (with skmeans))

#features	Method	a	b	c	d	e	f	g	h	i
k	NMF+T	0.482	0.220	0.369	0.251	0.301	0.338	0.357	0.363	0.340
(ours)	WNMF+T	0.437	0.217	0.356	0.223	0.234	0.290	0.315	0.332	0.285
	GNMF+T	**0.532**	0.259	0.405	0.288	0.325	0.394	0.411	0.420	0.381
$k \times 10$	NMF+T	0.221	0.121	0.201	0.170	0.181	0.235	0.249	0.207	0.204
(ours)	WNMF+T	0.325	0.131	0.277	0.159	0.179	0.161	0.215	0.251	0.177
	GNMF+T	0.527	**0.300**	**0.450**	**0.331**	**0.404**	**0.423**	**0.469**	**0.441**	**0.415**
$k \times 30$	NMF+T	0.270	0.120	0.275	0.196	0.231	0.245	0.259	0.235	0.237
(ours)	WNMF+T	0.186	0.082	0.247	0.093	0.072	0.092	0.163	0.127	0.072
	GNMF+T	0.507	0.269	0.434	0.289	0.339	0.388	0.422	0.424	0.365
other	MTrick	0.441	0.283	0.396	0.284	0.339	0.383	0.409	0.396	0.364
transfer	SDT	0.358	0.118	0.210	0.158	0.154	0.160	0.188	0.254	0.179
w/o	NMF	0.435	0.249	0.387	0.281	0.331	0.358	0.386	0.377	0.354
transfer	WNMF	0.315	0.215	0.364	0.235	0.273	0.262	0.356	0.326	0.312
	GNMF	0.393	0.161	0.345	0.243	0.247	0.308	0.332	0.326	0.276
(union)	NMF-All	0.316	0.122	0.206	0.097	0.079	0.159	0.145	0.223	0.133
	WNMF-All	0.207	0.076	0.214	0.108	0.085	0.135	0.139	0.233	0.128
	GNMF-All	0.429	0.150	0.262	0.172	0.131	0.219	0.253	0.368	0.202
baseline	skmeans	0.208	0.107	0.213	0.143	0.154	0.206	0.224	0.208	0.185
	Ncut	0.462	0.267	0.381	0.298	0.371	0.383	0.403	0.402	0.372

conducted additional experiments by applying the conventional NMF based clustering algorithms to the union of data instances in the source and the target domain. The results are shown in the rows "(union)" with suffix "-All" in Table 4 and Table 5.

By comparing the results of the corresponding methods (*e.g.*, the row for NMF (w/o transfer) which was applied only to the data instances in the target domain and the row for NMF-All), we may see that the performance degraded when NMF based methods were applied to the union of data instances in both tables. Thus, simply merging the data instances and increasing the number of instances does not contribute to improving performance. On the other hand, our methods (the rows "k (ours)" with +T) outperformed the conventional methods (with -All). These results indicate that the reported performance improvement in our method was obtained not due to the increase of the instances, but due to our transfer hypothesis and the method.

3.3   Discussions

The results in Section 3.2 indicate that the algorithm TNT in [9] improves the performance of document clustering, not because the available amount of instances is simply increased, but because it can exploit the knowledge learned in the source domain. Especially, it was effective and robust, especially in GNMF+T with eq. (19). On the other hand, the performance of SDT was rather low. The performance of MTrick was better than that of SDT, but the better performance of MTrick was partly due to the careful initialization of the matrices in the algorithm [14].

4   Concluding Remarks

This paper presented a theoretical analysis of our topic graph based transfer learning method (TNT) and proved the convergence of the learning algorithm in TNT. The method is based on a transfer hypothesis that two domains in transfer learning have similar feature spaces, and this hypothesis is formalized as a regularization term based on a graph structure. The performance of the method is evaluated through extensive experiments over document clustering problems. The results are encouraging, and show that our method can improve the performance of document clustering by exploiting the knowledge learned in the source domain. Since various extensions have been proposed for NMF algorithms, we plan to extend our approach to deal with other NMF based algorithms with different regularization terms in near future.

Acknowledgments. This work is partially supported by the grant-in-aid for scientific research (No. 24300049) funded by MEXT in Japan and the Murata Science Foundation.

References

1. Cai, D., He, X., Wu, X., Han, J.: Non-negative matrix factorization on manifold. In: Proc. ICDM 2008, pp. 63–72 (2008)
2. Cover, T., Thomas, J.: Elements of Information Theory. Wiley (2006)
3. Dai, W., Xue, G.-R., Yang, Q., Yu, Y.: Co-clustering based classification for out-of-domain documents. In: Proc. KDD 2007, pp. 210–219 (2007)
4. Dhillon, J., Modha, D.: Concept decompositions for lage sparse text data using clustering. Machine Learning 42, 143–175 (2001)
5. Ding, C., Li, T., Peng, W., Park, H.: Orthogonal nonnegative matrix tri-factorizations for clustering. In: Proc. KDD 2006, pp. 126–135 (2006)
6. Gao, J., Fan, W., Jiang, J., Han, J.: Knowledge transfer via multiple model local structure mapping. In: Proc. KDD 2008, pp. 283–291 (2008)
7. Lee, D.D., Seung, H.S.: Algorithms for non-negative matrix factorization. In: Proc. NIPS 2001, pp. 556–562 (2001)
8. Ling, X., Dai, W., Xue, G., Yang, Q., Yu, Y.: Spectral domain-transfer learning. In: Proc. KDD 2008, pp. 488–496 (2008)
9. Ogino, H., Yoshida, T.: Topic graph based non-negative matrix factorization for transfer learning. In: Kryszkiewicz, M., Rybinski, H., Skowron, A., Raś, Z.W. (eds.) ISMIS 2011. LNCS, vol. 6804, pp. 260–269. Springer, Heidelberg (2011)
10. Pan, S.J., Yang, Q.: A survey on transfer learning. IEEE Transactions on Knowledge and Data Engineering 22(10), 1345–1359 (2010)
11. Strehl, A., Ghosh, J.: Cluster ensembles — a knowledge reuse framework for combining multiple partitions. J. Machine Learning Research 3(3), 583–617 (2002)
12. von Luxburg, U.: A tutorial on spectral clustering. Statistics and Computing 17(4), 395–416 (2007)
13. Xu, W., Liu, X., Gong, Y.: Document clustering based on non-negative matrix factorization. In: Proc. SIGIR 2003, pp. 267–273 (2003)
14. Zhuang, F., Luo, P., Xiaong, H., He, Q., Xiong, Y., Shi, Z.: Exploiting associations between word clusters and document classes for cross-domain text categorization. In: Proc. SDM 2010, pp. 13–24 (2010)

Selective Weight Update for Neural Network – Its Backgrounds

Yoshitsugu Kakemoto[1] and Shinichi Nakasuka[2]

[1] The JSOL, Ltd, Tokyo, Harumi Center Building, 2-5-24 Harumi, Chuo-ku Japan
kakemoto.yoshitsugu@jsol.co.jp
[2] The University of Tokyo, Tokyo, 7-3-1 Hongo, Bunkyo-ku Japan
nakasuka@space.t.u-tokyo.ac.jp

Abstract. VSF–Network, Vibration Synchronizing Function Network, is a hybrid neural network combining Chaos Neural Network and hierarchical neural network. VSF–Network is designed for symbol learning. VSF–Network finds unknown parts of input data by comparing to stored pattern and it learns unknown patterns using unused part of the network. New patterns are learned incrementally and they are stored as sub-networks . Combinations of patterns are represented as combinations of the sub-networks. In this paper, the two theoretical backgrounds of VSF–Network are introduced. At the first, an incremental learning framework with Chaos Neural Networks is introduced. Next, the pattern recognition with the combined with symbols is introduced. From the viewpoints of9 differential topology and mixture distribution, the combined pattern recognition by VSF-Network is explained. Through an experiment, both the incremental learning capability and the pattern recognition with pattern combination are shown.

Index Terms: Incremental learning, Chaos Neural network, nonlinear dynamics, catastrophe theory.

1 Introduction

The purpose of our research is developing a model of symbol-generation with neural networks. We have reported our model and its performance in recent years[1,2]. In this paper, we show the backgrounds of VSF–Network, Vibration Synchronizing Function Network.

A human bing performs cognitive operations for pattersns obtained from abstracting data about external environment and use them as symbol. High level reasoning is the useful for robots to work autonomously in external environment. Symbol is a dominant background of the reasoning. We propose the symbol-generation process as the process that a system abstracts input data and learns patterns treated as symbol[1,2].

Models about symbol-generation have been proposed in the past years. Inamura[3] has proposed a model of stochastic behavior recognition and symbol-generation. The relations among patters are a key component of symbol generation. In the field of semantics, many theories about symbol have been also proposed. Chandler has proposed the model of the double articulation[4]. Semiotic codes belong to either single articulation, double articulation or non articulation. The double articulation enables a semiotic code to form an infinite number of meaningful combinations using a small number of low-level units.

T. Yoshida et al. (Eds.): AMT 2013, LNCS 8210, pp. 116–125, 2013.
© Springer International Publishing Switzerland 2013

Based on these studies, the generation process about symbol is summarized as follows.

1. At the first stage, patterns are learned by abstracting input data. The pattern is a prototype of a symbol.
2. A combination of patterns is learned by refining learned patterns.
3. The refined symbols and the combinations of them are maintained.

We focus on the first two steps to developing our model.

2 Incremental Learning by Chaos Neural Network

2.1 Incremental Learning

For neural networks, the learning of symbol is an instance of incremental learning[5] because neural networks learn incrementally new patterns keeping learned patterns. On the incremental learning, correlations among learned patterns take an important role. Lin and Yao[6] have proposed Negative Correlation Leaning model as a model of the incremental learning. In the proposed method, the neural network leans incrementally with increasing neurons based on the correlation between new patterns and learned patterns. If we can estimate the suitable number of neurons defined before learning, it is reasonable to increase the number of neurons according to progress of learning. However, there are redundant neurons in learned natural network. The over-learning occurs as increasing neurons. If we increase neurons, the problem changes for the worse. One way to solve the problem is reusing neurons and learning new patterns incrementally, if there are neurons that do not participate pattern recognition on a neural network. In VSF–Network, unknown patterns that have low correlation to learned patterns are learned by reusing a part of neurons in the neural network. VSF–Network is decided into sub-networks by the reusing. VSF–Network learns patterns incrementally by dividing the network into sub-networks, if it has redundant neurons at the first step of incremental learning.

2.2 Pattern Recognition by Chaos Neural Network

Researches about properties of the stored patterns in a neural network have been studied in the field of associative memory since the 1980s. In the associative memory by Chaos Neural Network[7], CNN, we can find chaotic retrieval dynamics in addition to normal dynamics on associative memory[8]. The chaotic dynamics can be find when input pattern is a partial pattern to stored patterns.

The i-th output x_i of neuron i in CNN that has M-input neurons and N-chaotic neurons is defined by (1).

$$x_i(t+1) = f\left[\xi_i(t+1) + \eta_i(t+1) + \zeta_i(t+1)\right] \tag{1}$$

In (1), an input term ξ_i, a feedback term η_i of internal status and an inhibitory ζ_i are respectively defined as follows.

$$\xi_i(t+1) = k_s\xi_i(t) + \sum_{j=1}^{N} v_{ij}A_j(t) \tag{2}$$

$$\eta_i(t+1) = k_n \eta_i(t) + \sum_{j=1}^{N} w_{ij} x_j(t) \tag{3}$$

$$\zeta_i(t+1) = k_r \zeta_i(t) - \alpha x_i(t) - \theta(1 - k_r) \tag{4}$$

In (2), (3) and (4), $A(t)$ is input at a time t, and v_{jk} is the connection weight between the $k-th$ element of input pattern and the chaos neuron j. w_{ij} is the connection weight between chaos neuron i and chaos neuron j. k_s, k_n, and k_r are a parameter for the each term. α is inhibitory strength parameter and θ_i is threshold for the inhibitory term. f is assumed to be Sigmoid function,

$$f_a(x) = \frac{1}{1 + e^{-ax}}.$$

2.3 Correlation between Associative Memory and Patterns

The dynamics of CNN is a dynamics of associative memory. Hopfield[9] shows that associative memory in the form of (5) has attractors and some of attractors reach equilibrium points, if W is a symmetric matrix and f is a monotonic increasing function.

$$x(t+1) = f(u(t)), \ u(t) = WX(t) - \alpha I \tag{5}$$

On (5), f is Sigmoid function, vector u is an inner status of neurons and W_{ij} is a connection weight between neuron i and j. α is a parameter to be 0 to diagonal element of W. ξ_μ is a column vector describing a pattern μ. We assume that W are updated with self correlations of patterns, if new patterns are provided. That is,

$$W = \sum_{\mu=1}^{p} \xi_\mu \xi_\mu^T - \alpha I. \tag{6}$$

If the initial value of (5) is data of learned pattern μ, it is $X(0) = \xi_\mu$ and a status of the associative memory for a pattern μ is $U_\mu(t) = WX_\mu(t-1) - \alpha I$. Statues of the attractor depends on WX and W depends on the collation between U_μ, so ξ_μ is $U_\mu \xi_\mu^T \simeq 1$. For other pattern $\nu(\neq \mu)$, we assume that

$$U_\mu \xi_{\nu,a}^T \simeq 1, \ U_\mu \xi_{\nu,c}^T \simeq O\left(\frac{1}{\sqrt{|N_c|}}\right). \tag{7}$$

In (7), $\xi_{\nu,a}$ is high correlated elements, $\xi_{\nu,c}$ is low correlated part of an input pattern ξ_μ and N_c is the set of elements of $\xi_{\nu,c}$. The status U_ν of the associative memory is,

$$U_\nu \propto WX_\nu = \sum_{\mu}\left(\xi_\mu \xi_\mu^T - \alpha I\right) X_\nu^T = \sum_{\mu}\left(\xi_\mu \xi_\mu^T\right) X_\nu^T - \alpha I X_\nu^T. \tag{8}$$

Because the second term of (8) is constant regardless of W, we take account only the first term of (8). Here, we introduce column partitioning for W. We say $W = [w_a, w_c]$ and (8) is

$$WX_\nu^T = w_a \xi_{\nu,a}^T + w_c \xi_{\nu,c}^T = \sum_{i \in N_a} w_i \xi_{(\nu,a),i} + \sum_{i \in N_c} w_i \xi_{(\nu,c),i}, \tag{9}$$

where N_a is the set of elements of $\xi_{\nu,a}$.

The result of (9) shows that the attractor reaches an equilibrium point ξ_μ, if $|N_a|$ is greater than $|N_c|$. Otherwise, if $|N_c|$ is greater than a certain value, an averaged status between ξ_μ and ξ_ν is an equilibrium point. Thus, an equilibrium point for a input ξ_ν depends on the ratio of $|N_a|$ and $|N_c|$.

2.4 Correlation on Chaos Neural Network

If a CNN works based on the dynamics (5), internal status u_i of CNN neuron i changes its values periodically and the values coincident with other neuron's values for the periods. This phenomenon is called synchronized oscillation.

The ξ_i term of CNN(1) is the self-interaction and η_i term is an interaction with other neuron j ($\neq i$), therefore CNN is a system that a status of neuron i diffuses to whole CNN through η term and each neuron interacts with others. For this reason, CNN is an instance of GCM, Globally Coupled Mapping[10], . The retracting observed on GCM is that some neurons show synchronized oscillation with other elements[11]. Patterns correlated with known pattern reach a equilibrium point corresponding to stored patterns as the result of the time evolution defined by (5). In contrast, other elements that have low correlation with the learned patterns do not reach any equilibrium point and they are in a precarious state. We analyze the relation between behavior of CNN neurons and the equilibrium stats.

CNN is a GCM defined by (10) on $\mathbb{R}^N (N \geq 1)$. In (10), f_a is assumed to be Sigmoid function.

$$F_{a,\varepsilon} : \mathbb{R}^N \to \mathbb{R}^N, x = (x_1, \cdots, x_N)^T \mapsto y = (y_1, \cdots, y_N))^T \tag{10}$$

$$y_i = (1 - \varepsilon) f_a(x_i) + \frac{\varepsilon}{N} \sum_{j=1, j \neq i}^{N} f_a(x_j) \ (1 \leq i \leq N)$$

The derivation of $F_{a,\varepsilon}$ is given by

$$F_a'(x) = (1 - \varepsilon) \begin{pmatrix} f_a'(x_i) & & 0 \\ & \ddots & \\ 0 & & f_a'(x_i) \end{pmatrix} + \frac{\varepsilon}{N} \begin{pmatrix} 1 \\ \vdots \\ 1 \end{pmatrix} (f_a'(x_1), \cdots, f_a'(x_N))$$

Komuro[11] studied dynamics on GCM and analyzed behavior of GCM elements. The invariant manifold H_σ is a sub-space H^N that coordinate permutation P_σ is invariant. H_σ is define by $H_\sigma = \{x \in \mathbb{R}^N ; x_i = x_{\sigma(i)}, 1 \leq i \leq N\}$. For any $\sigma \in S_N$ and $x \in H_\sigma$, a orthogonal complement $H_\sigma^\perp$ of H_σ is invariant to $F'(x)$. Here, σ is a cyclic permutation $\sigma = (i_{11}, \cdots, i_{1m_1}) \cdots (i_{k1}, \cdots, i_{km_k})$ where k is the number of groups and m is the number of elements in the group, and S_N is a symmetric group of degree N. The eigenvalue value of $F(x)$ toward the orthogonal complement $H_\sigma^\perp$ is

$$(1 - \varepsilon) f_a'(x_{i_{j1}}), \quad (j = 1, \cdots, k). \tag{11}$$

Based on the these definitions, we analyze dynamics of CNN as GCM. u_μ is a status of neuron i for a know pattern μ and u_ν is a status of neuron i for an unknown pattern

ν. Furthermore, $u_{\nu,a}$ is a part of u_{ν} that has high correlation with u_{μ} and $u_{\nu,c}$ is a part of u_{ν} that has low correlation with u_{μ}. For $u_{\nu,a}$ and $u_{\nu,c}$, we define permutation group σ_a and σ_c respectively. The eigenvalue of H_{σ_a} is $F(x)' u_{a,i} = (1 - \varepsilon) f_a'(x) u_{a,i}$, and the eigenvalue of H_{σ_c} is $F(x)' u_{c,i} = (1 - \varepsilon) f_a'(x) u_{c,i}$. From the definition $u_{\nu,a}$ and $u_{\nu,c}$, there are stable equilibrium points on H_{σ_a} but that is not alway the case for H_{σ_c}. Therefore, the eigenvalue of H_{σ_a} shows a stable status but the eigenvalue of H_{σ_c} dose not show a stable status.

For the retrieval process of CNN, the highly correlated elements retract to other elements and the low correlated elements do not retract. As the result, the low correlated elements show isolate oscillation. This fact shows that the low correlated group tends to show isolate oscillation comparing with highly correlated group. Therefore, role of a neuron to the patterns recognition is determined through the status of each elements based on their property as GCM.

3   Recognition of Combinational Patterns

On VSF–Network, a hierarchical neural network is divided into sub-networks with the selective weight update. The sub-network generated in this process corresponds to proto-symbol by the sub-network. The learning by VSF–Network is an estimation of a limited mixture model for a probability density function $p_k(x|a)$ with a parameter c_k,

$$f(x|a) = \sum_{k=1}^{K} c_k p_k(x|a_k) \tag{12}$$

where $p_k(x|a_k)$ is a probability density corresponding to each proto-symbol, that is a single pattern.

All patterns learned by VSF–Network do not always need for recognition. Multiple components or single component of limit mixture model are selected based on inputs. If a pattern for single pattern is given, VSF–Network retrieves a pattern for single pattern. The probability density learned by VSF–Network is a multi-modal distribution in some ranges of value and it is a single-modal distribution in other ranges. This fact means that VSF–Network is require for determining the final output based on the topological property of manifold spaned by a neural network. We apply Stochastic catastrophe theory[12,13] to implement a combinational pattern recognition on VSF-Network. Stochastic catastrophe theory privies an estimation of multi modal probability density through use of the bifurcation of solution at the singular points.

In general, there are singular points on manifold spanned by non-liner function. At neighborhood of singular point, bifurcation occurs. The number of value corresponding to a variable changes at bifurcation point. For example,the number of real root of 3-th degree equation $x^3 - cx = 0$ is 1 at the left side of the point $c = 0$ and it is 3 at the right site of the point $c - 0$. Thom[14] proposed catastrophe theory about the bifurcations at equilibrium points. Cobb[12,13] proposed to turn catastrophe theory into a stochastic catastrophe theory by adding to a dynamic system a Wiener process, $dW(t)$, with variance λ^2 and to treat the resulting equation as a SDE(13), stochastic differential equation,

$$dY = \frac{\partial V(Y;\alpha,\beta)}{\partial Y} dt + dW(t). \tag{13}$$

This SDE is then associated with a probability density that describes the distribution of the netrork's states on any moment in time, which may be expressed as

$$f(y) = \frac{\psi}{\sigma^2} \exp\left[\frac{\alpha(y-\lambda) + \frac{1}{2}\beta(y-\lambda)^2 - \frac{1}{4}(y-\lambda)^4}{\sigma^2}\right] \tag{14}$$

Here, ψ is a normalizing constant, and λ merely determines the origin of scale of the state variable. In this stochastic context, β is called the bifurcation factor, as it determines the number of modes of the density function, while α is called asymmetry factor as it determines the direction of the skew of the density. The density is symmetric if $\alpha = 0$ and becomes left or right skewed depending on the sign of α[13].

Grasman[15] proposed estimation method for the density function (14) based on methods proposed by Cobb[12,13]. If we have a set of measured dependent variables $Y_1, Y_2, \cdots Y_p$ to a first order approximation we say

$$y = w_0 + w_1 Y_1 + w_2 Y_2 + \cdots + w_p Y_p \tag{15}$$

where $w_1, w_2, \cdots w_p$ are the first order coefficients of a polynomial approximation to the transformation. Similarly, the parameters α and β are canonical variates in the sense that they are smooth transformations of actual control variates. For experimental parameters or measured independent variables $X_1, X_2, \cdots X_p$, we can write

$$\alpha = a_0 + a_1 X_1 + a_2 X_2 + \cdots + a_p X_p \tag{16}$$

$$\beta = b_0 + b_1 X_1 + b_2 X_2 + \cdots + b_p X_p \tag{17}$$

The core of Grasman's method is the fitting method that performs maximum likelihood estimation of all the parameters from (15) to (17) for observed dependent variables $Y_{i1}, Y_{i2} \cdots Y_{ip}$, and independent variables $X_{i1}, X_{i2} \cdots X_{ip}$, for subjects $i = 1 \cdots n$.

$$L(a, b, w; Y, X) = \sum_{i=0}^{n} \log \psi_i - \sum_{i=0}^{n} \left[\alpha_i y_i + \frac{1}{2}\beta_i y_i^2 - \frac{1}{4} y_i^4 (y-\lambda)^4\right] \tag{18}$$

Note that compared to (14), we have absorbed the location and scale parameters and into the coefficients $w_0, w_1, \cdots, w_p$.

The learning by VSF–Network is a minimizing L with respect to the parameters $w_0, \cdots, w_p, a_0, \cdots, a_p, b_0, \cdots, b_p$. An equilibrium points, as a function of the control parameters α and β, are solutions to the equation

$$\alpha + \beta y - y^3 = 0 \tag{19}$$

This equation has one solution if $\lambda = 27\alpha - 4\beta^3$, which is known as Cardan's discriminant, is greater than zero, and has three solutions if $\lambda < 0$. Because VSF–Network performs incremental learning, so it should update the parameters to $\lambda < 0$. λ is determined by the parameters α and β, so VSF–Network should learn unknown patterns to increase β without changing parameter corresponding to known patterns.

4 VSF–Network and Selective Weight Updating

We have proposed VSF–Network[1,2] to implement incremental learning of symbols and pattern recognition with their combination. VSF–Network is composed of BP–module that is a hierarchical network and CNN–module that is CNN. BP–module is trained with the selective weight update rule. CNN–module finds new parts of an input data and an used part of hidden layer neurons of BP–module.

4.1 Learning Procedure

VSF–Network works for the incremental learning only, so the learning of VSF–Network is assumed that the initial connection weights among layers have been learned before the its incremental learning. The learning of VSF–Network is performed as follows.

1. Data are provide to the input layer of BP–module.
2. The outputs of the hidden layer in BP–module are applied to CNN–module and they are used for the initial state of each neuron of CNN-module.
3. From the initial state, CNN–module performs the retrieval process based on the dynamics of (1) for times $t = 1, \cdots, T$. The synchronous rate defined by (20) is calculated.
4. The rest process of the forward path on BP–Module is performed. The error $E_{k,\mu}$ between the output $\hat{Y}_k$ from BP-module and the expected output Y^μ for the input data is calculated.
5. The connection weights among layers are updated based on the weight update rule defined by (21) and (22).

VSF–Network measure synchronous rate between chaos neuron i and j at a part of the times $t = 1, \cdots, T$. This measurement is implemented with a correlation integral(20) based on Heaviside function H.

$$C(r) = \frac{1}{n^2} \sum_{i,j=1 i \neq j}^{n} H(r - |x_i - x_j|) \tag{20}$$

4.2 Selecting Weights Updating

If inputs for CNN are statuses of hidden layer, we can find redundant neurons at hidden layer that do not have a role in pattern recognition based on the discussions of the previous sections. On CNN, redundant neurons that do not relate to pattern recognition show asynchronous oscillation while the network retrieves patterns. By updating only connection weights for neurons that show asynchronous oscillation, the network can learn new pattern without degradation of capability for learned patterns. By these ideas, we have been proposed weight updating method for neural network[1]. We can summarize properties of the selected weight updating rule as the following points.

– If multi-layer network has redundant neurons, then the weights are updated.
– The connection weights for neurons that show synchronous oscillation to other neurons are not updated and show asynchronous oscillation to other neurons are updated.

If there is a significant correlation between chaos neuron i and chaos neuron j, i and j output a similar value. To accentuate the effects of CNN, we apply the correlations among weights. The delta rule for multi-layer network is changed based on this selective weights update rule as follows.

$$\Delta W_{ij} = \begin{cases} \eta \frac{\partial E_{ij}}{\partial W_{ij}} & (\lambda_i \leq P) \\ 0 & (\lambda_i > P) \end{cases},$$

$$\frac{\partial E_{ij}}{\partial W_{ij}} = (1.0 - |cor_{ki,kj}|)^{-1} \sum_{j=1}^{n} \frac{\partial E_{jk}}{\partial W_{jk}} f'(H_i^\mu). \tag{21}$$

In (21), η is coefficient for update, E_{jk} is learning error, λ_i is degree of coincidence calculated with (20) and P is threshold for λ_i. ΔW_{ij} is the delta value for the weight between the i-th input layer neuron and the j-th hidden layer neuron. The update rule for ΔW_{jk} layer is,

$$\Delta W_{jk} = \begin{cases} \eta \frac{\partial E_{jk}}{\partial W_{jk}} & (\lambda_i \leq P) \\ 0 & (\lambda_i > P) \end{cases},$$

$$\frac{\partial E_{jk}}{\partial W_{jk}} = \left(\prod_{i=1}^{m} (1.0 - |cor_{ij}|) \right)^{-1} \sum_{j=1}^{n} E^\mu f'(O_k^\mu) \tag{22}$$

In (22), ΔW_{jk} is the delta value for the weight between the j-th hidden layer neuron and the k-th output layer neuron. $cor_{ki,kj}$ is correlation between neuron ki and kj in hidden layer.

5   Experiment and Result

In this section, we show a basic capability of VSF–Network through an experiment. In this paper,we show an experiment result about basic capability of VSF-Network. We show other results of experiments about various aspects of dynamics on VSF-Network in recent papers[1,2].

The task for the experiment is a learning of avoiding obstacles by a rover. The rover learns whether it can avoid an obstacle when the rover begins turning to the left from the point placed in. If the rover can avoid the obstacle, the output is $= 1$ otherwise the output is $= 0$. We have three conditions about obstacle setting. The condition 1 is T-Junction obstacle condition, the condition 2 is simple obstacle condition and the condition 3 is combined obstacle condition.

The procedure of the experiment are described as fellows.

- The initial step
  - The initial weight for BP–Module is learned by multi-layer network using 6000 records of from condition 1 of each task.
- The incremental learning step
  - We provide the different patterns from the patterns that are assigned at the previous step. The input data is 1500 records from the condition 2.
- The step for conforming learning performances.

- We compare MSE, Mean Squired Error, of each trial of the incremental learning to show the effect of the learnings.
- To confirm the combination form, the data combined the condition 1 and condition 2 are used in this step.

We show the result of incremental learning by VSF–Network for this task. Fig.1 shows the changes of MSE with the progress of the incremental learning for this task. For incremental learning of the task, the effect of VSF–network is observed. VSF-Network learns new patterns incrementally and its weights are not destroyed. The combined patterns are learned without learning with the progress of incremental learning. After a certain number of incremental learning, MSE of every incremental learning reaches a equilibrium status. VSF–Network incrementally learns by reusing neurons which are considered as inconsequential neurons for identification of learned patterns. The incremental learning stops when redundant neuron is lost.

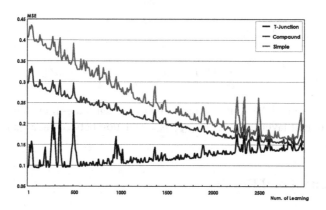

Fig. 1. Learning progress of VSF–Network

6 Conclusion

In this paper, we show the theoretical backgrounds of VSF–Network. The background consists of two parts. The first background is incremental learning and CNN. VSF–Network can identify new parts of input data and a part of neurons for reusing based on a dynamics of CNN. It learns only new parts of input data and it store in sub-networks. These sub-networks are generated automatically using unused parts of VSF–Netwok.

Another background is a mixture density estimation by stochastic catastrophe theory. The ability of VSF–Network for recognizing combined patterns that are learned by every sub-network can be explained by topological properties of solution space. Through the experiments, we show that VSF–Network can recognize the combined patterns only if it have learned parts of the patterns

The next step of our research concerns a detail consideration of the its dynamics. This points is related to the first background of VSF–Network and its capability for incremental learning.

References

1. Kakemoto, Y., Nakasuka, S.: The dynamics of incremental learning by vsf-network. In: Alippi, C., Polycarpou, M., Panayiotou, C., Ellinas, G. (eds.) ICANN 2009, Part I. LNCS, vol. 5768, pp. 688–697. Springer, Heidelberg (2009)
2. Kakemoto, Y., Nakasuka, S.: Neural assembly generation by selective connection weight updating. In: Proc. IjCNN 2010 (2010)
3. Inamura, T., Tanie, H., Nakamura, Y.: Proto-symbol development and manipulation in the geometry of stochastic model for motion generation and recognition. Technical Report NC2003-65, IEICE (2003)
4. Chandler, D.: Semiotics for Beginners. Routledge (1995)
5. Giraud-Carrier, C.: A note on the utility of incremental learning. AI Communications 13, 215–223 (2000)
6. Lin, M., Tang, K., Yao, X.: Incremental learning by negative correlation leaning. In: Proc. of IJCNN 2008 (2008)
7. Aihara, T., Tanabe, T., Toyoda, M.: Chaotic neural networks. Phys. Lett. 144A, 333–340 (1990)
8. Uchiyama, S., Fujisaki, H.: Chaotic itinerancy in the oscillator neural network without lyapunov functions. Chaos 14, 699–706 (2004)
9. Hopfield, J.: Neurons with graded response have collective computational properties like those of two-stage neurons. Proceedings of the National Academy of Sciences of U.S.A. 81, 13088–13092 (1984)
10. Kaneko, K.: Chaotic but regular posi-nega switch among coded attractors by cluster size variation. Phys. Rev. Lett. 63, 219 (1989)
11. Komuro, M.: A mechanism of chaotic itinerancy in globally coupled maps. In: Dynamical Systems (NDDS 2002) (2002)
12. Cobb, L., Ragade, R.: Applications of catastrophe theory in the behavioral and life. Behavioral Science 79(23), 291 (1978)
13. Cobb, L., Watson, B.: Statistical catastrophe theory: An overview. Mathematical Modellin 23(8), 1–27 (1980)
14. Thom, R.: Stability and Morphogenesis.: Essai D'une Theorie Generale Des Modeles. W. A. Benjamin, California (1973)
15. Grasman, R., van der Maas, H., Wagenmakers, E.: Journal of statistical software. Mathematical Modelling 32, 1–27 (2009)

Toward Robust and Fast Two-Dimensional Linear Discriminant Analysis

Tetsuya Yoshida and Yuu Yamada

Graduate School of Information Science and Technology,
Hokkaido University
N-14 W-9, Sapporo 060-0814, Japan
{yoshida,yamayuu}@meme.hokudai.ac.jp

Abstract. This paper presents an approach toward robust and fast Two-Dimensional Linear Discriminant Analysis (2DLDA). 2DLDA is an extension of Linear Discriminant Analysis (LDA) for 2-dimensional objects such as images. Linear transformation matrices are iteratively calculated based on the eigenvectors of asymmetric matrices in 2DLDA. However, repeated calculation of eigenvectors of asymmetric matrices may lead to unstable performance. We propose to use simultaneous diagonalization of scatter matrices so that eigenvectors can be stably calculated. Furthermore, for fast calculation, we propose to use approximate decomposition of a scatter matrix based on its several leading eigenvectors. Preliminary experiments are conducted to investigate the effectiveness of our approach. Results are encouraging, and indicate that our approach can achieve comparative performance with the original 2DLDA with reduced computation time.

1 Introduction

Inexpensive high resolution digital cameras makes it easy to take a lot of photos in daily life, and the photos are often uploaded into social networks to share with friends. Folksonomy or social tagging have been widely used for effective sharing of resources [8]. For example, del.icio.us is well known for social book-marking. For the handling of image files, pixiv has been used as a community site for image uploading. In addition, Flickr is a well-known photo-sharing site. Since tags or meta information may be available on images, exploitation of the available label information is useful to improve the performance of image categorization.

In order to exploit the label information on the images, this paper presents an approach toward robust and fast Two-Dimensional Linear Discriminant Analysis (2DLDA). We propose to use simultaneous diagonalization of scatter matrices to realize stable calculation of eigenvectors in 2DLDA. Furthermore, we propose to use approximate decomposition of a scatter matrix based on several leading eigenvectors. Preliminary experiments over the extended Yale Face Database B are conducted, and the performance of our approach is investigated through the comparison with other 2-dimensional methods.

The rest of this paper is organized as follows. Section 2 explains related work to clarify the context of this research. Section 3 explains the details of our approach.

T. Yoshida et al. (Eds.): AMT 2013, LNCS 8210, pp. 126–135, 2013.
© Springer International Publishing Switzerland 2013

Section 4 reports a preliminary evaluation of our approach and a comparison with other methods. Section 5 summarizes our contributions and suggests future directions.

2   Related Work

Since images are usually represented as 2-dimensional objects in computer, they can be naturally represented as matrices, where each element in a matrix corresponds to a pixel value in an image. On the other hand, most statistical methods [1] (e.g., PCA (Principal Component Analysis) or SVD (Singular Value Decomposition)) assume vector representation of objects, as in the bag-of-word model. Thus, when conducting statistical analysis of images under matrix representation, the images are first converted into high-dimensional vectors in previous approaches. However, converting images into vectors may lead to some computational problem due to the high-dimensionality of converted vectors.

In order to exploit matrix representation of images, several methods have been proposed under the name of "two-dimensional" analysis. Two-Dimensional PCA (2DPCA) [9] was proposed for faster computation of eigenvectors in terms of an "image" covariance matrix. This method was further extended to Two-Dimensional SVD (2DSVD) [2]. Both methods convert the original 2-dimensional images (matrices) into smaller matrices in terms of the eigenvalue analysis of the image covariance matrix. However, even when label information (e.g., a tag or category name) is provided for each image, these methods cannot exploit such information to improve the performance of image categorization. In order to exploit label information, Linear Discriminant Analysis (LDA) was extended to two-dimensional analysis in [10].

3   Robust and Fast Two-Dimensional Linear Discriminant Analysis

3.1   Preliminaries

A bold normal uppercase letter is used for a matrix, and a bold italic lowercase letter for a vector. For a matrix $\mathbf{A}$, $\mathbf{A}_{ij}$ stands for an element in $\mathbf{A}$, $\mathrm{tr}(\mathbf{A})$ stands for the trace of $\mathbf{A}$, and $\mathbf{A}^T$ stands for its transposition. $\mathbf{I}$ stands for a unit matrix. When a matrix $\mathbf{A}$ is not singular, its inverse matrix is represented as $\mathbf{A}^{-1}$. We assume that an image is represented as a 2-dimensional pixels in this paper. Thus, ith image is represented as a matrix $\mathbf{A}_i \in \mathbb{R}^{r \times c}$, for $i = 1, \ldots, n$, where n is the number of images. Also, we assume that n images are groupd into k classes $\Pi_1, \ldots, \Pi_k$, where Π_j has n_j images. Let $\bar{\mathbf{A}}_j = \frac{1}{n_j} \sum_{\mathbf{A} \in \Pi_j} \mathbf{A}$ be the mean of the jth class, $j = 1, \ldots, k$, and $\bar{\mathbf{A}} = \frac{1}{n} \sum_{j=1}^{k} \sum_{\mathbf{A} \in \Pi_j} \mathbf{A}$ be the global mean.

3.2   Two-Dimensional Linear Discriminant Analysis

In discriminat analysis, the separation between classes is maximized while the closeness within classes is minimized [1]. The former is quantified in terms of *within-class* scatter matrix $\mathbf{S}_w$, and the latter is quantified in terms of *between-class* scatter matrix $\mathbf{S}_b$. Linear Discriminant Analysis (LDA) tries to find the optimal linear transformation, which realizes the above objective, by converting the original data with the transformation. The matrix for the optimal linear transformation is obtained by solving a generalized eigenvalue problem for the scatter matrices.

Two-Dimensional Linear Discriminant Analysis (2DLDA) [10] is an extension of LDA for 2-dimensional images. 2DLDA tries to find two linear transformation matrices $\mathbf{U} \in \mathbb{R}^{r \times q_r}$ and $\mathbf{V} \in \mathbb{R}^{c \times q_c}$, which convert each image $\mathbf{A}_i$ to a (smaller) matrix $\mathbf{Y}_i = \mathbf{U}^T \mathbf{A}_i \mathbf{V}$. As in LDA, the matrices $\mathbf{U}$ and $\mathbf{V}$ are obtained by solving generalized eigenvalue problems for scatter matrices. Difference from the standard LDA is that, since rows and columns are dual in a matrix, two sets of scatter matrices are considered in 2DLDA, and matrices $\mathbf{U}$ and $\mathbf{V}$ are iteratively calculated.

The scatter matrices for obtaining $\mathbf{U}$ are defined by fixinig $\mathbf{V}$ as

$$\mathbf{S}_w^V = \mathrm{tr} \left(\sum_{j=1}^k \sum_{\mathbf{A} \in \Pi_j} (\mathbf{A} - \bar{\mathbf{A}}_j) \mathbf{V} \mathbf{V}^T (\mathbf{A} - \bar{\mathbf{A}}_j)^T \right) \tag{1}$$

$$\mathbf{S}_w^V = \mathrm{tr} \left(\sum_{j=1}^k n_j (\bar{\mathbf{A}}_j - \bar{\mathbf{A}}) \mathbf{V} \mathbf{V}^T (\bar{\mathbf{A}}_j - \bar{\mathbf{A}})^T \right) \tag{2}$$

$\mathbf{U}$ is obtained by solving the corresponding generalized eigenvalue problem: $\mathbf{S}_w^V \boldsymbol{u} = \lambda \mathbf{S}_b^V \boldsymbol{u}$, where $\boldsymbol{u}$ is a generalized eigenvector and λ is the corresponding generalized eigenvalue. By taking a specified number (denoted as q_r in this paper) of leading generalized eigenvectors (in descending order of eigenvalues), the qth column of $\mathbf{U}$ is set to the vector $\boldsymbol{u}_q$.

Similarly, the scatter matrices for obtaining $\mathbf{V}$ are defined by fixing $\mathbf{U}$ as

$$\mathbf{S}_w^U = \mathrm{tr} \left(\sum_{j=1}^k \sum_{\mathbf{A} \in \Pi_j} (\mathbf{A} - \bar{\mathbf{A}}_j)^T \mathbf{U} \mathbf{U}^T (\mathbf{A} - \bar{\mathbf{A}}_j) \right) \tag{3}$$

$$\mathbf{S}_w^U = \mathrm{tr} \left(\sum_{j=1}^k n_j (\bar{\mathbf{A}}_j - \bar{\mathbf{A}})^T \mathbf{U} \mathbf{U}^T (\bar{\mathbf{A}}_j - \bar{\mathbf{A}}) \right) \tag{4}$$

The columns of $\mathbf{V}$ are set to the leading a specified number (denoted as q_c in this paper) generalized eigenvectors of the corresponding generalized eigenvalue problem.

3.3 Robust Calculation of 2DLDA

One problem in 2DLDA is that, contrary to the well-known Laplacian eigenmap [7], it is necessary to calculate eigenvectors of *asymmetric* matrices. For example, matrix $\mathbf{U}$ is set to several leading eigenvectors of $(\mathbf{S}_w^V)^{-1}\mathbf{S}_b^V$, but matrix $(\mathbf{S}_w^V)^{-1}\mathbf{S}_b^V$ may be asymmetric. Since calculating eigenvectors of an asymmetric matrix may be numerically unstable, the performance of 2DLDA may be degraded.

In order to cope with the above problem, we propose to use the technique in [6] for the stable calculation of $\mathbf{U}$ and $\mathbf{V}$ in 2DLDA. Since this technique can be applied to both $\mathbf{U}$ and $\mathbf{V}$, we explain it by denoting $\mathbf{S}_w^V$ and $\mathbf{S}_b^V$ as $\mathbf{S}_w$ and $\mathbf{S}_b$, respectively.

The scatter matrix $\mathbf{S}_w$ is a symmetric matrix with positive eigenvalues by definition. Thus, it can be decomposed as

$$\mathbf{S}_w = \mathbf{H}\boldsymbol{\Lambda}\mathbf{H}^T \tag{5}$$

where $\mathbf{H}$ is orthogonal and $\boldsymbol{\Lambda}$ is diagonal. With $\mathbf{H}$ and $\boldsymbol{\Lambda}$, $(\mathbf{H}\boldsymbol{\Lambda}^{-1/2})^T$ $\mathbf{S}_w\mathbf{H}\boldsymbol{\Lambda}^{-1/2} = \mathbf{Z}\boldsymbol{\Sigma}\mathbf{Z}^T$, where $\mathbf{Z}^T\mathbf{Z} = \mathbf{I}$ and $\boldsymbol{\Sigma}$ is diagonal. By using matrix $\mathbf{H}\boldsymbol{\Lambda}^{1/2}\mathbf{Z}$, both $\mathbf{S}_w$ and $\mathbf{S}_b$ can be simultaneously diagonalized as

$$\mathbf{S}_w = \mathbf{H}\boldsymbol{\Lambda}^{1/2}\mathbf{Z}(\mathbf{H}\boldsymbol{\Lambda}^{1/2}\mathbf{Z})^T \tag{6}$$

$$\mathbf{S}_b = \mathbf{H}\boldsymbol{\Lambda}^{1/2}\mathbf{Z}\boldsymbol{\Sigma}(\mathbf{H}\boldsymbol{\Lambda}^{1/2}\mathbf{Z})^T \tag{7}$$

Since $\mathbf{S}_w^{-1} = \mathbf{H}\boldsymbol{\Lambda}^{-1}\mathbf{H}^T$, matrix $\mathbf{S}_w^{-1}\mathbf{S}_b$ can be decomposed as

$$\mathbf{S}_w^{-1}\mathbf{S}_b = \mathbf{H}\boldsymbol{\Lambda}^{-1/2}\mathbf{Z}\boldsymbol{\Sigma}\mathbf{Z}^T\boldsymbol{\Lambda}^{1/2}\mathbf{H}^T$$

$$= \boldsymbol{\Delta}\boldsymbol{\Sigma}\boldsymbol{\Delta}^{-1} \tag{8}$$

where $\boldsymbol{\Delta} = \mathbf{H}\boldsymbol{\Lambda}^{-1/2}\mathbf{Z}$. Thus, columns of $\boldsymbol{\Delta}$ correspond to the eigenvectors of $\mathbf{S}_w^{-1}\mathbf{S}_b$ [6]. By substituting $\mathbf{S}_w^V$ and $\mathbf{S}_b^V$ into $\mathbf{S}_w$ and $\mathbf{S}_b$, we can stably calculate matrix $\mathbf{U}$ as the eigenvectors of $(\mathbf{S}_w^V)^{-1}\mathbf{S}_b^V$. The same procedure can be used for calculating $\mathbf{V}$.

3.4 Fast Calculation of 2DLDA

In the robust calculation in Section 3.3, all the eigenvectors of $\mathbf{S}_w$ are used in spectral decomposition [6]. For fast calculation of 2DLDA, we propose to use approximate decomposition of $\mathbf{S}_w$ based on its several leading eigenvectors as

$$\mathbf{S}_w \simeq \tilde{\mathbf{H}}\tilde{\boldsymbol{\Lambda}}\tilde{\mathbf{H}}^T \tag{9}$$

where $\tilde{\mathbf{H}}^T\tilde{\mathbf{H}} = \mathbf{I}$ and $\tilde{\boldsymbol{\Lambda}}$ is diagonal. With the above $\tilde{\mathbf{H}}$ and $\tilde{\boldsymbol{\Lambda}}$, similar relations in the approximate decomposition hold, e.g., $(\tilde{\mathbf{H}}\tilde{\boldsymbol{\Lambda}}^{-1/2})^T\mathbf{S}_w\tilde{\mathbf{H}}\tilde{\boldsymbol{\Lambda}}^{-1/2} = \tilde{\mathbf{Z}}\tilde{\boldsymbol{\Sigma}}\tilde{\mathbf{Z}}^T$, where $\tilde{\mathbf{Z}}^T\tilde{\mathbf{Z}} = \mathbf{I}$ and $\tilde{\boldsymbol{\Sigma}}$ is diagonal. Similarly, $\mathbf{S}_w^{-1}\mathbf{S}_b$ can be approximately decomposed as

$$\mathbf{S}_w^{-1}\mathbf{S}_b \simeq \tilde{\boldsymbol{\Delta}}\tilde{\boldsymbol{\Sigma}}\tilde{\boldsymbol{\Delta}}^{-1} \tag{10}$$

where $\tilde{\boldsymbol{\Delta}} = \tilde{\mathbf{H}}\tilde{\boldsymbol{\Lambda}}^{-1/2}\tilde{\mathbf{Z}}$. Use of only several leading eigenvectors can contribute to fast calculation of linear transformation matrices in 2DLDA.

4 Evaluation

4.1 Experimental Setting

Datasets. Based on previous work [9], the effectiveness of the approaches in Section 3 was evaluated for image categorization. In the following experiments, we used the extended Yale Face Database B[1]. This dataset contains 16128 images of 28 human subjects under 9 poses and 64 illumination conditions. Each image in this dataset is represented as a 480 by 640 matrix.

Evaluation Measures. Cluster assignment was evaluated with respect to the Normalized Mutual Information (*NMI*) [5]. Let C, $\hat{C}$ stand for the random variables over the true and assigned clusters. *NMI* is defined as:

$$NMI = \frac{I(\hat{T};T)}{(H(\hat{T}) + H(T))/2} \quad (\in [0,1]) \tag{11}$$

where $H(\cdot)$ is Shannon Entropy, $I(;)$ is Mutual Information. *NMI* in eq.(11) corresponds to the accuracy of cluster assignment. The larger *NMI* is, the better the clustering assignment is.

Purity is defined based on the contingency table of the ground truth cluster C_j and the constructed cluster $\hat{C}_h$ as

$$Purity = \frac{1}{n} \sum_{j=1}^{k} \max_{h} |C_j \cap \hat{C}_h| \quad (\in [0,1]) \tag{12}$$

where n is the number of data. As shown in eq.(12), it is calculated based on the number of shared instances (in this paper, images) among C_j and $\hat{C}_h$. The larger Purity is, the more cohesive the constructed clusters are.

Methods. We evaluated the performance of the following state-of-the-art 2-dimensional methods:

- 2DLDA [10]
- app2DLDA (our approach)
- 2DPCA [9]
- 2DSVD [2]

For a given 2-dimensional image, which is represented as a matrix, the above methods construct a compressed (smaller size) matrix. Thus, image clustering was conducted by applying some off-the-shelf clustering algorithm to the constructed matrices. We used the standard kmeans algorithm [3] in the subsequent experiments. When applying kmeans to matrices, the matrices were converted into vectors with vec operator [4]. Also, we evaluated the performance of kmeans as a baseline for image clustering. Since kmeans is a partitioning based clustering

---

[1] http://vision.ucsd.edu/~leekc/ExtYaleDatabase/ExtYaleB.html

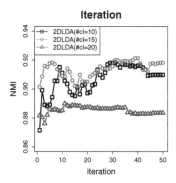

Fig. 1. Influence of the number of iterations

algorithm, we used the number of clusters (which corresponds to the number of humans in the dataset) in the subsequent experiments.

Evaluation Procedure. For a specified set of clusters, 10 images were randomly sampled for each cluster to construct a test set for the clusters. Furthermore, for each test set, clustering was repeated 10 times with different initial configuration. Thus, 100 runs were conducted for each set of clusters, and the averaged results are reported in the subsequent experiments. Since 2DLDA constructs linear transformation matrices based on the labeled images, for a specified set of clusters, 40 labeled images were randomly sampled for each cluster to construct a training set.

Parameters. 2DLDA has two kinds of hyper-parameters: the number of eigenvectors, and the number of iterations. The number of columns in matrix $\mathbf{U}$ is denoted as q_c, and that in $\mathbf{V}$ is denoted as q_r, respectively. The maximum number of eigenvectors with positive eigenvalues in LDA is the number of cluster minus one. Thus, in order to exploit the label information on images, both q_c and q_r were set to the number of cluster minus one. The number of eigenvectors is also a hyper-parameter in other 2-dimensional methods. Since 2DSVD tries to construct compressed images, q_c and q_r were set to 5 % of the pixel size in the original image. Thus, since the original images were 480 by 640 matrices, q_c and q_r were set to 24 and 32, respectively. In 2DPCA, q_c was set to 20.

In 2DLDA, linear transformation matrices are obtained by *iteratively* calculating several leading eigenvectors in the generalized eigenvalue problems. Thus, since the number of iterations may affect the performance of 2DLDA, we evaluated the influence of the number of iterations. The results are summarized in Fig. 1. In Fig. 1, the horizontal axis corresponds to the number of iterations, and the vertical axis corresponds to *NMI* in eq.(11). In the legend, black line with squares corresponds the result when the number of clusters was set to 10 (#cl = 10 in the legend); green line with circles, to 15 (#cl = 15); red line with triangles, to 20 (#cl = 20).

The results in Fig. 1 indicate that the clustering performance of 2DLDA is stable with respect to the number of iterations, and that only a small number of

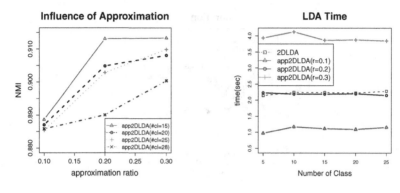

Fig. 2. Results of of approximate decomposition in eq.(9)

iterations for calculating matrices $\mathbf{U}$ and $\mathbf{V}$ is enough to improve its performance in 2DLDA. Thus, in order to reduce the computation time, we set the number of iterations to two in the subsequent experiments.

4.2 Evaluations of Approximate Decomposition

The influence of approximation in eq.(9) is summarized in the left-hand-side of Fig. 2. The horizontal axis corresponds to the approximation ratio (hereafter denoted as r), and the vertical axis corresponds to *NMI*. Here, the approximation ratio (r) is defined as the number of used leading eigenvectors divided by the number of clusters minus one (i.e., the number of maximum eigenvectors with positive eigenvalues in 2DLDA). Results for different number of clusters are shown in the figure.

The left-hand-side of Fig. 2 indicate that the clustering performance can be improved by increasing the number of used leading eigenvectors in the approximation. However, the performance improvement may be saturated around $r=0.2$. This result indicate that it is not necessary to use all the eigenvectors in 2DLDA to achieve competitive performance.

The right-hand-side in Fig. 2 shows the computation time in the decomposition of $\mathbf{S}_w^{-1}\mathbf{S}_b$. The running time was measured on a computer with Debian/GNU Linux, Intel Xeon W5590 (3.33GHz), 36 GB memory. As expected, using only some leading eigenvectors contributes to faster computation (e.g., app2DLDA ($r=0.1$)). However, app2DLDA ($r=0.3$) was slower than the original 2DLDA. This may be due to our current inefficient implementation for eigenvalue decomposition. Further investigation is needed to resolve this problem.

4.3 Comparisons with Two-Dimensional Methods

Comparisons of our approach with other 2-dimensional methods are summarized in Figs. 3 and 4. The horizontal axis in these figures corresponds to the number of classes. The vertical axis in Fig. 3 corresponds to *NMI* in eq.(11); and that

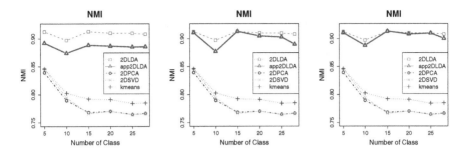

Fig. 3. Results of *NMI* (from left to right: *r*=0.1, 0.2, 0.3)

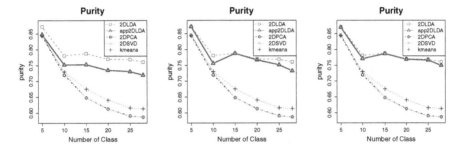

Fig. 4. Results of Purity (from left to right: *r*=0.1, 0.2, 0.3)

in Fig. 4, to Purity in eq.(12). In the legend, brown lines with squares corresponds 2DLDA; red lines with triangles, to app2DLDA; blue lines with circles, to 2DPCA; water-blue line with crosses, to 2DSVD; black lines, to kmeans. Also, the approximation ratio (r) in Figs. 3 and 4 was set to 0.1, 0.2 and 0.3 (from left to right), respectively.

The results in Figs. 3 and 4 indicate that LDA based methods outperform other 2-dimensional methods (2DPCA and 2DSVD) and kmeans. This would be because LDA based methods can exploit the labeled images when constructing compressed (smaller sized) matrices via linear transformation. Also, when the approximation ratio (r) is equal to or greater than 0.2, the performance of the proposed approach (app2DLDA) is almost equivalent to that of the original 2DLDA, albeit our approach requires less computation than 2DLDA (as shown in the right-hand-side of Fig. 2). On the other hand, the performance of 2DPCA and that of 2DSVD were almost the same, and were inferior to kmeans when the number of clusters was increased. This might be because the number of images per cluster was relatively small in the experiments.

Detailed comparisons of 2DLDA and our approach are shown in Figs. 5. The vertical axis in the left-hand-side of Figs. 5 corresponds to *NMI*, and that in the right-hand-side of Figs. 5 corresponds to Purity. The brown lines with squares corresponds 2DLDA; the red lines with triangles, to app2DLDA ($r = 0.1$); the blue lines with circles, to app2DLDA ($r = 0.2$); the water-blue line with crosses, to app2DLDA ($r = 0.3$).

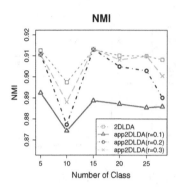
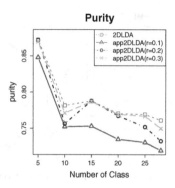

Fig. 5. Results

The results in Figs. 5 also indicate that the performance of our approach is almost equivalent to that of the original 2DLDA. Except for when the approximation ratio (r) was set to 0.1, our approach showed competitive performance with the original 2DLDA.

4.4   Discussion

Although matrices **U** and **V** are iteratively calculated in 2DLDA [10], we confirmed that a small number of iterations is enough for image categorization (see Fig. 1). Under this setting, the results of experiments indicate that approximate decomposition in our approach can achieve competitive performance with the original 2DLDA, and can outperform both 2DPCA and 2DSVD in image categorization. However, the results also revealed that much work need to be conducted for efficient calculation of leading eigenvectors in the approximation. We plan to hack the available linear algebra packages for faster computation in near future.

5   Concluding Remarks

This paper presented an approach toward robust and fast Two-Dimensional Linear Discriminant Analysis (2DLDA). Since the repeated calculation of eigenvectors of asymmetric matrices may lead to unstable performance, we proposed to use simultaneous diagonalization of scatter matrices so that eigenvectors can be stably calculated. Furthermore, we proposed to use approximate decomposition of a scatter matrix based on its several leading eigenvectors. Results of preliminary experiments indicate that approximate decomposition in our approach can achieve competitive performance with the original 2DLDA, and that the proposed approach can outperform other 2-dimensional methods in image categorization. We plan to conduct more in-depth analysis of our approach, especially the computational issue, and extend it based on the analysis in near future.

Acknowledgments. This work is partially supported by the grant-in-aid for scientific research (No. 24300049) funded by MEXT in Japan and the Murata Science Foundation.

References

1. Anderson, T.W.: An Introduction to Multivariate Statistical Analysis. Wiley-Interscience (2003)
2. Ding, C., Ye, J.: Two-dimensional singular value decomposition (2dsvd) for 2d maps and images. In: Proc. of SDM 2005, pp. 32–43 (2005)
3. Hartigan, J., Wong, M.: Algorithm as136: A k-means clustering algorithm. Journal of Applied Statistics 28, 100–108 (1979)
4. Harville, D.A.: Matrix Algebra From a Statistican's Perspective. Splinger (2008)
5. Strehl, A., Ghosh, J.: Cluster ensembles — a knowledge reuse framework for combining multiple partitions. J. Machine Learning Research 3(3), 583–617 (2002)
6. Swets, D.L., Weng, J.J.: Using discriminat eigenfeatures for image retrieval. IEEE Transactions on Pattern Analysis and Machine Intelligence 18(8), 831–836 (1996)
7. von Luxburg, U.: A tutorial on spectral clustering. Statistics and Computing 17(4), 395–416 (2007)
8. Voss, J.: Tagging, folksonomy & co – renaissance of manual indexing? In: Proc. 10th International Symposium for Information Science, pp. 234–254 (2007)
9. Yang, J., Zhang, D., Frangi, A.F., Yu Yang, J.: Two-dimensional pca: A new approach to appearance-based face representation and recognition. IEEE Transactions on Pattern Analysis and Machine Intelligence 26(1), 131–137 (2004)
10. Ye, J., Janardan, R., Li, Q.: Two-dimensional linear discriminant analysis. In: Proc. of NIPS 2004, pp. 1569–1576 (2004)

Research on the Algorithm of Semi-supervised Robust Facial Expression Recognition

Bin Jiang, Kebin Jia*, and Zhonghua Sun

School of Electronic Information & Control Engineering,
Beijing University of Technology, Beijing 100124, China
kebinj@bjut.edu.cn

Abstract. Under the condition of multi-databases, a novel algorithm of facial expression recognition was proposed to improve the robustness of traditional semi-supervised methods dealing with individual differences in facial expression recognition. First, the regions of interest of facial expression images were determined by face detection and facial expression features were extracted using Linear Discriminant Analysis. Then Transfer Learning Adaptive Boosting (TrAdaBoost) algorithm was improved as semi-supervised learning method for multi-classification. The results show that the proposed method has stronger robustness than the traditional methods, and improves the facial expression recognition rate from multiple databases.

Keywords: Facial Expression Recognition, Semi-Supervised Learning, TrAdaBoost.

1 Introduction

Facial expression recognition can be help for obtaining human's psychology and emotion, and providing technical support to enhance research in the field of human-computer interaction and multimedia information processing[1]. It relies to a large extent on the number of labeled images for training classifier. Large training set can better reflect the real distribution of samples and consequently obtaining good generalization error. However, image labeling is very difficult, expensive and time consuming. Semi-supervised learning [2] makes use of a lot of unlabeled samples combined with training set. These unlabeled samples are helpful to construct the exact model for facial expression classification. So semi-supervised learning becomes an important method to solve the problem with lack of labeled data and to improve the practicability of the facial expression recognition.

* Supported by the National Natural Science Foundation of China under Grant No.30970780 and Ph.D. Programs Foundation of Ministry of Education of China under Grant No. 20091103110005.

T. Yoshida et al. (Eds.): AMT 2013, LNCS 8210, pp. 136–145, 2013.
© Springer International Publishing Switzerland 2013

In recent years, many semi-supervised learning algorithms were proposed for facial expression recognition. In 2010, Hady et al [3] proposed a facial expression learning framework to combine Tri-Class SVMs with Co-Training method. In 2011, Chen et al [4] took all three semi-supervised assumptions, including smoothness, cluster, and manifold assumptions, together into account during boosting learning. Experiment demonstrates that the algorithm yields favorite results for facial expression recognition tasks in comparison to other state-of-the-art semi-supervised learning algorithms. Although these algorithms have good performance on facial expression recognition, an important problem may be overlooked by the most semi-supervised learning algorithms. That is, facial expressions depend on human face. The differences caused by individual will bring great difficulties to facial expression recognition [5]. Traditional semi-supervised learning methods for facial expression recognition only use samples from the same database, not only remove the differences caused by individual, but also meet the semi-supervised assumption that the distributions of the training and testing data are the same [6]. However, the facial expression images don't always have the same ethnic, gender or cultural background in different facial expression databases. Semi-supervised learning assumption which is the training and testing data has the same distribution, may not hold [7]. So in case of multi-databases the performance of current facial expression recognition systems may decrease significantly.

According to the above problem, TrAdaBoost (Transfer Learning Adaptive Boosting) [8] algorithm leverages the source domain data to construct a high-quality classification model for the target domain data, and improves the recognition rate and robustness on the classification of multi-databases. But TrAdaBoost does not belong to semi-supervised learning, and not need labeled data in the training stage. Meanwhile, TrAdaBoost is only suitable for two-class classification, and does not meet the requirements of semi-supervised facial expression recognition. So we propose a novel algorithm called Semi-Supervised Learning Adaptive Boosting. On one hand, the proposed algorithm retains the ability of TrAdaBoost and uses knowledge transfer to reduce the interference of individual differences in multi-databases facial expression recognition. On the other hand, the proposed algorithm can realize multi-classification. Compared with traditional semi-supervised facial expression recognition algorithms, Semi-supervised learning Adaptive Boosting is effective and robust.

2   Our Approach

2.1   The Principle of TrAdaBoost

Because our algorithm and TrAdaBoost is closely related, the principle of TrAdaBoost needs to be analyzed at first. Let X be data and $Y = \{0, 1\}$ be the set of category labels. The training data set $T \subseteq \{X \times Y\}$ is partitioned into two labeled sets T_d and T_s. T_d represents the diff-distribution training data that $T_d = \{(x_i^d, c(x_i^d))\}$, where $x_i^d \in X_d$ $(i = 1, 2, ..., n)$, and $c(x_i^d)$ returns the label for the data instance x_i^d. T_s represents the same-distribution training data that

$T_s = \{(x_i^s, c(x_i^s))\}$, where $x_i^s \in X_s$ $(i = 1, 2, ..., m)$. n and m are the sizes of T_d and T_s, respectively. The training set $T = \{(x_i, c(x_i))\}$ is defined as follows:

$$x_i = \begin{cases} x_i^d, & i = 1, ..., n \\ x_i^s, & i = n+1, ..., n+m \end{cases}$$

The test data set is denoted by $S = \{(x_i^t)\}$, where $x_i^t \in X_s (i = 1, 2, ..., q)$. Here, q is the size of test data set. TrAdaBoost can transfer the knowledge from X_d to X_s, and construct an effective classification model for S.

The formal descriptions of TrAdaBoost algorithm are given as follows:

Transfer Learning Adaptive Boosting Algorithm

Input the labeled training data set T, the unlabeled test data set S,

a base learning algorithm **Learner**, and the maximum number of iterations N.

Initialize the initial weight vector, that $w^1 = (w_1^1, ..., w_1^{n+m})$

and the parameter is computed that $\beta = 1/(1 + \sqrt{2 \times \ln n/N})$.

For $t = 1, 2, ..., N$

    1. Set $p^t = w^t/(\sum_{i=1}^{n+m} w_i^t)$.

    2. Call **Learner** to classify dataset with the distribution p^t .

    Then, get back a hypothesis $h_t : X \to Y$.

    3. Calculate the error of h_t on T_s:

$$\epsilon_t = \sum_{i=n+1}^{n+m} \frac{w_i^t \times |h_t(x_i) - c(x_i)|}{\sum_{i=n+1}^{n+m} w_i^t}$$

    4. Set $\beta = \epsilon_t/(1 - \epsilon_t)$, and ϵ_t is required to be less than $1/2$.

    5. Update the new weight vector:

$$w_i^{t+1} = \begin{cases} w_i^t \times \beta^{|h_t(x_i) - c(x_i)|}, & i = 1, ..., n \\ w_i^t \times \beta^{-|h_t(x_i) - c(x_i)|}, & i = n+1, ..., n+m \end{cases}$$

Output the hypothesis:

$$h_f(x) = \begin{cases} 1, & \prod_{t=\lceil N/2 \rceil}^{N} \beta_t^{-h_t(x)} \geq \prod_{t=\lceil N/2 \rceil}^{N} \beta_t^{-\frac{1}{2}} \\ 0, & otherwise \end{cases}$$

Where $\lceil N/2 \rceil$ denotes ceiling function, which returns the next highest integer that is greater than or equal to N/2. Just like AdaBoost [9] algorithm, TrAdaBoost also aims to boost the accuracy of a weak learner by carefully adjusting the weights of training data and learn a classifier accordingly. Further more, TrAdaBoost assumes that, due to the difference in distributions between training and testing data, some of training data may be useful in learning for testing data but some of them may not and could even be harmful. It attempts to iteratively reweight training data to reduce the effect of the bad training data while encourage the good training data to contribute more for testing data classification. However, TrAdaBoost is a binary classifier, and can't solve the multi-class learning problem. Meanwhile, TrAdaBoost belongs to supervised learning, and not a semi-supervised learning method. So we propose a novel algorithm called Semi-Supervised Learning Adaptive Boosting.

2.2   The Principle of Our Approach

The formal descriptions of semi-supervised learning adaptive boosting algorithm are given as follows:

Semi-Supervised Learning Adaptive Boosting Algorithm

Input training data set T, the labels of training data $y_i \in Y = \{1, 2, ..., l\}$,

$i = 1, 2, ..., n + m$, the unlabeled test data set S, a base learning algorithm

Learner, and the maximum number of iterations N.

Initialize nearest neighbor classifier is applied to assign the labels of unlabeled

training data and the initial weight vector, that $w^1 = (w_1^1, ..., w_1^{n+m})$ and the

parameter is computed that $\beta = 1/(1 + \sqrt{2 \times \ln n / N})$.

For $t = 1, 2, ..., N$

1. Set $p^t = w^t / (\sum_{i=1}^{n+m} w_i^t)$.

2. Call **Learner** to classify dataset with the distribution p^t.

   Then, get back a hypothesis $h_t : X \to Y$.

3. Calculate the error of h_t on T_s, where, for any predicate π, we define

   $[\pi]$ to be 1, if π holds and 0 otherwise.

$$\epsilon_t = \sum_{i=n+1}^{n+m} \frac{w_i^t [h_t(x_i) \neq y_i]}{\sum_{i=n+1}^{n+m} w_i^t}$$

4. Set $\beta = \epsilon_t / (1 - \epsilon_t)$, and ϵ_t is required to be less than $1/2$.

5. Update the new weight vector:

$$w_i^{t+1} = \begin{cases} w_i^t \times \beta^{[h_t(x_i) \neq y_i]}, & i = 1, ..., n \\ w_i^t \times \beta^{-[h_t(x_i) \neq y_i]}, & i = n+1, ..., n+m \end{cases}$$

Output the hypothesis:

$$h_f(x) = \arg\max_{y \in Y} \sum_{t=1}^{N} (\log \frac{1}{\beta_t}) [h_t(x) = y]$$

Our approach has retained the ability to transfer knowledge and improved on the TrAdaBoost in two ways.

On one hand, TrAdaBoost belongs to supervised learning method, and the labels of training data are needed for the experiment. But semi-supervised facial expression recognition requires large amount of unlabeled data together with the labeled data to build classifiers. We randomly choose some training samples as labeled data for every facial expression, and the rest of training samples will be unlabeled data. If we choose all training data as labeled data, our algorithm will not belong to semi-supervised learning. But if we don't supply any label information for training data, the classification accuracy of our algorithm will be decreased, and the performance of knowledge transfer will also be affected. A trade-off is a situation that we choose proportion of certain training samples as labeled data for every facial expression, and the rest of training samples become unlabeled data. So the requirement of semi-supervised learning and performance of our algorithm will be taken into consideration. Nearest neighbor classifier is working on computing the label of unlabeled training samples in order to eliminate unwanted effect.

On the other hand, TrAdaBoost can only deal with binary classification. However, facial expression recognition needs multi-classification. Because TrAdaBoost is an extension of the AdaBoost algorithm, we learn from an extension of the AdaBoost algorithm for multi-class learning, called AdaBoost.M1 [9]. A lot of changes are made as follows: (1) Learner should be a multi-classifier; (2) the equation of ϵ_t is changed; (3) we revise the equation to update the new weight vector; (4) the equation of hypothesis is inherited from AdaBoost.M1 algorithm. In the extremely, if training and testing data follow the same distribution, our approach will be equivalent to AdaBoost.M1.

3   Simulations and Experiments

3.1   Experimental Setting

In this experiment, we evaluated our algorithm on Japan Female Facial Expression Database [10] (JAFFE), Cohn-Kanade AU-Coded Database [11] (CK) and Radboud Faces Database [12] (RaFD). As human emotions are commonly categorized into six discriminative basic classes (angry, disgust, fear, happy, sad and surprise) in facial expression recognition, we extracted 180 images from JAFFE, 240 peak emotional images from CK, and 342 frontal images from RaFD database. These images contains of six basic facial expressions, and each emotion has the same number of images in order to avoid data imbalanced.

Our experimental procedure is categorized into three stages: preprocessing, feature extraction and expression classification. For the purpose of comparisons with the methods in Refs. 9, 13 and 14, we crop these original images of size 256×256, 640×490, and 681×1024 into 168×120 by removing the back-ground influences. Several representative samples are shown in Fig.1.

Since the illumination condition is varied to the images in experimental databases, we apply histogram equalization to eliminate lighting effects.

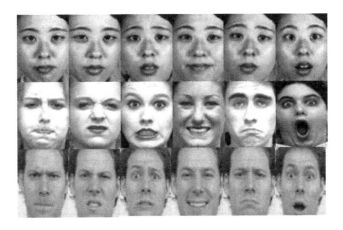

Fig. 1. Sample Images

After the preprocessing of experiment, we apply principal component analysis to reduce the feature dimensionality into a 40 dimensional sub-space. Then linear discriminant analysis (LDA) is used to extract important features from each image. We adopt k nearest neighbor (k-NN) to identify six facial expressions in the experimental databases, knn is also the base classifier in AdaBoost.M1, RegBoost and our proposed method. It can be called by Learner. Let $k = 11$ in our experiments.

3.2   Comparisons and Analysis

In order to test the performance of the proposed algorithm, Label Propagation (LP) [13], AdaBoost.M1 and RegBoost[14] algorithm have been selected as baseline algorithms. LP is a classical algorithm in semi-supervised learning, and RegBoost is an effective semi-supervised learning algorithm for facial expression recognition.

Firstly, we choose 180 samples of JAFFE database as source domain T_d, and 240 samples of CK database are selected as target domain T_s. 180 samples, which come from JAFFE database, belong to training data set, so that $n = 180$. The rest training samples are randomly chosen from CK database at a certain rate for each facial expression, and the rate is from 10% to 50%. So that $m = r * n$. The rest samples in T_s are used as test data set S, so that $q = 240 - r * n$. We perform the same procedure of training and testing repeatedly for 10 times, and average all the 10 recognition rates to obtain the final performance of experimental methods. Note that, the number of labeled and unlabeled samples in training data set is 1:1. The number of basic classifier and iteration is ten.

Table 1 shows the performance comparisons among our proposed algorithm and baseline algorithms. We can see that our proposed algorithm is better than baseline algorithms in the most cases. Then we choose 240 samples of CK database as source domain T_d, and 180 samples of JAFFE database are selected as target domain T_s. The results are shown in Table 2.

Table 1. Mean Error Rates of JAFFE and CK Databases

	10%	20%	30%	40%	50%
LP	63.20%	61.67%	60.86%	59.05%	57.20%
AdaBoost.M1	47.16%	42.60%	38.71%	36.13%	36.40%
RegBoost	46.58%	42.21%	38.23%	35.77%	36.00%
Proposed	46.53%	42.25%	38.23%	35.48%	35.73%

Table 2. Mean Error Rates of CK and JAFFE Databases

	10%	20%	30%	40%	50%
LP	74.17%	66.14%	62.13%	54.76%	47.50%
AdaBoost.M1	62.63%	61.06%	58.06%	56.55%	55.50%
RegBoost	65.13%	68.33%	60.28%	66.65%	57.83%
Proposed	61.99%	56.59%	57.31%	55.95%	55.17%

In Table 2, the recognition rates are low. Because there are only Japanese females in JAFFE database, we dont need to take into consideration of individual differences. However, there are many people in CK database, for example, European, Asian and African. Even though they have the same facial expression, the intensity and appearance of facial expression are different. Individual differences will add to our difficulties in the experiment. But our algorithm still achieves higher recognition rates than baseline algorithms, when the rate is equal to 10%, 20% or 30%.

Secondly, we choose 180 samples of JAFFE database as source domain T_d, and 342 samples of RaFD database are selected as target domain T_s. The experimental setting is same to the experiments on JAFFE and CK databases. The results are shown in Table 3. Then we choose 342 samples of RaFD database as source domain T_d, and 180 samples of JAFFE database are selected as target domain T_s. The results are shown in Table 4.

Table 3 and Table 4 show that our algorithm has better performance on the JAFFE and RaFD database. Because the people in the two databases are relatively homogeneous, we can avoid interference of individual differences.

Finally, we choose 240 samples of CK database as source domain T_d, and 342 samples of RaFD database are selected as target domain T_s. Then we choose 342 samples of RaFD database as source domain T_d, and 240 samples of CK database are selected as target domain T_s. The results are shown in Table 5 and Table 6.

Table 3. Mean Error Rates of JAFFE and RaFD Databases

	10%	20%	30%	40%	50%
LP	63.95%	58.04%	56.63%	50.96%	49.60%
AdaBoost.M1	47.38%	41.11%	37.60%	32.89%	32.18%
RegBoost	46.76%	40.33%	36.81%	32.07%	31.03%
Proposed	46.76%	40.26%	36.42%	32.00%	30.95%

Table 4. Mean Error Rates of RaFD and JAFFE Databases

	10%	20%	30%	40%	50%
LP	69.58%	60.00%	58.21%	57.14%	58.33%
AdaBoost.M1	48.82%	42.37%	37.56%	30.24%	30.00%
RegBoost	48.82%	42.37%	37.56%	30.24%	30.00%
Proposed	47.43%	41.05%	36.15%	29.52%	26.67%

Table 5. Mean Error Rates of CK and RaFD Databases

	10%	20%	30%	40%	50%
LP	68.90%	64.29%	59.67%	58.05%	55.18%
AdaBoost.M1	56.51%	50.99%	40.70%	37.68%	36.76%
RegBoost	56.35%	50.99%	40.70%	37.68%	36.76%
Proposed	56.16%	50.31%	40.04%	36.95%	35.99%

When CK database is used as source domain, the recognition rates of Table 5 are higher than Table 2. Because there are people of mainly European in CK and RaFD databases, in Table 2 we use CK and JAFFE databases, the ethnic of people is different. The problem of individual differences is more serious in Table 2. When RaFD database is used as source domain, the recognition rates of AdaBoost.M1 and RegBoost are same in Table 4 and Table 6. Because a large quantity of useful information are supplied by RaFD database, and RegBoost algorithm is based on the framework of AdaBoost algorithm. Both of them make up the difference between two algorithms.

Table 6. Mean Error Rates of RaFD and CK Databases

	10%	20%	30%	40%	50%
LP	66.47%	61.09%	52.46%	53.04%	52.27%
AdaBoost.M1	49.51%	42.93%	35.94%	33.14%	29.39%
RegBoost	49.51%	42.93%	35.94%	33.14%	29.39%
Proposed	48.68%	42.47%	35.72%	32.65%	28.64%

The experimental results show that semi-supervised learning adaptive boosting algorithm has stronger robustness than the baseline methods, and improves the facial expression recognition rate from multiple databases.

4   Conclusion

In this paper, a novel algorithm called Semi-Supervised Learning Adaptive Boosting was proposed. Our proposed method has employed the knowledge transfer to construct a high-quality classification model, and to overcome problem of the labeled and unlabeled data, which are drawn from different feature distribution to some extent. Compared with traditional semi-supervised learning methods for facial expression recognition, simulation experiment shows that the proposed algorithm achieves the highest classification accuracy. Simulation experimental results indicate that, the proposed algorithm is efficient.

References

1. Jiang, B., Jia, K.B., Yang, G.S.: Research Advance of Facial Expression Recognition. Computer Science 38, 25–31 (2011)
2. Luo, S.W.: The Perception Computing of Visual Information. Science Press, Beijing (2010)
3. Abdel Hady, M.F., Schels, M., Schwenker, F., Palm, G.: Semi-supervised facial expressions annotation using co-training with fast probabilistic tri-class SVMs. In: Diamantaras, K., Duch, W., Iliadis, L.S. (eds.) ICANN 2010, Part II. LNCS, vol. 6353, pp. 70–75. Springer, Heidelberg (2010)
4. Chen, K., Wang, S.H.: Semi-Supervised Learning via Regularized Boosting Working on Multiple Semi-Supervised Assumptions. IEEE Transactions on Pattern Analysis and Machine Intelligence 33, 129–143 (2011)
5. Tan, H.C., Zhang, Y.J.: Person-Independent Facial Expression Recognition Based on Person-Similarity Weighted Distance. Jounal of Electronics & Information Technology 29, 455–459 (2007)
6. Zhou, Z.H., Yang, Q.: Machine learning and its applications. Tsinghua University Press, Beijing (2011)
7. Yan Jr., H.B., Ang, M.H., Poo, A.N.: Cross-Dataset Facial Expression Recognition. In: IEEE International Conference on Robotics and Automation, pp. 5985–5990. IEEE Press, Shanghai (2011)

8. Dai, W.Y., Yang, Q., Xue, G.R., et al.: Boosting for Transfer Learning. In: 24th International Conference on Machine Learning, pp. 193–200. Omni Press, Corvallis (2007)

9. Freund, Y., Schapire, R.: A decision-theoretic generalization of on-line learning and an application to boosting. Journal of Computer and System Sciences 55, 119–139 (1997)

10. Lyons, M., Akamatsu, S., Kamachi, M., et al.: Coding facial expressions with Gabor wavelets. In: 3rd IEEE International Conference on Automatic Face and Gesture Recognition, pp. 200–205. IEEE Press, Nara (1998)

11. Kanade, T., Cohn, J., Tian, Y.L.: Comprehensive database for facial expression analysis. In: 4th IEEE International Conference on Automatic Face and Gesture Recognition, pp. 46–53. IEEE Press, Grenoble (2000)

12. Langner, O., Dotsch, R., Bijlstra, G., et al.: Presentation and validation of the Radboud Faces Database. Cognition & Emotion 24, 1377–1388 (2010)

13. Zhu, X.J., Ghahramani, Z.: Learning from labeled and unlabeled data with label propagation. Technical Report, Carnegie Mellon University (2002)

14. Chen, K., Wang, S.H.: Regularized Boost for Semi-Supervised Learning. In: 21st Conference on Neural Information Processing Systems, pp. 281–288. MIT Press, Cambridge (2007)

Identification of K-Tolerance Regulatory Modules in Time Series Gene Expression Data Using a Biclustering Algorithm

Tustanah Phukhachee and Songrit Maneewongvatana

Department of Computer Engineering, King Mongkut's University of Technology Thonburi,
Bangkok, Thailand
s54450015@st.kmutt.ac.th, songrit@cpe.kmutt.ac.th

Abstract. Nowadays, biclustering problem is still an intractable problem. But in time series expression data, the clusters can be limited those with contiguous columns. This restriction makes biclustering problem to be tractable problem. However existing contiguous column biclustering algorithm can only find the biclusters which have the same value for each column in biclusters without error tolerance. This characteristic leads the algorithm to overlook some patterns in its clustering process. We propose a suffix tree based algorithm that allows biclusters to have inconsistencies in at most k contiguous column. This can reveals previously undiscoverable biclusters. Our algorithm still has tractable run time with this additional feature.

Keywords: biclustering, error tolerance, regulatory modules, suffix tree, time series gene expression data.

1   Introduction

Clustering microarray data is one of many challenging problems nowadays. Since the DNA chips techniques enable simultaneous measurements of the expression level for wide range of genes on given experiment conditions [1]. These wide ranges of the data and complexity of clustering microarray data problem make the problem to be difficult.

Biclustering is the technique that simultaneously clusters the microarray data on both genes and conditions. The data in each cluster exhibit highly correlated behaviors between the subgroups of genes and conditions. It has showed many advantages in identification the local expression patterns and has been extensively studied and surveyed [2-4]. Biclustering problem has many approaches proposed to date. Most of the biclustering approaches presented are heuristic and thus do not guarantee to find the optimal solution. Some other cases used exhaustive search which have limitation on the input size in order to obtain reasonable runtimes. The dealing with exact value of original expression data matrix is one difficulty of biclustering problems. Thus, finding coherent behaviors regardless of the exact values have

T. Yoshida et al. (Eds.): AMT 2013, LNCS 8210, pp. 146–155, 2013.
© Springer International Publishing Switzerland 2013

becomes great interest. These circumstances lead to the introduction of many new methods based on discrete matrix [5-18]. Unfortunately, all these variations of the problem remain NP-hard.

Although the problem remains NP-hard, there exist restrictions to biclustering problem which lead it to be a tractable problem. For example, if the expressions data are organized that expression level of various snapshots of the same condition are represented as group of time-sorted contiguous columns and biclusters are limited to contiguous snapshots. This study focuses on such setting, which uses to identify coherent expression patterns shared by group of genes in consecutive time points.

Gene expression values in each cluster discovered by existing biclustering algorithms are consistent with the criteria defined when clustering was performed. For example, if the criterion is to find the cluster of elements with the same value, all elements in resulting cluster all have the same value. However, it may useful to allow cluster with few elements that do not conform to the criteria in order to obtain larger cluster. This paper, we focus on such clustering. We define a k-tolerance bicluster to be a bicluster with at most k contiguous columns can exhibit inconsistency on the gene expression pattern. By allowing these additional tolerances, the biclustering algorithm can discover more related and hidden patterns than the standard techniques used nowadays.

In this work, we propose an algorithm to find and report all k-tolerance maximal contiguous column coherent biclusters (k-CCC-Biclusters) within tractable time. Maximal k-CCC-Bicluster is a k-CCC-Bicluster which no existed k-CCC-Biclusters that can be the superset of it. Our algorithm is based on contiguous column coherent biclustering algorithm (CCC-Biclustering algorithm) [19] and is improved by considering the error tolerance of the input expression data in contiguous time after the suffix tree creation.

The content in this paper is organized as follows: in Section 2, we survey related work. In Section 3, notations used throughout this paper are defined. In Section 4, the k-tolerance contiguous column coherent biclustering algorithm is proposed. Then we present our experimental results in Section 5. Lastly, we discussed the optimal value of k and concluded our work in Section 6.

2   Related Work

At present, large number of biclustering algorithms were proposed to solve the general case of biclustering [3-4]. We can group biclustering algorithms into 3 main groups: heuristic algorithms, exhaustive algorithms, and condition-based algorithms. Our work is focuses on condition-based algorithms.

The large majority of biclustering algorithms use heuristic approaches to identify biclusters [3]. The Coupled Two-Way Clustering (CTWC) [20] which uses only subsets of rows or columns that are identified as stable clusters in previous clustering iteration are candidates for the next iteration. This heuristic leads to avoid all possible combinations which reduce search time of the algorithm with the cost of accuracy of the algorithm.

Cheng and Church [2] introduced the first biclustering algorithm applied to gene expression data. This algorithm is also heuristic algorithm. The algorithm starts with iteratively removal of a row or a column that gives the maximum decrease of similarity score, mean squared residue, H, until no further decrease of H. Then the algorithm iteratively adds a row or a column which gives minimum increase of H back to construct the bigger similarity biclusters with H lower than some threshold, δ. Lastly, it reports the bicluster and marks the newly found bicluster in order to find the next one. This heuristic method makes the algorithm very fast. However with this techniques and the marking make the discovery of highly overlapping biclusters unlikely, since elements of already identified biclusters have been marked by random noise [3].

On the contrary, the exhaustive algorithms such as brute force or exhaustive enumeration methods can find the optimal set of biclusters [3]. However, due to their high complexity, they can be applicable only when the input size is small [3]. This led the practical exhaustive algorithms to be condition-based algorithm instead.

Although there is some condition-based biclustering algorithm which addressed the consecutive columns, the first was purposed by Ji and Tan [6]. The exact complexity of this algorithm is hard to estimate by their description. Still this algorithm considered to be in linear time. Yet, there exist another proposal by Madeira et al. which addressed the problem of finding maximal contiguous column coherent biclusters (CCC-Biclusters) [19]. This algorithm can find and report all biclusters in linear time and in the size of the expression matrix. Although our context is based on this algorithm, however in our work we also proposed the improvement to allow the maximal k-CCC-Biclusters to clustered the error tolerance data altogether. This will increase flexibilities and open the new path in finding the larger significant biclusters from the data matrix which will improve the efficiency and accuracy to the standard CCC-Biclustering algorithm.

3   Definitions

Our definitions of CCC-Biclusters and suffix tree are based on [19]. This leads us to introduce some definitions about k-tolerance data that relate to our work.

We denote gene expression matrix with $|R|$ rows and $|C|$ columns as M where R is the set of genes and C is the set of its conditions. The expression of the genes i under condition j is represented by M_{ij}.

In this work, the input data used was already discretized to be in regulatory modules with 3 symbols as $\{D, N, U\}$ which refer to DownRegulated, NoChange, and UpRegulated respectively.

Our definitions used throughout this work are defined as follows:

Definition 1: bicluster and trivial bicluster. Bicluster $B = (I, J)$ is the submatrix M_{IJ} defined by $I \subseteq R$, subset of rows, and $J \subseteq C$, subset of columns. Trivial bi-cluster is a bicluster with only 1 row and 1 column.

Definition 2: k-CCC-Bicluster. CCC-Bicluster M_{IJ} is k-tolerance CCC-Bicluster, k-CCC-Bicluster, if the CCC-Bicluster has at most k contiguous internal columns that have some data inconsistency.

Definition 3: maximal *k*-CCC-Bicluster. *k*-CCC-Bicluster is maximal if no other *k*-CCC-Bicluster exists that can be its superset, for all other *k*-CCC-Bicluster M_{ST}, if $I \subseteq S$ and $J \subseteq T$ then $I = S$ and $J = T$.

Definition 4: string depth and string label. String depth is the length of string to the specific position. String depth of node v in suffix tree T is the length of the symbols from the root node to node v in T. We denote this as $P(v)$ and denote the path which contain all these symbols from the root to v as string label.

Definition 5: number of leaves. For each internal node v in suffix tree T, we denote the number of leaves in the subtree with v as root of the subtree by $L(v)$.

Definition 6: suffix link. For the node v in T with label $x\alpha$ where x is a single character and α is a string (possibly empty), if there is another node u with label α then there will be a link from v to u. We call it the suffix link. The special case is when α is empty then $x\alpha$ has a suffix link from v to the root node.

Definition 7: MaxNode. MaxNode is an internal node v of the suffix tree T which satisfies one of these following conditions:

(a) Node v does not have incoming suffix links, or

(b) Node v has only incoming suffix links from node u such that for every node u, $L(u) < L(v)$.

4   Algorithms

4.1   Preprocessing Step

Our algorithm assumed that the gene expression input data has already been discretized. Therefore, first part of our work is to apply the alphabet transformation technique which was introduced by Madeira et al. [19] to our discrete input data as follow.

In alphabet transformation process, for each string S^o in set of strings $\{S_1^o, ..., S_{|R|}^o\}$, we append each character, C^o, in S^o by its position in the column, We now obtain the set of string $\{S_1', ..., S_{|R|}'\}$ with $C_i' = C_i^o i$ where C_i^o is the ith position of S^o.

We then append terminator symbols to each string. This symbol must not appear anywhere else (usually symbol $\$$ was used). For each string S' in set of strings $\{S_1', ..., S_{|R|}'\}$, we inserted the special symbols $\$x$ to the end of the string where x is the row of the string. Now we obtain set of string $\{S_1'', ..., S_{|R|}''\}$ with $S_x'' = S_x' \$x$. Therefore, our last column now is termination symbol and our columns size is increased by 1.

4.2   Suffix Trees and CCC-Biclusters

In this part of context, we state the connection of maximal CCC-Biclusters of the input data matrix M_{IJ} and the node in generalized suffix tree T built from the set of strings $\{S_1'', ..., S_{|R|}''\}$ which taken from the preprocessing step.

(a) Each internal node in suffix tree T corresponds to one row-maximal, right-maximal CCC-Bicluster in matrix M with at least two rows. Since an internal node v in T has the common substring length $P(v)$ for each of its leaves. Therefore, each internal node v defines a CCC-Bicluster that has $P(v)$ columns and $L(v)$ rows. Every right-maximal, row-maximal CCC-Bicluster with at least two rows corresponds to internal node in T.

(b) An internal node corresponds to a maximal CCC-Bicluster if and only if there is no suffix link from any node with the same value of $L(v)$ pointing to it. Since if there is an incoming suffix link from an internal node u to node v with $L(u) = L(v)$ then bicluster corresponds to v be already included in bicluster corresponds to u, $v \subset u$. So v is not maximal CCC-Bicluster.

(c) An internal node in T corresponds to a left-maximal CCC-Bicluster if and only if it satisfies *Definition 7*. Since an internal node v can be maximal CCC-Bicluster from the *(b)* or in fact that $L(u) < L(v)$ which lead v and u to be separate biclusters.

These facts lead to the theorem from [19] which was defined as follows:

Theorem 1: Every maximal CCC-Bicluster with at least two rows corresponds to an internal node in the generalized suffix tree T that satisfies *Definition 7*, and each of these internal nodes defines a maximal CCC-Bicluster with at least two rows. We define these nodes as N.

4.3   Postprocessing Step: K-Tolerance Clustering Process

As stated in part 4.2.*(a)*, each internal node of the suffix tree correspond to each bicluster of the input data. We will use this characteristic of the suffix tree in order to allow error tolerance to be clustered and be including to our result.

(a) The result biclusters which include error tolerance are in form of *FxE* denoted as front part (F), tolerance part (x) and extended part (E) respectively.

(b) Since each internal node v in the suffix tree T has the common longest substring, every internal node in the suffix tree corresponds to the front part, F, of bicluster result with error tolerance included.

(c) In many cases there are no x and E which lead the result to consist of only F part. These results correspond to the normal standard biclusters.

(d) Tolerance part is the part that we allow expression data to be inconsistence among the k columns which have to assigned before performing algorithm.

(e) Each E can be found by performing suffix tree over the substring, Z, in child of each internal node. These internal nodes corresponding to its own F, NF. These Z will exclude the first k characters which corresponding to x.

With all above facts considered, it leads us to the theorem which also defined using the facts in part 4.2 as follows:

Theorem 2: Every maximal CCC-Bicluster with at least two rows corresponds to an internal node NF in the generalized suffix tree T that satisfies *Definition 7*, and each of these internal nodes defines a maximal CCC-Bicluster with at least two rows. We define these nodes as *NN*.

4.4   K-CCC-Biclustering: A Biclustering Algorithm for Finding and Report All Maximal K-CCC-Biclusters

Theorem 1 implies that there is an algorithm which can find and report all maximal CCC-Biclusters of discretized and transformed gene expression matrix M in time linear and size of the input matrix since it corresponds to a suffix tree which has all these properties. *Theorem 2* also implies the same thing with E part.

From *Theorem 1* and *Theorem 2* by combining string of N, F, with x and string of NF, E, lead the algorithm to find bicluster with error tolerance included.

Algorithm1 performed the alphabet transformation to the discrete input data as described in the part 4.1. After that our algorithm starts to build a generalized suffix tree from the set of strings $\{S_1^{''},...,S_{|R|}^{''}\}$ which obtained from the preprocessing step. We now check each internal node whether the conditions in *Theorem 1* are met. Nodes that do not meet the required conditions are marked as "invalid." Then we build generalized suffix tree for each N with string of its child node, $\{S_1^C,...,S_{|R|}^C\}$ exclude first characters, then exclude 2, ..., k characters. We then check each internal node whether the conditions in *Theorem 2* are met $P(w)$ is string depth of internal node in *Theorem 2*. Nodes that do not meet the required conditions are marked as "invalid." When finish we combine the F and E of correspond valid node together. Next, if no duplicate have been reported and $F+x+E$ is maximal. We report our k-CCC-Biclusters which corresponded to the combination of valid internal nodes and with special symbol x for characters that were excluded. Lastly, checking if v was already be the part of k-CCC-Bicluster or all reported k-CCC-Biclusters have the different leaves than v. If it is, we report the k-CCC-Bicluster corresponding to only F part.

Algorithm1: k-CCC-Biclustering

```
Input: Discretized gene expression matrix M
  Perform alphabet transformation and obtain{S₁",...,S|R|"}.
  Build a generalized suffix tree T for{S₁",...,S|R|"}.
  For each internal node v∈T do
    Mark node v as "Valid."
    Compute the string depth P(v).
  For each internal node v∈T do
    Compute the number of leaves L(v) in the subtree
    rooted at v.
  For each internal node v∈T do
    If there is a suffix link from v to a node u and
    L(u)=L(v)do
      Mark node u as "Invalid."
  For each internal node v∈T do
```

```
If v marked as "Valid" do
   For i∈{1,2,…,k}do
      Build a generalized suffix tree T_n from {S_1^C,…,S_{|R|}^C}
      excluding i first characters.
      For each internal node w∈T do
         Mark node w as "Valid."
         Compute the string depth P(w).
      For each internal node w∈T do
         Compute the number of leaves L(w) in the subtree
         rooted at w.
      For each internal node w∈T do
         If there is a suffix link from w to a node y
         and L(y)=L(w) then
            Mark node y as "Invalid."
      For each internal node w∈T do
         If w marked as "Valid" do
            If no duplicate of w have been reported and
            F+x+E is maximal do
               Report the k-CCC-Bicluster corresponding
               to F+x+E.
   If v has not reported as k-CCC-Bicluster or all
   reported bicluster have rows different than v do
      Report the k-CCC-Bicluster corresponding to v.
```

4.5   Complexity Analysis of K-CCC-Biclustering

Our k-CCC-Biclustering algorithm corresponds to suffix tree and performed by depth-first searches as CCC-Biclustering algorithm. Since our algorithm extends each internal node with suffix tree so our algorithm is a solution based algorithm which complexity varies for each input. From the observation of our random synthetic data, the increasing rate of internal node, IN, the worst case is roughly $|C|^{1.5}$. With this internal node considered we concluded that the complexity of our algorithm is $O(k|R|^2|IN|^2)$ or $O(k|R|^2|C|^3)$.

However our algorithm only extends the valid internal node. For most case only small portion of genes are forming an internal node which usually can be discarded. This characteristic leads the complexity of our algorithm to be mostly $O(k|R||C|^3)$.

5   Experimental Results

In this section we present and discuss our experimental results.

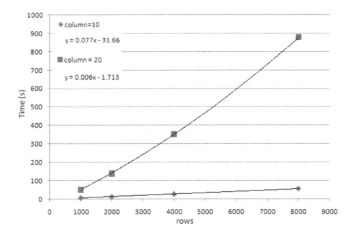

Fig. 1. Time results and trend lines of k-CCC-Biculstering algorithm consider row increasing ($k = 2$)

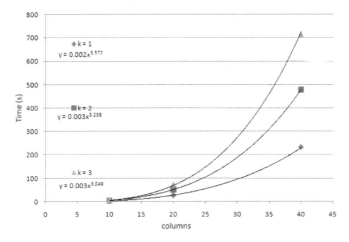

Fig. 2. Time results and trend lines of k-CCC-Biculstering algorithm consider column increasing ($k = 2$)

From the result of our experiments Figs. 1 and 2, we conclude that our analysis is correct. Since the rate of internal node increase from rows is linearly so the complexity when increasing row is at $O(|R|^2)$ with possibility to be nearly $O(|R|)$. On the other hand connection between the increasing rate of internal node and column are not linear. This leads us to consider the connection between them which is varying since our algorithm is solution-based algorithm. In our case the worst case is about $|C|^{1.5}$, which lead our complexity to be at $O(|C|^2)$ when consider the column increasing. Our ex-perimental results also stated that k is in linear time complexity with our algorithm.

When considering all factors together, this leads the complexity of our algorithm from the experiments to be at $O(k|R||C|^3)$ which still tractable time.

6   Conclusions

From our experiment we conclude that our algorithm can discover biclusters which allows k-tolerance errors in $O(k|R|^2|C|^3)$.

In many cases, our algorithm can also considered to be linearly time on rows with complexity of $O(k|R||C|^3)$ as explained in analysis section.

For many field of works, the column in time-series gene expression data can be small constant value. This type of works can lead the k-CCC-Biclustering algorithm to be nearly linear in $O(k|M|)$ time, $|M|$ is the size of input matrix. These works are the ideal to increase the flexibility of the result with our algorithm.

In the error tolerance algorithms, the characteristic which allows erroneous data to be clustered in the same group of biclusters provided more flexibility to the algorithm. However this flexibility is provided with the risk of the fault positive biclusters increasing. Since our algorithm is at its best with small constants columns. The value of also has to be small so the result will still be significant. From our experiments, k also has much impact to the number of the result bicluster. This leads to the risk of fault positive bicluster increasing. With this fact considered, we specify that k in k-CCC-Biclustering algorithm is the best when it is equal to 1.

Our experiments also suggest that there are more previously indiscoverable clusters can be found by using this algorithm. This can open new paths to find the biclusters which may be hidden in the expression data matrix. Erroneous in gene expression data can be the result of many factors such as environment setting, user, preprocessing step or even due to some process in collecting data. These factors can leads to some minor error in gene expression data which usually be overlooked by standard contiguous column coherent biclustering algorithm.

References

1. McLachlan, G.J., Do, K., Ambroise, C.: Analysing Microarray Gene Expression Data. John Wiley & Sons, New Jersey (2004)
2. Cheng, Y., Church, G.M.: Biclustering of Expression Data. In: Proc. 8th Int'l Conf. Intelligent Systems for Molecular Biology (ISMB 2000), pp. 93–103. ACM Press, New York (2000)
3. Madeira, S.C., Oliveira, A.L.: Biclustering Algorithms for Biological Data Analsis: a survey. IEEE/ACM Trans. Computational Biology and Bioinformatics 1(1), 24–45 (2004)
4. Van Mechelen, I., Bock, H.H., De Boeck, P.: Two-Mode Clustering Methods: A Structured Overview. Statistical Methods in Medical Research 13(5), 979–981 (2004)
5. Ben-Dor, A., Chor, B., Karp, R., Yakhini, Z.: Discovering Local Structure in Gene Expression Data: The Order-Preserving Submatrix Problem. In: Proc. 6th Int'l Conf. Computational Biology (RECOMB 2002), pp. 49–57 (2002)
6. Ji, L., Tan, K.: Identifying Time-Lagged Gene Clusters Using Gene Expression Data. Bioinformatics 21(4), 509–516 (2005)
7. Koyuturk, M., Szpankowski, W., Grama, A.: Biclustering Gene-Feature Matrices for Statistically Significant Dense Patterns. In: Proc. 8th Int'l Conf. Research in Computational Molecular Biology (RECOMB 2004), pp. 480–484 (2004)

8. Liu, J., Wang, W., Yang, J.: Biclustering in Gene Expression Data by Tendency. In: Proc. 3rd Int'l IEEE CS Computational Systems Bioinformatics Conf. (CSB 2004), pp. 182–193 (2004)

9. Liu, J., Wang, W., Yang, J.: A Framework for Ontology-Driven Subspace Clustering. In: Proc. ACM SIGKDD 2004, pp. 623–628 (2004)

10. Liu, J., Wang, W., Yang, J.: Gene Ontology Friendly Biclustering of Expression Profiles. In: Proc. 3rd IEEE CS Computational Systems Bioinformatics Conf. (CSB 2004), pp. 436–447 (2004)

11. Liu, J., Wang, W., Yang, J.: Mining Sequential Patterns from Large Data Sets. Advances in Database Systems, vol. 18. Kluwer Academic Publishers (2005)

12. Lonardi, S., Szpankowski, W., Yang, Q.: Finding Biclusters by Random Projections. In: Sahinalp, S.C., Muthukrishnan, S.M., Dogrusoz, U. (eds.) CPM 2004. LNCS, vol. 3109, pp. 102–116. Springer, Heidelberg (2004)

13. Madeira, S.C., Oliveira, A.L.: A Linear Time Algorithm for Biclustering Time Series Expression Data. In: Casadio, R., Myers, G. (eds.) WABI 2005. LNCS (LNBI), vol. 3692, pp. 39–52. Springer, Heidelberg (2005)

14. Murali, T.M., Kasif, S.: Extracting Conserved Gene Expression Motifs from Gene Expression Data. In: Proc. 8th Pacific Symp. Biocomputing (PSB 2003), vol. 8, pp. 77–88 (2003)

15. Prelic, A., Bleuler, S., Zimmermann, P., Wille, A., Bühlmann, P., Gruissem, W., Hennig, L., Thiele, L., Zitzler, E.: A Systematic Comparison and Evaluation of Biclustering Methods for Gene Expression Data. Bioinformatics 22(10), 1282–1283 (2006)

16. Sheng, Q., Moreau, Y., De Moor, B.: Biclustering Microarray Data by Gibbs Sampling. Bioinformatics 19(2), 196–205 (2003)

17. Tanay, A., Sharan, R., Shamir, R.: Discovering Statiscally Significant Biclusters in Gene Expression Data. Bioinformatics 18(1), 136–144 (2002)

18. Wu, C., Fu, Y., Murali, T.M., Kasif, S.: Gene Expression Module Discovery Using Gibbs Sampling. Genome Informatics 15(1), 239–248 (2004)

19. Madeira, S.C., Teixeira, M.C., Sá-Correia, L., Oliveira, A.L.: Identification of Regulatory Modules in Time Series Gene Expression Data Using a Linear Time Biclustering Algorithm. IEEE/ACM Transaction Computational Biology and Bioinformatics 7(1), 153–165 (2010)

20. Getz, G., Levine, E., Domany, E.: Coupled Two-Way Clustering Analysis of Gene Microarray Data. Proc. Natural Academy of Sciences Us, 12079–12084 (2000)

21. Madeira, S.C., Gonçalves, J.P., Oliveira, A.L.: Efficient Biclustering Algorithms for identifying transacriptional regulation relationships using time series gene expression data. INESC_ID Tec. Rep. 22 (2007)

22. Peeters, R.: The Maximum Edge Biclique Problem Is NP-Complete. Discrete Applied Math. 131(3), 651–654 (2003)

Ranking Cricket Teams through Runs and Wickets

Ali Daud[1] and Faqir Muhammad[2]

[1] Department of Computer Science & Software Engineering,
IIU, Islamabad 44000, Pakistan
[2] Department of Business Administration, AU, Islamabad 44000, Pakistan
ali.daud@iiu.edu.pk, aioufsd@yahoo.com

Abstract. Teams are ranked to show their authority over each other. The International Cricket Council (ICC) ranks the cricket teams using an ad-hoc points system entirely based on the winning and losing of matches. In this paper, adoptions of PageRank and h-index are proposed for ranking teams to overcome the weakness of ICC ad-hoc point system. The intuition is to get more points for a team winning from a stronger team than winning from a weaker team by considering the number of runs and wickets also in addition to just winning and losing matches. The results show that proposed ranking methods provide quite promising insights of one day and test team rankings.

Keywords: indexing, ranking, cricket teams, runs and wickets.

1 Introduction

Sports ratings are performed for showing the standings of different teams and players by analyzing the results of competitions or matches. The team with the highest points is usually ranked number 1. Traditional sports rankings are based on win, loss or tie ratios or polls which are subjective rating of the teams, such as, ICC cricket rankings are based on many ad-hoc rules [14]. Borooah and Mangan [2], criticized that current point system based ranking schemes are opaque, so the methods used by ICC for rankings of cricket teams and players still needs to be investigated properly to provide better ranking methods.

State-of-the-art indexing and ranking algorithms such as h-index [8] and PageRank [12] can be adopted to rank cricket teams. The number of runs and wickets from which matches are won can be thought of as citations. The intuition is that the more the average number of runs or wickets from which matches are won by a team the higher the team-index (our proposed method) he has. PageRank is an iterative algorithm [12] which was used to rank web pages on the basis of inlinks and the importance of those pages that are providing those inlinks. The nodes in a graph can be taken as teams and the links between them are the matches played between them. If a team A wins from another team B, team B will provide an inlink to team A. In this way a directed graph is built which can be used to rank teams by considering graph weightage in addition the simply considering the number of won or lost matches.

We propose Team-index (t-index) T-index considers only the number of runs and wickets from which matches are won, while the strength and weakness of the teams

T. Yoshida et al. (Eds.): AMT 2013, LNCS 8210, pp. 156–165, 2013.
© Springer International Publishing Switzerland 2013

from which a team wins is ignored. Consequently, to consider the strength and weakness of teams from which a team wins we propose TeamRank (TR) which is an adoption of PageRank [12]. The intuition is that the more a team wins matches from the stronger teams the higher it will be ranked, while it ignores the number of runs and wickets from which matches are won. Consequently, we propose weighted TeamRank (WTR) which also considers the weight of number of runs and wickets from which a team wins from other teams. The results and discussions prove that our proposed methods are useful and should be used to rank cricket teams.

The contributions in this work are as follows, (1) proposal of graph and non-graph weightage based ranking algorithms for cricket teams ranking, (2) addition of number of runs and wickets instead of simply using match won or lost information and (3) hybridization of non-graph weightage and graph weightage based ranking algorithms to provide a unified solution.

2   Cricket Teams Ranking

In this section, before describing our proposed (1) Team-Index, (2) Simple Team Rank, (3) Weighted Team Rank, and (4) Unified Weighted Team Rank methods, we briefly introduce related ICC ranking method [14] for ranking cricket teams.

2.1   ICC Cricket Teams Ranking System

The international governing body of cricket is international Cricket Council (ICC). ICC awards championship trophies to the teams with highest ratings in both ODI and test matches. Test cricket matches can last up to five days with each day broken into three sections punctuated by lunch and tea breaks. ODI cricket matches are the faster alternative, typically completed in one day, and with a maximum of 50 over's permitted per team. An "over" is defined as a set of six balls bowled consecutively. The ICC employs ratings formulas for both leagues to determine a champion [14].

2.1.1   Test Matches

i.   Add one point to a team for winning a match, after a series between two teams; add a half-point to both teams for drawing a match. Add a bonus point to the team that won the series; add a half-point to each team if the series ended in a draw.

ii.   Multiply the team's series result by 50 points more than the opponent's rating, if the ratings gap between the two teams was less than 40 points at the start of the series. Then add that total to the opponent's series result multiplied by 50 points less than the opponent's rating.

iii. Multiply the stronger team's series result by 10 points more than their own rating (if the ratings gap was equal to or more than 40 points), then add that total to the opponent's series result multiplied by 90 points less than the team's own rating. The weaker team multiplies its series result by 90 points more than their own rating, and then adds that total to the opponent's series result multiplied by 10 points less than the team's rating.

iv.   Add the new point's totals to the team's points total before the series began. Remove points from matches that no longer fall within the past three years. Update the number of matches by adding one more than the number of games in a series. For example, if a series lasted two matches, you would add three matches to the total.

v.   Divide the updated points total by the updated match total. This represents the team's rating, and comparisons of ratings will yield the team's ranking [14].

2.1.2 ODI Matches

i.   Add one point to a team for winning the match, after a series between two teams, and a half-point to each team for a draw.

ii.   Score 50 points more than the opponent's rating for the winner if the gap between the two teams at the outset of the match was less than 40 points. Score 50 points fewer than the opponent's rating. In case of a tie, each team scores the opponent's rating.

iii.   Score 10 points more than the stronger team's rating in a win or 90 points fewer than its rating in a loss (if the gap between teams' ratings was more than or equal to 40 points). The weaker team scores 90 points more than its rating for a win or 10 points fewer than its rating for a loss. For ties, the stronger team scores 40 points fewer than its rating and the weaker team scores 40 points more than its rating.

iv.   Add the new point totals to the existing point total for each team before the series started. Update the match numbers, as well. Throw out all points and matches that no longer fall within the last three years.

v.   Divide the new points total by the new matches' total. This will provide the rating for each team, and ratings comparisons will order the teams into rankings [14].

2.2   Team-Index (T-Index)

Our first proposed method T-index is an adoption of h-index [8]. In T-index like h-index teams are referred to as author's and papers as number of runs and wickets from which matches are won by the team. The idea is that if a team wins matches from more number of runs and wickets the higher the t-index the team will have.

We take an example of a team A which played 15 matches in total from which 7 matches are won by runs and 8 matches are won by wickets. Here, it is necessary to mention that in cricket game one team bat first and score runs while other team bowls. So in case team batting first wins it wins from runs while in case team batting second wins it wins from wickets. This is why teams can win from runs and wickets both in different matches. Table 1(a) and Table 1(b) for team A are used to calculate its T-index as follows.

Table 1(a). Matches won from wickets

T1	
No. of Matches	No. of Wickets
1	10
2	9
3	8
4	8
5	7
6	2
7	3
8	1
Total	48

Table 1(b). Matches won from runs

T2	
No. of Matches	No. of Runs
1	138
2	99
3	86
4	84
5	52
6	6
7	5
Total	470

$$\text{T} - \text{index} = \frac{\sqrt{T1 + T2}}{2} = \frac{\sqrt{3.46 + 10.83}}{2} \qquad (1)$$

$$= \frac{14.88}{2} = 7.14$$

$$T1 = \frac{\sqrt{\text{Total no. of wickets}}}{2} = \frac{\sqrt{48}}{2} = \frac{6.928}{2} = 3.46$$

$$T2 = \frac{\sqrt{\text{Total no. of runs}}}{2} = \frac{\sqrt{470}}{2} = \frac{21.68}{2} = 10.83$$

2.3 TeamRank (TR)

Our second proposed method TR is an adoption of page rank algorithm. PageRank [12] is considered as one of the most important graph based page ranking algorithms on the web. TR of a team should be high if the team wins many matches from other teams and those teams are strong (those teams also had won many matches). TR is calculated by using the following formula.

$$TR(A) = \frac{1\text{-}d}{N} + d[\frac{TR(T_i)}{CT_i} + \cdots + \frac{TR(T_n)}{CT_n}] \qquad (2)$$

Where, $TR(A)$ is the TeamRank of Team A, $TR(T_i)$ is the TeamRank of Teams T_i which link (lose matches) to Team A, CT_i is the number of outlinks (matches lost) by team T_i, d is a damping factor which can be set between 0 and 1, and N is total number of teams. Here inlinks refer to the matches won say a team has won 10 matches then inlinks will be 10 and outlinks refer to the matches lost from another team say a team lost 14 matches then outlinks will be 14.

2.4   Weighted Team Rank (WTR)

Our third proposed method is a weight based enhancement in TR. Assigns larger rank values to stronger teams instead of dividing the rank value of a team evenly among it's outlink matches. Instead of only considering the number of matches we do in TR we also consider the number of runs and wickets from which the matches are lost. The idea is that if a team lost matches from more runs and wickets will contribute less to the rank of the team being ranked.

An example is provided to show how the parameter of runs and wickets impact the ranking of teams. Suppose we have two teams A and B with same number of lost matches 10. If the sum of the runs from which team A lost those 10 matches is 200 and sum of the wickets is 30 and the sum of the runs from which team B lost those matches is 100 and sum of the wickets is 15. It will have different rank scores of teams. As in the following example we can see that when both Team A and B has lost same number of matches they have same scores 0.1 while contributing to some other teams from which they have lost matches. But when runs and wickets are considered we can see that Team A has rank score of 0.056 while Team B has rank score 0.061 which is higher as compared to Team A. We can see that team has lost matches from more runs or wickets as compares to Team B, as Team A's lost matches runs sum is 200 and sum of wickets is 30 while Team B's lost matches runs sum is 100 and sum of wickets is 15. Finally we can say that if a team's loses matches from more runs and wickets it contributes less score to the team from which the team has lost matches. WTR is calculated by using Eq. 3.

Table 2(a). Team Rank scores contribution when runs and wickets are ignored

Team A	Team B
$\frac{1}{10} = 0.1$	$\frac{1}{10} = 0.1$

Table 2(b). Team Rank scores contribution when runs and wickets are considered

Team A	Team B
$\frac{1}{\frac{60(10)+20(200)+20(30)}{60+200+30}} = 0.056$	$\frac{1}{\frac{60(10)+20(100)+20(15)}{60+100+15}} = 0.061$

$$WTR(A) = \frac{1-d}{N} + d \sum \frac{WTR(T_i)}{WC(T_i)} \qquad (3)$$

Where, $WTR(A)$ is the weighted TeamRank of Team A, $WTR(T_i)$ is the TeamRank of Teams T_i which have lost matches to Team A, $WC(T_i)$ is the number of outlinks (matches lost) by team T_i, in order to calculate $WC(T_i)$ (weighted outlinks) we use weighted arithmetic mean formula given in Eq. 4. One can try different weights for matches, runs and wickets such as (50(matches), 25(runs), 25(wickets)) or (40(matches), 30(runs), 30(wickets)). We used 60% weigtage for match result as it is important to win or lose as compared to number of runs or wickets whose weightage is 20% and 20%, respectively used in this work.

$$Weighted\ Outlinks = \frac{60(matches) + 20(runs) + 20(wickets)}{matches + runs + wickets} \qquad (4)$$

2.5   Unified Weighted Team Rank (UWTR)

Our proposed fourth method UWTR is hybridization of t-index and WTR methods. UWTR combines the power of a team in terms of winning number of matches in terms of runs and wickets without considering graph weightage (t-index) and power of a team in terms of the power of the teams from which those matches are won with considering graph weightage. UWTR is calculated by using Eq. 5, where, WT_j is the T-index of the team for which we are calculating the rank and $\sum_{t=1}^{N} WT_i$ is the sum of t-index of all teams.

$$UWTR = \frac{1-d}{N}\left(\frac{WT_j}{\sum_{t=1}^{N} WT_i}\right) + d\sum \frac{WTR(T_i)}{WC(T_i)} \qquad (5)$$

3   Experimental

3.1   Dataset

The dataset for experiments is taken from the cricinfo web site [15] from 2000 to March 2012 and ICC cricket rankings point system of ODI and test matches is taken as existing method. There are ten teams which has been given the test status by ICC. Teams are categorized into strong and weak teams on the basis of opinions about their performance. Strong teams are Australia, India, Pakistan, South Africa, Sri Lanka, West Indies, England, New Zealand and weak teams are Zimbabwe, Bangladesh.

Here it is necessary to mention that there exists no Gold Standard dataset with which the ranking results of existing and our proposed methods can be compared to find accuracy in terms of precision, recall or f-measure. Consequently, the results of our proposed methods are subjectively compared with the existing ICC rankings to show their effectiveness. The subjective discussions are performed with the help of several cricket team players of our university.

3.2   Results and Discussions

3.2.1 One Day International Matches

One day international team's rankings are provided in Table 3. T-index is used to rank teams by considering the number of runs and wickets from which the teams have won matches. The top 3 teams ranked by ICC ranking are Australia, South Africa and India, respectively. The top ranked team for t-index is also Australia with clear difference of score in comparison to other teams at number 2 and 3, which is same as ICC cricket ranking. While the second team ranked is South Africa and third team is

Sri Lanka which is not same as ICC rankings. The Indian team is ranked 5th by t-index due to winning from less number of runs and wickets from other teams as compared to Sri Lanka which have won matches from more wickets and runs as compared to India so is ranked 3rd.

Australia again stands first by our second method TR, which is same position given in ICC rankings. But by point system India is ranked 3rd while by TR method India is ranked at number 2nd. By analyzing data we have found that Indian team has won more matches against strong teams like Australia, Pakistan, South Africa and Sri Lanka as compared to South Africa. So the inlinks weights by winning from stronger teams are more for India as compared to South Africa. Pakistan and Sri Lanka are also ranked higher by TR as compared to South Africa due to their most winnings from stronger teams.

By applying our proposed third WTR method we got different results as from simple team rank method. By this method South Africa ranked first, Sri Lanka is second and Australia is ranked third because we are calculating weight-age of each team against other team by considering number of runs and wickets from which matches are won. South Africa has won more matches from weaker teams resulted in winning matches from large number of runs and wickets so is ranked number one.

As unified team rank is the combination of two techniques which are team index and weighted team rank, for ODI matches Australia hold first position, South Africa and Sri Lanka on second and third and so on. One can see that rankings provided by t-index and UWTR are same for all teams which shows that considering runs and wickets are both useful though similar results are obtained when graph based strength or weakness of teams is considered. For our proposed all methods England team is ranked 6th or 7th due to winning from less runs and wickets.

Table 3. Teams Ranking W.R.T ODIS

ICC RANKING				
	TEAM	MATCH	POINTS	RATING
1	Australia	49	6030	123
2	South Africa	30	3549	118
3	India	55	6409	117
4	England	40	4469	112
5	Sri Lanka	55	6111	111
6	Pakistan	48	4989	104
7	New Zealand	31	2667	86
8	West Indies	33	2814	85
9	Bangladesh	36	2408	67
10	Zimbabwe	33	1511	46

TEAM INDEX		TEAM RANK		WEIGHTED TEAM RANK		UNIFIED TEAM RANK	
Australia	3.80789	Australia	0.133857	South Africa	0.0512701	Australia	0.00630054
South Africa	3.53553	India	0.121694	Sri Lanka	0.0512691	South Africa	0.00586111
Sri Lanka	3.53550	Pakistan	0.119422	Australia	0.0512602	Sri Lanka	0.00586098
Pakistan	3.4641	Sri Lanka	0.11424	India	0.0512553	Pakistan	0.00573821
India	3.3541	South Africa	0.110733	New Zealand	0.0512466	India	0.00556624
New Zealand	3.08221	New Zealand	0.0981	England	0.0512424	New Zealand	0.00512497
England	2.95804	England	0.0946053	Pakistan	0.051204	England	0.00492394
West Indies	2.78388	West Indies	0.081723	West Indies	0.0511923	West Indies	0.00463663
Bangladesh	2.17945	Zimbabwe	0.0636812	Bangladesh	0.0509535	Bangladesh	0.00363257
Zimbabwe	2.12132	Bangladesh	0.0619443	Zimbabwe	0.0508243	Zimbabwe	0.00352188

3.2.2 Test Matches

Test teams rankings are provided in Table 4. Test match is played for five days in which each team can play two innings. The teams can win from wickets and runs and sometime even one innings plus some wickets or runs. T-index is used to rank test teams by considering the number of runs and wickets from which the teams have won matches. The top 3 teams ranked by ICC ranking are England, South Africa and Australia, respectively. The top ranked teams for t-index are Australia, England and Sri Lanka in which Australia with clear difference of score in comparison to teams at 2nd and 3rd number, which is different from ICC cricket rankings in which England is ranked number 1. The south African team is ranked 6th by t-index due to winning from less number of runs and wickets from other teams as compared to Sri Lanka which have won matches from more wickets and runs as compared to South Africa so is ranked 3rd.

Table 4. Teams Ranking W.R.T Tests

ICC RANKING - TEST MATCH				
	TEAM	MATCH	POINTS	RATING
1	England	48	5614	117
2	South Africa	32	3709	116
3	Australia	46	5153	112
4	India	46	5103	111
5	Pakistan	35	3781	108
6	Sri Lanka	38	3780	99
7	West Indies	38	3212	85
8	New Zealand	28	2366	85
9	Bangladesh	18	135	8

TEAM INDEX		TEAM RANK		WEIGHTED TEAM RANK		UNIFIED TEAM RANK	
Australia	3.7081	Australia	0.158223	India	0.0515155	Australia	0.00692822
England	3.31662	England	0.136533	England	0.0515	England	0.0062163
Sri Lanka	3.20156	South Africa	0.122923	Australia	0.0514915	Sri Lanka	0.00598051
Pakistan	3	India	0.11662	South Africa	0.0514199	Pakistan	0.0056282
India	2.95804	Sri Lanka	0.0968203	Pakistan	0.0514129	India	0.00556456
South Africa	2.95804	Pakistan	0.0959763	New Zealand	0.0513838	South Africa	0.00555422
New Zealand	2.69258	New Zealand	0.08725	Sri Lanka	0.0513102	New Zealand	0.00506361
West Indies	2.34521	West Indies	0.0746774	West Indies	0.0511389	West Indies	0.00440602
Zimbabwe	1.65831	Zimbabwe	0.0587397	Zimbabwe	0.0503121	Zimbabwe	0.00305885
Bangladesh	1.5	Bangladesh	0.0522367	Bangladesh	0.0502923	Bangladesh	0.00276505

Australia again stands first by our second method TR, which is different from ICC rankings in which England is ranked first. But by point system South Africa is ranked 2nd while by TR method South Africa is ranked at number 6 and Sri Lanka is ranked at number 3. The data analysis explains that Sri Lanka team has won more matches against strong teams like Australia, India, Pakistan and South Africa as compared to South Africa. Even Indian and Pakistani team is ranked higher due to winning matches from strong teams. So the in link weights by winning from stronger teams are more for Australia and Sri Lanka as compared to South Africa.

By applying our proposed third WTR method we got different results as from simple team rank method and WTR results for ODI matches. By this method India is ranked first, England is ranked second and Australia is ranked third because we are calculating weight-age of each team against other team by considering number of runs and wickets from which matches are won. India has won more matches from weaker teams resulted in winning matches from large number of runs and wickets so is ranked number one.

As unified team rank is the combination of two techniques which are team index and weighted team rank for test matches. Australia holds first position, England and Sri Lanka on second and third, respectively and so on. One can see that rankings provided by t-index and UWTR are same for all teams which shows that considering runs and wickets are both useful though similar results are obtained when graph based strength or weakness of teams is considered. For our proposed all methods Australian team is ranked number one for both ODI and Test matches due to winning from more runs and wickets as well as from stronger teams. The results for WTR method are a bit different though in which other teams are ranked on top.

4   Related Work

H-Index [8] and G-index [6] was proposed for scientist's productivity indexing in co-author networks. Both h-index and g-index ignored number of years in which the scientist has published papers, so Burrell [4] proposed m-quotient by including career length in existing indexing h-index.

PageRank [12] was first used for ranking web pages. It provides query independent importance of web pages. Consequently, for results dependent on query based importance topic-sensitive PageRank is proposed [7]. The main idea was to rank web pages on the basis of same topic web pages linking to them and their importance on that topic. The problem of treating all links equally when rank scores are being calculated is raised and Weighted PageRank Algorithm was proposed [13]. It takes both inlinks and outlinks importance into account and distributes rank scores based on the popularity of web pages and showed better performance. Bundit et al., [10] highlighted the time factor importance in order to find authoritative web pages and proposed Time-Weighted PageRank. PageRank was applied to many domains other than ranking of web pages, such as; a Personalized PageRank [9] is proposed to analyze protein interaction networks.

A few social network analysis researchers' interest is also attracted by crickets' popularity. Bailey and Clarke [1], investigated the inefficiencies occurred in market in player head to head betting for 2003 cricket world cup. Bracewell and Ruggiero [3] have shown interest in performance monitoring of an individual batsman's performance in different matches by using a parametric control chart. Duchet al., [5] have used social network analysis based network approach which is applied for quantifying individual soccer players performance. An initial effort is made to apply PageRank [12] to teams and captains ranking in cricket [11]. Unfortunately, they have not considered h-index based researcher productivity methods for ranking cricket teams and also ignored the number of runs and wickets parameters for both graph and non-graph based weightage methods.

5   Conclusions

This work concludes that number of runs and wickets from which team wins are important and affect teams ranking. The weightage factor is also important when two teams' wins similar number of matched from similar kind of opponents. The hybridization of h-index and PageRank based methods for ranking cricket teams is also effective as it considers graph, non graph weightage as well as number of runs and wickets for both. Similar methods can be applied to T20 matches.

References

1. Bailey, M.J., Clarke, S.R.: Market inefficiencies in player head to head betting on the 2003 cricket world cup. Economics, Management and Optimization in Sport 11, 185–201 (2004)
2. Borooah, V.K., Mangan, J.E.: The "Bradman class": An exploration of some issues in the evaluation of batsmen for test matches, 1877 - 2006. Journal of Quantitative Analysis in Sports 6(3) (2010)
3. Bracewell, P.J., Ruggiero, K.: A parametric control chart for monitoring individual batting performances in cricket. Journal of Quantitative Analysis in Sports 5, 1–16 (2009)
4. Burrell, Q.L.: Hirsch's h-index: a stochastic model. Journal of Informetrics 1(1), 16–25
5. Duch, J., Waitzman, J.S., Amaral, L.A.N.: Quantifying the performance of individual players in a team activity. PLoS One 5, e10937 (2010)
6. Egghe, L.: Theory and Practice of the g-index. Jointly published by Akadémiai Kiadó, Budapest and Springer, Dordrecht; Scientometrics 69(1), 131–152 (2006)
7. Haveliwala, T.H.: Topic-sensitive PageRank. In: Proc. of the International Conference on World Wide Web (WWW), pp. 517–526 (2002)
8. Hirsch, J.E.: An index to quantify an individual research output. Proc. of the National Academy of Sciences of the United States of America 102, 16569–16572 (2005)
9. Ivn, G., Grolmusz, V.: When the Web Meets the Cell: Using Personalized Page Rank for Analyzing Protein Interaction Networks. Bioinformatics 27(3), 405–407 (2011)
10. Manaskasemsak, B., Rungsawang, A., Yamana, H.: Time-weighted web authoritative ranking. Information Retrieval Journal 14(2), 133–157 (2011)
11. Mukherjee, S.: Identifying the greatest team and captain: A complex network approach to cricket matches. Physica A: Statistical Mechanics and its Applications 391(23), 6066–6076 (2012)
12. Page, L., Brin, S., Motwani, R., Winograd, T.: The pagerank citation ranking: Bringing order to the web. Technical report, Stanford Digital Libraries SIDL-WP-1999-0120 (1998)
13. Xing, W., Ghorbani, A.: Weighted PageRank Algorithm. In: Proc. of 2nd Annual Conference on Communication Networks and Services Research, pp. 305–314 (2004)
14. http://www.ehow.com/how_6916968_calculate-icc-rankings.html
15. http://www.cricinfo.com

Information and Rough Set Theory Based Feature Selection Techniques

Liam Cervante and Xiaoying Gao

School of Engineering and Computer Science
Victoria University of Wellington, P.O. Box 600, Wellington, New Zealand
{liam.cervante,xgao}@ecs.vuw.ac.nz

Abstract. Feature selection is a well known and studied technique that aims to solve "the curse of dimensionality" and improve performance by removing irrelevant and redundant features. This paper highlights some well known approaches to filter feature selection, information theory and rough set theory, and compares a recent fitness function with some traditional methods. The contributions of this paper are two-fold. First, new results confirm previous research and show that the recent fitness function can also perform favorably when compared to rough set theory. Secondly, the measure of redundancy that is used in traditional information theory is shown to damage the performance when a similar approach is applied to the recent fitness function.

1   Introduction

In many situations a large number of features are introduced in order to describe the target objects in the universe. This large number of features allows for different concepts and patterns to be identified and can help with numerous problems, such as classification. Often, however, too many features can contribute to "the curse of dimensionality", a major obstacle in classification. In addition to this, the presence of noisy or highly correlated features can decrease performance. Feature selection is an important and well known technique for solving the above problems [1]. Feature selection can be described as follows: given a set of n features, G, find a set of m features, F, such that $m < n$ and $F \subset G$. F should be representative of G, and should have eliminated any irrelevant or redundant features hence increasing efficiency and enhancing classification accuracy.

Any feature selection algorithm has two key aspects: the search strategy and the evaluation criterion (fitness function). The evaluation criterion measures how good the selected features are, this information can then be used for a number of things. For example, it can be used to guide the search, by choosing the next feature or highlighting a good path to explore, or to decide when to stop. Evaluation criterion can be categorized into wrapper approaches and filter approaches. Wrapper approaches embed learning algorithms, such as a Naive Bayes classifier, into the evaluation criterion, while filter approaches use mathematical models as estimates of goodness. Wrapper approaches usually achieve better results than

T. Yoshida et al. (Eds.): AMT 2013, LNCS 8210, pp. 166–176, 2013.
© Springer International Publishing Switzerland 2013

filter approaches, but the cost of the learning algorithm makes them computationally expensive and can lead to a loss of generality as the algorithm will pick features that perform well for that classifier and that particular training set [2].

Filter approaches rely on the mathematical model used to estimate the goodness of the selected features rather than the actual measure used in wrapper approaches. The performance of the algorithm is then dependent on how good an estimate the mathematical model provides. Many different models have been proven effective, including information measures [3] and rough set theory [4]. A filter approach that achieves a good estimate has the ability to perform favorably when compared to wrapper approaches. If it is a good estimate it will likely select good features, and possibly match the wrapper approach, but do so using fewer resources. This means that a filter approach could achieve a more complete search than a wrapper approach in the same amount of time, the potential for finding better subsets is then increased.

The second aspect of any feature selection algorithm is the search strategy. To perform a complete search of every possible feature subset would be unfeasible. A dataset with instances described by only 30 features has 2^{30} possible feature subsets, each of which would need to be evaluated. To overcome this issue more complex search strategies have been devised. Greedy algorithms exist but these have the problem of getting stuck in local minima and maxima [1]. Evolutionary techniques can also be used to perform the search. Increasing the complexity of the search strategy aims to be able to perform a more complete search.

Previous work presented a new information theory function that performed well compared to traditional information theory when using particle swarm optimization as the search technique [5]. This paper focuses on comparing and evaluating that new filter based evaluation criterion on a larger scale. Comparison with traditional information theory [3,6] is performed again and rough set based [7,8] techniques are introduced and compared with both the information theory techniques. Comparison between information theory and rough set theory lacks landmark research, the recent technique has also only been compared with information theory and not rough set theory.

The contributions of this paper are two fold. Firstly, comparison between the information theory and rough set theory approaches show that the recent fitness function produces favorable results when compared to both alternatives. Secondly, the results show that previous results could have been improved further by not considering the measure of redundancy seen in both the information theory approaches.

2   Background

2.1   Information Theory

Information theory as developed by Shannon [6] presents a way to quantify the level of uncertainty in random variables. From this we can derive the amount of information gained and shared between random variables.

Entropy, in information theory, can be described as the level of uncertainty in a random variable. Let X be a random variable with discrete values, the entropy of X, $H(X)$, is:

$$H(X) = - \sum_{x \in X} p(x) \log_2 p(x) \tag{1}$$

where $p(x) = P(X = x)$, the probability density function of X.

The joint entropy of two random variables, X and Y, can be described as:

$$H(X,Y) = H(Y,X) = - \sum_{x \in X,\, y \in Y} p(x,y) \log_2 p(x,y) \tag{2}$$

The joint entropy quantifies the degree of uncertainty in two random variables.

Gaining knowledge of a certain variable can often reveal information about others, this is measured by the conditional entropy. Assume that the variable Y is known then the conditional entropy of X given Y, $H(X|Y)$, is:

$$H(X|Y) = - \sum_{x \in X,\, y \in Y} p(x,y) \log_2 \frac{p(x)}{p(x,y)} \tag{3}$$

The conditional entropy can also be calculated using the entropy and joint entropy:

$$H(X|Y) = H(X,Y) - H(Y)$$

Finally, the information shared between two variables is defined as mutual information. The amount of information is shared between variables X and Y, the mutual information, $I(X;Y)$, is defined as:

$$I(X;Y) = - \sum_{x \in X,\, y \in Y} p(x,y) \log_2 \frac{p(x,y)}{p(x)p(y)} \tag{4}$$

As with the conditional entropy, shown above, mutual information can be defined using the other measures of entropy:

$$\begin{aligned} I(X;Y) &= H(X) - H(X|Y) \\ &= H(Y) - H(Y|X) \\ &= H(X) + H(Y) - H(X,Y) \end{aligned}$$

Information theory can be used to build a filter based evaluation criterion that can be used to calculate the information shared between features, for redundancy, and between the class value and the features, for relevance.

2.2   Rough Set Theory

Rough set theory, as developed by Pawlak [7], provides a formal approximation of a conventional set. The rough set is described by the lower and upper approximations of the conventional set. Let $\mathbf{U}$ be the universe, the set of instances, and let $\mathbf{A}$ be the set of attributes that describe the instances. Also, let $a(x)$ specify the value of attribute $a \in \mathbf{A}$ in instance $x \in \mathbf{U}$.

For any $P \subseteq \mathbf{A}$ we can define the indiscernible equivalence relation $IND(P)$:

$$IND(P) = \{(x, y) \in \mathbf{U}^2 \mid \forall a \in P. \; a(x) = a(y)\} \tag{5}$$

If $(x, y) \in IND(P)$ we say that x and y are indiscernible according to P. We can use the above relation to define equivalence classes, these are denoted $[x]_P$. This means that $y \in [x]_P \Leftrightarrow (x, y) \in IND(P)$.

Let $X \subseteq \mathbf{U}$ be the set we want to represent with P. We can define the upper and lower bounds of X according to P:

$$\underline{P}X = \{x \mid [x]_P \subseteq X\} \tag{6}$$

$$\overline{P}X = \{x \mid [x]_P \cap X \neq \emptyset\} \tag{7}$$

The rough set is then the tuple: $\langle \underline{P}X, \overline{P}X \rangle$. We can quantify the *accuracy* of the rough set using:

$$\alpha_P(X) = \frac{|\underline{P}X|}{|\overline{P}X|}$$

Which effectively measures how well the attributes in P separate the target set, X, from the rest of $\mathbf{U}$. If P is poorly chosen then few instances will be in the lower bound while many will be in the upper bound.

As with information theory, rough set theory can be used as filter approach to evaluation. Well selected features will separate the classes that instances can be assigned to.

3 Fitness Functions

This section presents evaluation criterion based on information theory and rough set theory and explains how they can be applied to the data. Fitness functions can be used to compare potential features, all three of the following functions work by adding the potential feature to the test and evaluating the new set together. As with the rough set theory the feature selection framework uses a set of instances that make up the universe $\mathbf{U}$, a set of features, F, that describe the instances, and each instance has a class value, c.

3.1 Paired Mutual Information

Peng et al. present an filter based fitness function based on information theory [3]. Peng et al. use mutual information to estimate the relevance and redundancy of the features as they are selected, the fitness function attempts to maximize the relevance and minimize the redundancy. Each feature, and the class label, needs to be treated as a random variable and the probability density functions can be calculated using a training set. For example, consider a particular feature,

$f \in F$, that can take three values: $\{0, 1, 2\}$. The number of times f takes the value 0 can be used to calculate $P(X = 0)$ and hence calculate the entropy and mutual information.

The fitness of a set of features, $G \subseteq F$, is calculated as:

$$Fitness(G) = D(G) - R(G) \tag{8}$$

$$D(G) = \frac{1}{|G|} \sum_{f \in G} I(f; c) \tag{9}$$

$$R(G) = \frac{1}{|G|^2} \sum_{f \in G, g \in G} I(f; g) \tag{10}$$

Equation (9) quantifies the average mutual information shared by each selected feature and the class label. Equation (10) quantifies the average mutual information shared by each of the selected features with every other feature, hence providing a measure of redundancy. This paper provides a comparison between the fitness function with (MI-R-P) and without (MI-NR-P) the measure of redundancy and shows that it is important to improving the accuracy.

3.2   Group Mutual Information

Where the function presented above considers pairs of features, our previous research presented a second information theory criterion that attempts to evaluate the features as a group [5]. The fitness function requires the joint entropy to be calculated over larger sets of random variables:

$$H(X_1, ..., X_n) = - \sum_{x_1 \in X_1} \cdots \sum_{x_n \in X_n} p(x_1, ..., x_n) \log_2 p(x_1, ..., x_n) \tag{11}$$

As above the fitness function attempts to maximize the relevance and minimize the redundancy.

$$Fitness(G) = D(G) - R(G) \tag{12}$$

$$D(G) = IG(c \mid G) \tag{13}$$

$$R(G) = \frac{1}{|G|} \sum_{f \in G} IG(f \mid \{G/f\}) \tag{14}$$

The criterion here attempts to quantify the amount of information gained about one random variable, or feature, given knowledge of a set of others. It can be calculated by:

$$IG(c \mid G) = H(c) - H(c|X)$$
$$= H(c) - (H(c \cup X) - H(X))$$
$$= H(c) + H(X) - H(c \cup X)$$

Cervante et al. use this function as a way to evaluate sets of features generated by a search using particle swarm optimization. As with the paired fitness function, the group evaluation is done using both with the measure of redundancy (MI-R-G) and without (MI-NR-G), the experimental procedure also involved a simpler process than particle swarm optimization to focus comparison on the fitness functions. As the group fitness function considers the selected features as a group rather than in a paired average it should provide a better estimate.

3.3   Probabilistic Rough Set Approximations

Rough set theory provides a natural way to evaluate feature sets, given the definitions. We can partition the universe using the class labels, each partition becomes a target set. The lower bound of each target set then measures the number of instances that have been completely separated from instances of other classes. Assume the universe, $\mathbf{U}$, has been partitioned into target sets: $\{U_1, ..., U_n\}$. An evaluation criterion for a subset of features, $G \in F$, is then:

$$Fitness(G, \mathbf{U}) = \frac{\sum_{U_i \in \mathbf{U}} |\underline{G}U_i|}{|\mathbf{U}|} \tag{15}$$

The evaluation criterion measures the number of instances that have been separated from instances of other classes by the features, a score of 1.0 means that G completely divides the classes.

It is possible that a minority of instances could share identical attributes but have different class labels. In practice this could happen due to minorities that are exceptions to common patterns or even human error in inputting the data. This being the case, having even one instance labeled differently means that it becomes impossible to find a satisfying assignment because of one mistake. To overcome this problem we can introduce probabilistic rough set approximations [8]. Rather than having a strict lower bound, we can relax it with varying degree using a value α. For a given target set X and a set of features G, we define the function μ to be:

$$\mu_G(x) = \frac{|[x]_G \cap X|}{|[x]_G|} \tag{16}$$

So, μ quantifies the proportion of $[x]_G$ that is also in the target set. Using this we can define the lower approximation of X according to G:

$$\underline{apr}_G X = \{x \mid \mu_G(x) \geq \alpha\} \tag{17}$$

Note that when α is set to 1.0 this calculation becomes the same as the lower bound calculation above, since every instance in $[x]_G$ must also be in X making it only true when it is a subset.

Using this approximation we update the fitness function to be:

$$Fitness(G, \mathbf{U}) = \frac{\sum_{U_i \in \mathbf{U}} |\underline{apr}_G U_i|}{|\mathbf{U}|} \tag{18}$$

We can set α to be 1.0 to mimic the strict lower bound definition and use a variety of values for α to hopefully improve performance. In experimental conditions, three conditions for α were considered, $\{1.0, 0.75, 0.5\}$, with $\alpha = 1.0$ as the baseline.

Table 1. Summary of results at peak accuracy

Criterion	Number of Features	Training Accuracy	Testing Accuracy
MI-NR-P	7	0.897	0.919
MI-R-P	4	0.939	0.945
MI-NR-G	4	0.939	0.945
MI-R-G	8	0.920	0.925
RS-1.0	12	0.932	0.922
RS-0.75	4	0.939	0.945
RS-0.5	4	0.939	0.945

4   Experimental Results

In order to evaluate the different criteria we tested them using a dataset from the UCI repository: Chess (King-Rook vs. King-Pawn). The dataset is split 52:48 into two classes. There are a total of 3196 instances and 36 attributes. We ran a simple forward selection algorithm and tested the accuracy on the training and testing set, derived from the dataset, as each feature was added to show how well each criterion estimated the goodness of potential features. A naive bayes classifier was used and the accuracy on the training set when using all features was 0.869 and on the testing set was 0.870.

8 different criteria were tested, the crisp rough set with $\alpha = 1.0$ (RS-1.0) and the two mutual information functions with no measure of redundancy (MI-NR-P and MI-NR-G for the pair and group criterion respectively) were tested, these are considered the baseline. Next we demonstrated the effect of adding the measure of redundancy in the mutual information criterion (MI-R-P and MI-R-G) and the effect of relaxing the value for alpha (RS-0.75 and RS-0.5). Each criterion peaked at a considerably lower number of features before the performance began degrading as more features were added. The results for the best performing features and how many were included can be seen in Table 1.

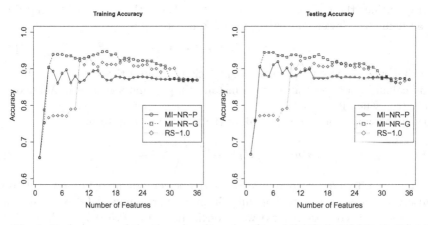

Fig. 1. Performance of the three baseline classifiers: MI-NR-P, MI-NR-G, RS-1.0

The following figures show the change in performance as more features are added. Figure 1. shows a comparison of the three baseline functions. Both the information theory criterion gain a large increase in accuracy very quickly, which the rough set function does not. However, after 12 features are selected the rough set function overtakes and gets better accuracy than the paired mutual information one achieves. Our group fitness function stays ahead of both.

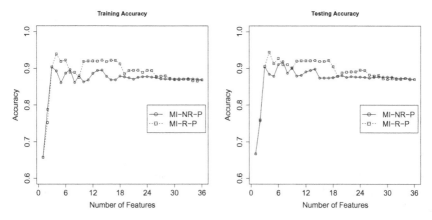

Fig. 2. Performance comparison between using and not using a measure of redundancy in the MI-P criterion

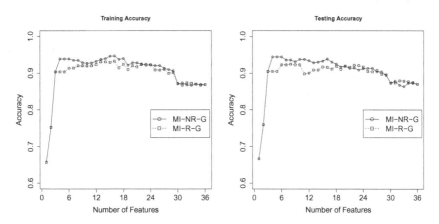

Fig. 3. Performance comparison between using and not using a measure of redundancy in the MI-G criterion

The remaining figures show attempts at further improving the accuracy. Figure 2. shows the introduction of a measure of redundancy to the pair criterion, making it the same as the fitness function as presented by Peng et al. Figure 3. shows the measure of redundancy introduced into our group fitness function. Interestingly, the added redundancy measure decreases the performance of

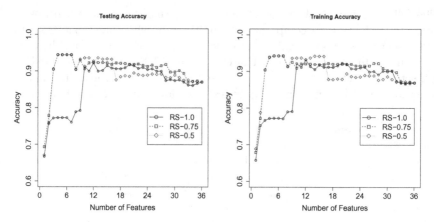

Fig. 4. Performance comparison for varying values of alpha, $\alpha = \{1.0, 0.75, 0.5\}$

the grouped fitness function while it improves the paired fitness. Finally, Figure 4. shows the changes seen with differing values for alpha in the probabilistic rough set approach. Both additional values for alpha show an increase in performance initially but when $\alpha = 0.5$ we see a large drop in performance after 17 features have been selected. With $\alpha = 0.75$ the performance remains strong for much longer.

5 Discussion

We can see that in this case it is possible to find the highest performing subset using all three criterion. MI-R-P, MI-NR-G, and the two probabilistic rough set approaches, RS-0.75 and RS-0.5, reach an accuracy of 0.939 after 4 features have been selected. In addition to this, the group function, MI-NR-G, and the two rough set approaches maintain this accuracy even as more features are being selected and only begin to drop when 7 features or more are selected. The paired function, MI-R-P, drops straight away but then levels, and after 7 features matches the rough set approaches. Our group fitness function, MI-NR-G, maintains a higher accuracy for much longer suggesting that it is finding better features than the alternate approaches, even though the accuracy isn't increased the danger of overfitting is lessened.

The reason of the accuracy drop when adding the redundancy measure to the group criterion could be cause it becomes too concerned with avoiding redundancy, too much weight is given to that consideration. The function already considers the fitness of the group as a whole so adding redundancy could be unnecessary. This is in contrast to the paired function, where the relevance measure does not consider the group and so it is important to add the measure of redundancy. In our earlier paper [5] that first introduced the group evaluation measure only the function that considers redundancy as well as relevance is used, accuracy could have been improved further by not considering the measure of redundancy.

Finally, it is clear that using probabilistic rough sets, RS-0.75 and RS-0.5, instead of the strict lower bound, RS-1.0, can select better features. However, when alpha was set very low, $\alpha = 0.5$, overfitting (searching for too long) becomes a significant risk, after 50% of the features had been selected the performance dropped to less than the performance of the strict definition. Using the strict lower bound likely means that it is difficult for the classifier to generalize while a low value for alpha leads to over generalization. The performance of the rough set theory function is then dependent on the choice of alpha. The optimum value of alpha could differ across multiple situations, the information theory functions do not have this problem as they have no global variables that need to be defined.

6 Conclusion

In conclusion, the results show that our previous group mutual information based approach can achieve a high accuracy and choose features to maintain it for longer. In addition to this, the redundancy measure used in the paired fitness function actually hurt the performance of the group fitness function. Finally, a comparison between these two information theory based approach and that of a rough set theory based approach was undertaken.

This paper presented the mathematical reasoning behind information theory and rough set theory, before showing two traditional evaluation criterion derived from the reasoning. Our group approach was then compared with the traditional approaches, and it is highlighted that previous work could have been improved by not considering the redundancy measure in the recent approach. This is in contrast to the traditional information theory approach in which considerable improvement is gained by including the measure of redundancy.

Results also show that while adding more features decreases the accuracy in the training and testing sets, the new approach can maintain a higher degree of accuracy for longer. This suggests that it is still finding good potential features where the others are not. Previous work compared the group fitness function favorably with the pair fitness function when particle swarm optimization was used as the search technique, future work could consider including a comparison with the rough set theory functions using a more complex search technique.

References

1. Guyon, I., Elisseeff, A.: An introduction to variable and feature selection. J. Mach. Learn. Res. 3, 1157–1182 (2003)
2. Kohavi, R., John, G.H.: Wrappers for feature subset selection. Artif. Intell. 97(1-2), 273–324 (1997)
3. Peng, H.P.H., Long, F.L.F., Ding, C.: Feature selection based on mutual information criteria of max-dependency, max-relevance, and min-redundancy. IEEE Transactions on Pattern Analysis and Machine Intelligence 27(8), 1226–1238 (2005)
4. Zhong, N., Dong, J., Ohsuga, S.: Using rough sets with heuristics for feature selection. J. Intell. Inf. Syst. 16(3), 199–214 (2001)

5. Cervante, L., Bing, X., Zhang, M.: Binary particle swarm optimisation for feature selection: A filter based approach. In: Proceedings of 2012 IEEE Congress on Evolutionary Computation, pp. 881–888. IEEE Press (2012)
6. Shannon, C.E., Weaver, W.: A Mathematical Theory of Communication. University of Illinois Press, Champaign (1963)
7. Pawlak, Z.: Rough sets. International Journal of Parallel Programming 11, 341–356 (1982), 10.1007/BF01001956
8. Yao, Y.: Probabilistic rough set approximations. Int. J. Approx. Reasoning 49(2), 255–271 (2008)

Developing Transferable Clickstream Analytic Models Using Sequential Pattern Evaluation Indices

Hidenao Abe

Department of Information Systems, Faculty of Information and Communications,
Bunkyo University 1100 Namegaya, Chigasaki, Kanagawa, 253-8550, Japan
hidenao@shonan.bunkyo.ac.jp

Abstract. In this paper, a method for constructing transferable "web" and "click-stream" prediction models based on sequential pattern evaluation indices is proposed. To predict end points, click streams are assumed as sequential data. Further, a sequential pattern generation method is applied to extract features of each click stream data. Based on these features, a classification learning algorithm is applied to construct click stream end point prediction models. In this study, the evaluation indices for sequential patterns are introduced to abstract each click-stream data for transferring the constructed predictive models between different periods. This method is applied to a benchmark clickstream dataset to predict the end points. The results show that the method can obtain more accurate predictive models with a decision tree learner and a classification rule learner. Subsequently, the evaluation of the availability for transferring the predictive morels between different periods is discussed.

Keywords: Sequential Pattern Mining, Clickstream Analysis, Transfer Learning.

1 Introduction

Considering the accumulations of massive log data on Web servers and other information systems, recent years have been seen an increase in the demand for efficient analytics for data streams, called "big-data analytics". To conduct efficient analyses of the big-data, data scientists need to combine conventional data mining techniques. In particular, for clickstream mining on web-sites, the issue with the analysis of big-data have been studied to identify an efficient method using only one type of model such as stochastic/probabilistic model[7].

The Two research issues considered in this study are based on the limitation of using the conventional web clickstream mining methods; the use of one type of model, and the unavailability of transfer of models among different periods. To solve the first issue, a combination of sequential pattern generation and classification model learning based on sequential pattern evaluation indices is proposed in this paper. In this combinatorial method, a set of sequential patterns is extracted from stored sequential data, and the construction techniques of predictive models are developed for the Web click-stream prediction, which predicts a visitor's action. These construction techniques can be used for the Web clickstream data acquisition times. For the second issue, a transfer

T. Yoshida et al. (Eds.): AMT 2013, LNCS 8210, pp. 177–186, 2013.
© Springer International Publishing Switzerland 2013

learning approach[8] is applied to a web clickstream prediction model for a sequential benchmark dataset. In order to construct predictive models that can be used during a different period and can be transferred, the clickstream instances change into the sequential pattern appearance information on sequential data as the values of a feature set which evaluates sequential patterns, and builds predictive models using classification learning algorithms.

In the remaining of this paper, two criteria of sequential pattern appearances and four evaluation index definitions are described in Section 2. The method for building web clickstream analytic models is described in Section 3. An experiment is performed by applying the proposed method to a web server log data from UCI KDD Archive [4] is described in Section 4. Finally, the conclusion and feature work are described in Section 5.

2   Generation of Sequential Patterns and Sequential Pattern Evaluation Indices

After the first frequent sequential pattern generation algorithm [2], many efficient algorithms [5] have been developed for generating sequential patterns from sequential datasets. Each sequential data consists of sequenced nominal values called "item", which are collected at particular start and end points. In such sequential data, a pair of items holds a half-ordered relationship that is defined by temporal difference, placement, and so on. From a sequential dataset S, sequential pattern generation algorithms generate sequential patterns that are a set of sub-sequences $\{s\}$ satisfying the criteria of each algorithm.

2.1   Sequential Pattern Generating Algorithms

There are two groups of algorithms developed in different contexts for the sequential pattern mining approach. The algorithms in one group (AprioriAll [2], PrefixSpan [9], and other frequent sequential pattern mining algorithms [5]) obtain sub-sequences with gaps between items. However, the algorithms in the other group (n-gram [12]) obtain sub-sequences without gaps.

Frequent sequential pattern extraction is applied to a sequential dataset D, which consists of sequential data s. Each sequential data s_i is comprises ordered items described as $s_i =< item_1, \cdots, item_{mi} > (item_k \in I)$, where the items included are in the domain of items $I = \{item_1, item_2, \cdots, item_m\}$. Each algorithm was implemented to search for possible sub-sequence in S with frequent sequential patterns sp, which appear more frequently than the given minimum support min_supp, achieving lower computational cost and higher memory efficiency. The frequencies of the frequent sequential patterns $|sp|$ always exceed the threshold, which is equated as $|sp| > min\_supp$. The frequent sequential patterns are often useful for characterizing the issue of the domain, which is described as the sequential dataset.

Researchers in the field of natural language processing have developed many sequential pattern extraction algorithms in text datasets as a different approach to extracting sequential patterns. Automatic term extraction is one of the important processes for identifying jargon, new combination of words, idioms, and so forth. In [6], a score called FLR was proposed to automatically identify combined nouns as terms. As described in [6], some measurements are used to identify such terms by using particular indexes on each algorithm.

For applying the FLR score based automatic term extraction method to a clickstream dataset, each sequence is assumed as one document and each page visit p_k as one noun. Under these assumptions, the method is applied to the sequential dataset S, which consists of each sequential data $s_i = <p_k> (1 \leq k \leq m_i)$. To find the characteristic sub-sequences with $FLR(Cs) > threshold$, the scores of the candidates of sub-sequence $Cs = <p_l> (1 \leq l \leq L < m_i)$ are calculated by using the following definition:

$$FLR(Cs) = f(Cs) \times (\prod_{l=1}^{L} (FL(p_l) + 1)(FR(p_l) + 1))^{\frac{1}{2L}}$$

where $f(Cs)$ denotes the frequency of a candidate Cs isolated, and $FL(p_l)$ and $FR(p_l)$ denote the frequencies of different orders on the former and the latter of each page visit p_l in bi-grams in each candidate sub-sequence Cs. As for the default threshold of the FLR score, $threshold = 1.0$ is set, denoted as $FLR(Cs) > 1.0$.

2.2 Defining Sequential Pattern Evaluation Indices

In this study, the measurements for sequential patterns are assumed as their evaluation indices. For the frequencies of the sub-sequences in a sequential dataset, two frequent criteria are used. One frequency basis is the document frequency (DF), which is the frequency of documents containing target words/phrases t in a document D, denoted as $DF(t, D)$. The other frequency basis is the term frequency (TF), which is the number of all the appearances of target words/phrases t in a document D, denoted as $TF(t, D)$.

In addition to the abovementioned frequency bases, the support, odds, and self-information criterion (entropy) index frames are combined with the frequency basis criteria. Term frequency and inversed document frequency (TFIDF [13]) is the most utilized importance index for textual data and sequential datasets on various domains that use the two frequency bases. Table 1 shows the definitions of the seven evaluation indices that are used in the rest of this paper.

3 Construction of Predictive Models Based on Sequential Pattern Evaluation Indices by Considering Transferring Models

In this section, a method to construct predictive models based on the sequential pattern evaluation indices by converting a sequential dataset to a relational dataset is described.

3.1 A Method for Constructing Clickstream Prediction Models Based on Sequential Patterns and Their Evaluation Indices

Usually, sequential data consists of various length of sequence of items. This is a major limitation for the application of conventional pattern generation methods such as rule

Table 1. Definitions of the sequential pattern evaluation indices including two frequency bases and the combinations of four index frames

	Basis of frequency counting for sp	
	Document frequency $DF = \|D_{\in sp}\|$	Term frequency $TF = \sum_i freq(sp, d_i)$
support	$DF / \|D\|$	$TF / \sum_{sp \in D} TF$
Odds	$DF / (DF - \|D\|)$	$TF / (TF - \sum_{sp \in D} TF)$
Self-entropy	$(DF / \|D\|) \log_2(DF / \|D\|)$	$(TF / \sum_{sp \in D} TF) \log_2(TF / \sum_{sp \in D} TF)$
TFIDF	$TF * \log(\|D\| / DF)$	

induction, and other methods for a relational dataset. Many sequential pattern generation algorithms have been developed for mining valuable information from sequential dataset. However, they tend to generate a large number of sequential patterns, and domain experts have to interpret their meanings by using certain evaluation indices. This pattern evaluation process is highly time-consuming, and it should be supported with a more efficient method.

In contrast, sequential patterns can be evaluated by using some evaluation indices based on their frequency in the on validation dataset. Each evaluation index was developed to quantify a particular aspect of sequential patterns. The combination of evaluation indices adequately for the given problem helps to reduce the cost of the manual evaluation process of the sequential patterns. Moreover, this prediction method needs results of each sequential data as class labels. Subsequently, the method constructs classification models to predict the class labels as the predictive models. The constructed prediction models should be more efficient than some objective criterion such as percentage of the majority class labels for the prediction model learning.

To solve abovementioned issue, an overview of a method for constructing sequential data label prediction models based on sequential patterns and their evaluation indices is shown in Figure 1.

This method is divided two major sub-processes. The first process is the calculation of sequential pattern evaluation indices for the generated sequential patterns. In this process, we generate sequential patterns with one sequential pattern generation algorithm at least once.

Subsequently, by using the calculated evaluation indices values for each sequential pattern, the method selects particular sequential patterns for each sequential data to convert it to one numerical vector. If a sequential data s_i includes a sequential pattern sp, which is selected by a conflict resolution strategy, the sequential data is converted to the numerical vector $d_i = < v_{ij} >$. Each value v_{ij} for $index_j$ is calculated as follows:

$$v_{ij} = Index_j(sp, D_{period})(sp \in s_i)$$

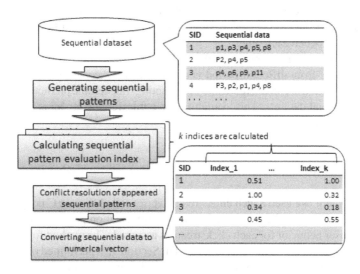

Fig. 1. An overview of the method for constructing predictive models of sequences based on sequential patterns and their evaluation indices

In this method, each representative sequential pattern sp for calculating index value is selected with certain criterion such as the maximum value for each index.

After converting the sequential dataset to the relational dataset, the values of each index are normalized for comparison with other relational datasets from different periods as follows.

$$Normalize(v_{ij}) = \frac{v_{ij} - min(v_{.j})}{max(v_{.j}) - min(v_{.j})}$$

where $min(v_{.j})$ denotes minimum value for $index_j$ within the dataset, and $max(v_{.j})$ denotes maximum value for $index_j$.

Applying this conversion process to entire sequential data in a sequential dataset S, a converted relational dataset D is obtained. On assigning class labels to corresponding instances in this relational dataset, any conventional classification learning algorithms can be applied to construct predictive models for predicting the labels of each sequence.

Because the sequential dataset can be obtained for generating nominal values when there are a start point and an end point, this method can be applied to various sequential datasets from each target domain. In this study, we apply the proposed method to a web clickstream dataset.

3.2 Transferring the Predictive Models Based on Sequential Pattern Evaluation Indices to Other Datasets

The proposed method assumes that a user can apply the same sequential pattern evaluation indices even for the sequential dataset from different periods. In the conversion process, if the same sequential evaluation indices are used for different sequential datasets, they generate relational datasets with the same features trivially. This assumption indicates that the constructed predictive models used on the different dataset can be transferred between each other.

In the transfer learning approach, the type of transfer with the same features for different datasets from different period is defined in [8]. According to this definition, the proposed method is considered while constructing transferable predictive models for different sequential datasets. This availability as a transferable model is also evaluated in the next section.

4   Experiment

In this experiment, we used a web clickstream dataset called 'Entree Chicago' [1]. This dataset consists of web clickstream of restaurant information pages within a site, and the result of user behavior for each clickstream is assigned. The method for constructing web clickstream prediction models, described in Section 3, is performed on this dataset.

The method converts each set of the sequential dataset to the relational datasets, which consists of values of the seven sequential pattern evaluation indices for the evaluation for predicting clickstream results. Subsequently, a constructed predictive model was applied to the other datasets at different periods for evaluating its availability as a transferable model.

4.1   Generating Sequential Patterns on Periodical Web Clickstream Dataset

Firstly the automatic term extraction method based on FLR scores is applied to the Entree Chicago datasets. Each web clickstream s is consisted by one sequence of pairs of one page view and one user action as one word $< p_i >$, denoted as $s =< p_1, p_2, \cdots, p_m >$ as one document. The web site hosts 675 pages corresponding to restaurants in eight cities in the U.S. Further, nine types of user actions were determined for each page view. Thus, the sequential datasets for each quarter of the year D_{period} could be obtained from the web server logs.

Table 2 shows the size of each sequential dataset $|D_{period}|$ and the number of sequential patterns extracted by the FLR score method.

According to the method illustrated in Figure 1, each sequential dataset on each period was converted to the relational dataset for predictive model construction using the seven sequential pattern evaluation indices and FLR score.

4.2   Evaluating the Availability for Predicting Web Clickstream Results Based on Sequential Pattern Evaluation Indices

After converting the sequential datasets on each period to the relational datasets RD_{period}, two class labels are assigned according to the determined end point in the original datasets. The class label "Give-up" means that a user quits browsing the website for some reason. In contrast, the label "Stay-in" is assigned when a user looks at a page on the website. The distributions of the class labels are shown in Table 3.

Table 2. Size of the sequential datasets and the number of the extracted sequential patterns using the automatic term extraction method based on FLR scores

period	$||D_{period}||$	# of Seq. Patterns
1996.Q3	754	731
1996.Q4	1785	1630
1997.Q1	5004	4277
1997.Q2	4001	3441
1997.Q3	3850	3354
1997.Q4	3533	2827
1998.Q1	5295	4396
1998.Q2	5501	4482
1998.Q3	4847	4017
1998.Q4	7955	6360
1999.Q1	6837	5491
1999.Q2	1298	1062

A 10-fold cross validation with 100 iterations was performed to evaluate the effect of the characteristic order entry sequences for the classification of the clickstream results using the following five classification learning algorithms: C4.5 [11], PART [3], SVM [10], and k-NN($k = 5$), implemented in Weka [15].

Table 4 shows the averaged accuracies and their significant advantages/disadvantages tested by a corrected t-test with a significant level $p = 0.05$.

The result shows that the predictive models constructed as decision trees by C4.5 predicted most class labels "Stay-in" and outperform the others. This indicates that predictive model construction based on sequential pattern evaluation indices is important to predict web clickstream results on this dataset.

On observing the details of the constructed predictive models of the web clickstream, Figure 2 shows the decision tree constructed by using the whole dataset of 1996.Q4, which consists of the eight sequential pattern evaluation indices. For the dataset of 1996.Q4, C4.5 achieves higher accuracy than only predicting "Stay-in" on the 100 times iterated 10-fold cross validation.

Although the predictive models with the sequential pattern evaluation indices do not mention some particular sequential patterns directly, the users can get the list of sequential patterns in each period by using the sequential pattern evaluation indices and their thresholds immediately.

4.3 Evaluating Availability for Transferring the Web Clickstream Predictive Models to Other Period

In order to evaluate the availability to transfer the web clickstream prediction models to other datasets from different periods, one predictive model as shown in Figure 2 was applied to other dataset from 1997.Q1 to 1999.Q2. The criterion for evaluating the availability of transfer was set up as the percentages of the majority class label for each dataset in this evaluation.

Table 3. Class distributions for the Entree Chicago datasets

period	RD_{period}	# of Stay-in	# of Give-up	% Maj. Class
1996.Q3	754	675	79	89.5
1996.Q4	1785	1572	213	88.1
1997.Q1	5004	4370	634	87.3
1997.Q2	4001	3539	462	88.5
1997.Q3	3850	3476	374	90.3
1997.Q4	3533	3108	425	88.0
1998.Q1	5295	4667	628	88.1
1998.Q2	5501	4848	653	88.1
1998.Q3	4847	4270	577	88.1
1998.Q4	7955	6913	1042	86.9
1999.Q1	6837	5976	861	87.4
1999.Q2	1298	1158	140	89.2

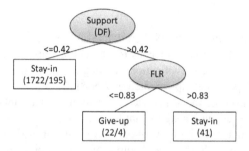

Fig. 2. Decision tree model constructed for the 1997 Q4 dataset. The oval denotes a tested attribute of the branch, and the rectangle denotes a leaf node for predicting class labels.

Table 5 shows the percentages of the majority class label and the correctly predicted ratio by considering each dataset as the test dataset.

The transfer application result in Table 5 shows that the predictive model predicted the "Give-up" and "Stay-in" class labels more accurately than only predicting "Stay-in" for the dataset on 1998.Q2. This indicates that the two evaluation indices and their thresholds mentioned previously work efficiently to predict the clickstream results in the different dataset from the different period. Moreover, we can determine the search region that consists of the two evaluation indices and the mentioned threshold for generating sequential patterns related to a particular class label within the next step.

If we construct the predictive model using the sequential patterns directly, it is very difficult because there is so small number of compatible sequential patterns between different periods. For example, the average number of compatible sequential pattern of 1996.Q4 is 252.7 to the other periods.

Table 4. Averaged accuracies(%) and their significant advantages/disadvantages. The significant advantage and disadvantage is marked as "v" and "*", respectively, under the significance level $p = 0.05$.

Dataset	% Maj. class	C4.5(J4.8)	PART	SVM(SMO)	5-NN
1996.Q3	89.5	89.5	89.49	89.5	88.1 *
1996.Q4	88.1	88.8 v	88.63	88.1	88.7
1997.Q1	87.3	87.9 v	88.03 v	87.3	87.5
1997.Q2	88.5	89.2 v	89.08 v	88.5	88.3
1997.Q3	90.3	91.0 v	90.93 v	90.3	90.6
1997.Q4	88.0	88.6 v	88.67 v	88.0	88.1
1998.Q1	88.1	89.4 v	89.25 v	88.1	88.0
1998.Q2	88.1	90.1 v	89.97 v	88.1	89.3 v
1998.Q3	88.1	89.7 v	89.39 v	88.1	88.7
1998.Q4	86.9	89.2 v	89.09 v	86.9	88.5 v
1999.Q1	87.4	88.5 v	88.54 v	87.4	88.0
1999.Q2	89.2	89.6	89.26	89.2	89.3
Average		89.3	89.2	88.3	88.6
[v/ /*]		[10/2/0]	[9/3/0]	[0/12/0]	[2/9/1]

Table 5. Transferring result of 1996.Q4 decision tree model to other dataset

RD_{period}(Tested)	% Maj. class	% Correctly Predicted
1997.Q1	87.3	84.7
1997.Q2	88.5	85.6
1997.Q3	90.3	90.3
1997.Q4	88.0	87.5
1998.Q1	88.1	88.1
1998.Q2	88.1	88.2
1998.Q3	88.1	88.1
1998.Q4	86.9	86.9
1999.Q1	87.4	87.4
1999.Q2	89.2	80.9

5   Conclusion

In this paper, a method to construct a clickstream prediction model based on sequential pattern generation and evaluation indices is described. Sequential datasets were converted to relational datasets, and subsequently, classification learning models constructed from the relational datasets were applied to predict user behaviors from each clickstream data.

In the experiment performed by using a benchmark dataset for web clickstream mining, most of the predictive models constructed by the four representative learning algorithms outperformed the ratios of the majority class label. The decision trees significantly improved the performance of the predictive models by using the values of the eight sequential pattern evaluation indices.

In the application of one predictive decision tree model learned from one dataset to the others from different period, using the transfer learning manner, the model achieved a better accuracy ratio in at least one different period. This indicates that both the evaluation indices and their adequate thresholds are essential for obtaining another set of sequential patterns in the predicted period.

As the future work, the proposed method should be comparable with conventional sparse tolerant learning algorithms such as stochastic models on the accuracy for predicting clickstream results. In addition, other types of sequential pattern evaluation indices from various fields such as frequent itemset evaluation indices [14] and automatic term extraction indices as described in [6] will be introduced in this framework. Many features of sub-sequences can be quantified by using these indices together.

References

1. Entree chicago recommendation data,
   http://kdd.ics.uci.edu/databases/entree/entree.html
2. Agrawal, R., Srikant, R.: Mining sequential patterns. In: Yu, P.S., Chen, A.L.P. (eds.) ICDE, pp. 3–14. IEEE Computer Society (1995)
3. Frank, E., Wang, Y., Inglis, S., Holmes, G., Witten, I.H.: Using model trees for classification. Machine Learning 32(1), 63–76 (1998)
4. Hettich, S., Bay, S.D.: The uci kdd archive, http://kdd.ics.uci.edu/
5. Mabroukeh, N.R., Ezeife, C.I.: A taxonomy of sequential pattern mining algorithms. ACM Comput. Surv. 43, 3:1–3:41 (2010),
   http://doi.acm.org/10.1145/1824795.1824798
6. Nakagawa, H.: Automatic term recognition based on statistics of compound nouns. Terminology 6(2), 195–210 (2000)
7. Padmanabhan, B.: Web Clickstream Data and Pattern Discovery, vol. 3, pp. 99–116 (2009)
8. Pan, S.J., Yang, Q.: A survey on transfer learning. IEEE Transactions on Knowledge and Data Engineering 22(10), 1345–1359 (2010)
9. Pei, J., Han, J., Mortazavi-Asl, B., Pinto, H., Chen, Q., Dayal, U., Hsu, M.C.: Prefixspan: Mining sequential patterns efficiently by prefix-projected pattern growth. In: Proc. of the 17th International Conference on Data Engineering, pp. 215–224. IEEE Computer Society, Los Alamitos (2001)
10. Platt, J.: Fast training of support vector machines using sequential minimal optimization. In: Burges, C., Smola, A. (eds.) Advances in Kernel Methods - Support Vector Learning, pp. 185–208. MIT Press (1999)
11. Quinlan, J.R.: Programs for Machine Learning. Morgan Kaufmann Publishers (1993)
12. Shannon, C.E.: A mathematical theory of communication. The Bell System Technical Journal 27, 379–423, 623–656 (1948)
13. Sparck Jones, K.: A statistical interpretation of term specificity and its application in retrieval. Document Retrieval Systems, 132–142 (1988)
14. Tan, P.N., Kumar, V., Srivastava, J.: Selecting the right interestingness measure for association patterns. In: Proceedings of International Conference on Knowledge Discovery and Data Mining, KDD 2002, pp. 32–41 (2002)
15. Witten, I.H., Frank, E.: Data Mining: Practical Machine Learning Tools and Techniques with Java Implementations. Morgan Kaufmann (2000)

Customer Rating Prediction Using Hypergraph Kernel Based Classification

Fatemeh Kaveh-Yazdy, Xiangjie Kong, Jie Li, Fengqi Li, and Feng Xia

School of Software, Dalian University of Technology, Dalian 116620, China
{f.kaveh,jie.jack.lee}@gmail.com, {xjkong,f.xia}@ieee.org,
lifengqi@dlut.edu.cn

Abstract. Recommender systems in online marketing websites like Amazon.com and CDNow.com suggest relevant services and favorite products to customers. In this paper, we proposed a novel hypergraph-based kernel computation combined with k nearest neighbor (kNN) to predict ratings of users. In this method, we change regular definition style of hypergraph diffusion kernel. Our comparative studies show that our method performs better than typical kNN, which is simple and appropriate for online recommending applications.

Keywords: Recommender System, Rating Prediction, Classification, Hypergraph Kernel.

1 Introduction

Recommender systems successfully contribute to the personalized marketing in recent years. Generally, recommender systems are software frameworks, which are exploited in business information systems. Business databases include key information of customers, characteristics of sold items or provided services and local information of sales departments [1–3]. Unfortunately, users'sentiments are hidden in commercial databases. Subsequently, business information systems need to employ recommender systems to uncover these patterns.

One of the major problems in recommender systems is volume data that should be processed to reveal users' behaviors. Mathematical basis of kernels comes back to inner product space which is sometimes referred to as a Hilbert space. Kernels try to find a newer representation form for data in Hilbert space to convert the nonlinear patterns to linear ones.

Let $\delta = (x_1, x_2, ..., x_n)$ be a set of objects and each x_i denotes an object from $\mathcal{X}$, then we have $\delta \subseteq \mathcal{X}$. For each $x_i \in \mathcal{X}$, there is a new representation in feature space $\mathcal{F}$. Mapping function Φ transforms all samples of δ object set to new feature space, in the new form of $\phi(\delta) = (\phi(x_1), \phi(x_2), ..., \phi(x_n))$.

$$\begin{aligned} \Phi : \mathcal{X} &\to \mathcal{H} \\ x &\mapsto \phi(x). \end{aligned} \tag{1}$$

T. Yoshida et al. (Eds.): AMT 2013, LNCS 8210, pp. 187–192, 2013.
© Springer International Publishing Switzerland 2013

Based on this basis kernel functions are defined. Let K denotes a positive definite kernel function of the form:

$$K : \mathcal{X} \times \mathcal{X} \to \mathbb{R}$$
$$x, x' \mapsto k(x, x'). \tag{2}$$

That is, a function which returns a real number for two given patterns x and x'. A kernel function can be defined as a comparison function between pairs of patterns. Consequently, kernels can show the similarity of two patterns; and regarding this characteristic, kernel are assumed to be symmetric functions [4].

Fortunately, kernel methods can be applied to any data structure by defining a suitable kernel function [5]. Pair wise definition of kernels analogically links them to graphs concepts and this similarity motivates us to use a hypergraph-based kernel in rating prediction.

Hypergraph was first introduced by Berge [6]. A hypergraph is a structure as $\mathcal{H}(\mathcal{V}, \mathcal{E})$, where $\mathcal{V}$ denote a finite set of objects i.e. $\mathcal{V} = \{v_1, v_2, ..., v_n\}$ and let $\mathcal{E} = \{e_1, e_2, ..., e_m\}$ is a family of subsets of $\mathcal{V}$ such that
1. $e_i \neq \emptyset$.
2. $\bigcup_i e_i = \mathcal{V}$.

The elements $v_1, v_2, ..., v_n$ are called vertices and $e_1, e_2, ..., e_m$ are called hyperedges, respectively. The simplest hyperedge containing only two vertices is a graph edge. Typical Adjacency matrices cannot be adopted for hypergraph; hence hypergraphs are represented by a $|\mathcal{V}| \times |\mathcal{E}|$ incidence matrix H, i.e.

$$\mathcal{H}(\mathbf{v_i}, \mathbf{e_j}) = \begin{cases} 1 & \text{if } v_i \in e_j \\ 0 & \text{otherwise} \end{cases} \tag{3}$$

Accordingly, vertex and hyperedge degree matrices can be defined. Let D_v denotes the diagonal matrix contains number of incidental hyperedges for vertices.

$$\mathbf{D_v} = \{d_{ii}\}_{i \in \{1,2,...,|\mathcal{V}|\}} = \{\sum_{j=1}^{|\mathcal{E}|} \mathbf{H}(v_i, e_j)\}. \tag{4}$$

Similarly, hyperedge degree matrix is filled with number of vertex members of hyperedges denoted by D_e, i.e.,

$$\mathbf{D_e} = \{d_{jj}\}_{j \in \{1,2,...,|\mathcal{E}|\}} = \{\sum_{i=1}^{|\mathcal{V}|} \mathbf{H}(v_i, e_j)\}. \tag{5}$$

Let Laplacian of a hypergraph be L and defined as,

$$\mathbf{L} = \mathbf{D_v} - \mathbf{H}\mathbf{H^T} \tag{6}$$

where H and D_v denote incidence and vertex degree matrices, respectively. Laplacian matrix is usually adopted in normalized form, which formed by dividing values of L by square root of D_v entries.

$$\tilde{\mathbf{L}} = \mathbf{D}_v^{-\frac{1}{2}} \mathbf{L} \, \mathbf{D}_v^{-\frac{1}{2}} \tag{7}$$

Conceptually, kernel matrices are then defined as the limits of matrix power series of the form with suitable parameters λ_i. [5]

$$\mathbf{K} = \sum_{i=1}^{m} \lambda_i \, (-\tilde{\mathbf{L}})^i \tag{8}$$

Kernel matrix can be adopted in classification and clustering processes and K nearest neighbor (kNN) is the most widely used classification algorithm that can be used in combination with a kernel. Kernel combination can be an appropriate solution for kNN memory optimization with respect to its ability to isolate the learning algorithm from training instances. Yu et al. [7] generalized the distance concept to inner product space and kernels. They substituted the square of norm distance in Euclidean space with Hilbertian space equivalence, as

$$d^2(\phi(x), \phi(y)) = \mathbf{K}(x, x) - 2\mathbf{K}(x, y) + \mathbf{K}(y, y) \tag{9}$$

2   The Proposed Method

We have established a novel algorithm for rating prediction which uses a hypergrph-based kernel in kNN framework to predict user's ratings. In this way, a hypergraph $\mathcal{G}$ for a rating dataset can be defined. Users are vertices of this hypergraph and there exists one hyperedge for each member of product-rating pair set.

Let r be the number of rating values and p denotes the number of products, therefore the count of defined hyperedges is $r \times p$. When customer C_i vote for product P_k, a rating value R_l, we assign 1 to hyperedge of (P_k, R_l) pair in the $C_i th$ row. Our Computed hypergraph kernel is used in k-nearest neighbor framework based on the suggested square distance method of Yu et al. [7]. Yu et al. proposed an extension for distance based kernels in kNN. However, we employed their method without any changes in a hypergraph-based kernel that is not formed based on a distance measure.

3   Experiments

We conduct an experiment on two rating datasets: a) MovieLens [1], and b) Rating of sweets[2]. Detailed information about this dataset is shown in Table 1. Our experiments consists a 5-fold cross-validation of proposed hypergraph-based kernel kNN method in comparison with typical kNN on two above datasets.

---

[1] http://www.movielens.org
[2] http://sweetrs.org/en/

Table 1. Detailed information of datasets which were used in experiments

Dataset	Users	Items	Ratings	Range of Ratings
MovieLens	943	1682	100000	[1,5]
Rating of Sweets	404	47	17903	[0,5]

3.1 Experimental Setup

The selected dataset is a subset of MovieLens collected data and rating of sweets dataset, provided by MLdata [3], is a subset of SweetRS customer rating. The incidence, Laplacian and kernel matrices are computed for the incidence matrix. In kernel equation, where the λ_i parameters are required to weight (normalized) Laplacian elements of power series, the first m eigenvalues of $-L$ matrix are used as λ_is. Table 2 shows numerical characteristics of the diffusion kernels of hypergraphs of datasets.

Table 2. Numerical characteristics of Kernels

Dataset	Users	Items	Item-rating Pairs	Size of K	m
MovieLens	943	1682	1682×5	943×943	60
Rating of Sweets	355	47	47×5	355×355	25

Experiments are performed on whole datasets to measure accuracy of our method in comparison with typical k nearest neighbor implementation. Typical kNN algorithm is implemented with city block distance metric. For each method (i.e., kernel-based and typical kNN), we estimate distance of neighbors of test set members from training set records and select select k nearest neighbors. The mean absolute error metric was used in experimental studies as accuracy criterion. The absolute error of each fold is computed as shown in equation (10).

$$E_{Abs} = \frac{\sum_{i=1}^{m} |R_i - \hat{R}_i|}{m} \tag{10}$$

Let m be the size of test set. R_i and $\hat{R}_i$ denote actual and predicted ratings, respectively. The overall error calculates as the average of the individual error measures of folds.

3.2 Results and Discussion

Accuracy of k nearest neighbor method is affected by the number of neighbors that are involved in labeling. Therefore, we repeat the test procedure with different numbers of neighbors. Tables 3 and 4 show the overall mean absolute error values of 5-fold cross-validation tests.

---

[3] http://mldata.org/repository/tags/data/sweets/

The error curves show that using kernel based nearest neighbor can decrease error values. This improvement comes back to the ability of our defined kernel in differentiate between users with similar but different patterns of rating. Users with same set of ratings (say $U_1 = \langle 1,1,0,0 \rangle$ and $U_2 = \langle 0,0,1,1 \rangle$) for different products have different rows/columns in incident matrix of our method. Subsequently, they would be differentiable; while, these users could not be differentiated using distance based kernels. Our proposed style of kernel definition does not assign weights but forms different rows for users like U_1 and U_2. Accordingly, our proposed provide more accurate similarity measure which helps kNN classifier to select nearest neighbors smarter than distance based typical kNN.

Table 3. The overall mean absolute error values of 5-fold cross-validation tests using kNN label prediction

Dataset	k=7	k=8	k=9	k=10	k=11	k=12	k=13	k=14	k=15
Movilens	0.445	0.455	0.463	0.471	0.480	0.484	0.487	0.489	0.487
SweetRS	0.324	0.326	0.32	0.317	0.312	0.312	0.310	0.304	0.300

Table 4. The overall mean absolute error values of 5-fold cross-validation tests using kernel based kNN label prediction

Dataset	k=7	k=8	k=9	k=10	k=11	k=12	k=13	k=14	k=15
Movilens	0.419	0.431	0.440	0.446	0.452	0.456	0.458	0.461	0.463
SweetRS	0.278	0.274	0.273	0.272	0.273	0.272	0.274	0.273	0.273

In addition, the accuracy improvement was achieved across all datasets, with different number of neighbors. Essentially, increasing number of neighbors can improve the accuracy of kNN, although this improvement in these experiments is not as significant as the improvement, provided by kernelised kNN algorithm. Although, increasing number of neighbors of typical kNN on SweetRS dataset has negative effect on the accuracy. However, this decreasing is about 2% that could be described by the dispersity of samples in SweetRS dataset.

4   Conclusion

We proposed a new method for customer rating prediction. In this method, we defined a hypergraph for a rating problem, in which customers are vertices and product-rating pairs form hyperedges. In this way, the problem of rating prediction is restated to a vertex classification. Ratings that are expressed by a customer for a product can be interpreted as customer's class label. After employing kernel based kNN, class of each customer is assigned to it. This method uses a hypergraph-based matrix with generalized k nearest neighbor. Hypergraph kernels were used in graph/network based data mining cases; however defining a kernel for a recommender system as we done is a new approach and may opens new doors to rating prediction problems. Results show that mean absolute error for our method is less than a typical kNN implementation.

Size of kernel in this method depends on number of human subjects which participate in voting/rating. The advantage of this type of kernel definition is its independency from count of products.Simplicity and low computational complexity are other advantages of our method that helps to get involve more neighbors in prediction.

Acknowledgement. This work was partially supported by the Natural Science Foundation of China under Grant No. 61203165, Liaoning Provincial Natural Science Foundation of China under Grant No. 201202032, and the Fundamental Research Funds for the Central Universities (DUT12JR10).

References

1. Debnath, S., Ganguly, N., Mitra, P.: Feature weighting in content based recommendation system using social network analysis. In: Proceedings of the 17th International Conference on World Wide Web, pp. 1041–1042. ACM (2008)
2. Lucas, J., Luz, N., Moreno, M., Anacleto, R., Almeida Figueiredo, A., Martins, C.: A hybrid recommendation approach for a tourism system. Expert Syst. Appl. 40(9), 3532–3550 (2013)
3. Julashokri, M., Fathian, M., Gholamian, M.: Improving customer's profile in recommender systems using time context and group preferences. In: The 5th International Conference on Computer Sciences and Convergence Information Technology (ICCIT), pp. 125–129 (2010)
4. Scholkopf, B., Smola, A.: Learning with kernels. MIT Press (2001)
5. Cook, D., Holder, L.: Mining Graph Data. Wiley-Interscience (2007)
6. Berge, C.: Graphs and Hypergraphs. North-Holland Publishing Company (1973)
7. Yu, K., Ji, L., Zhang, X.: Kernel nearest-neighbor algorithm. Neural Processing Letter 15(2), 147–156 (2002)

Preference Structure and Similarity Measure in Tag-Based Recommender Systems

Xi Yuan, Jia-jin Huang, and Ning Zhong

International WIC Institute, Beijing University of Technology, China
Department of Life Science and Informatics, Maebashi Institute of Technology,
Maebashi 371-0816, Japan
yuanxi_cathy@126.com, zhong@maebashi-it.ac.jp

Abstract. Social tagging systems extend recommender systems from the pair (user, item) to (user, item, tag). This paper discusses the framework of similarity measure on (user, item, tag) from qualitative and quantitative perspectives. The qualitative measure makes use of the preference structure relation on (user, item, tag), and the quantitative measure makes use of reflection on (user, item, tag). The k nearest neighbors and reverse k' nearest neighbors are used to generate recommendations.

1 Introduction

The development of the Internet and communication technology brings the information overload problem. The overload information makes Web users hard to find useful information that meets their needs and the information producers also hard to provide their information to the potential customers. Recommender systems [13] provide automated and personalized guide to the Web users based on historical information stored in the system. Collaborative filtering (CF) [4] recommendation and content-based filtering (CBF) recommendation [11] are widely used methods in recommender systems. The common techniques of these two methods for reducing the size of the neighborhood are to select only top-k similar neighbors which have great influence on target item into account, or select reverse top-k' similar neighbors which are affected by target item.

With the new generation of Web 2.0, a new and particular family of Web applications, namely social tag, has been emerging. Social tags [14] play a major role to Web users, who are not only allowed to publish and tag resources, but also freely create and share tags with free style, to realize the personalized classification of the Web resources with tags. The effect of users to both tags and resources makes the tag be a useful tool used in recommender system. The goal of the tag recommender system is to recognize the result set, which fits the users' interest preferences, to recommend. There exists three kinds of recommend methods in the tag-based recommender system [21]: Recommend the friend user set to the target user; recommend the item set to the target user may be interested in; and recommend the tag set to the target user may tagged in. Our research direction is to recommend items, which have three main studies:

T. Yoshida et al. (Eds.): AMT 2013, LNCS 8210, pp. 193–202, 2013.
© Springer International Publishing Switzerland 2013

network-based methods [22–24], tensor-based methods [10, 12, 16] and topic-based methods [5, 7]. Network-based methods produce recommendation based on a user-item-tag tripartite graph which consists of three integrated bipartite graphs. Tensor-based methods denote a user-item-tag 3-dimensional feature matrix to recommend. Topic-based methods recommend by assessing the implicit among user, item and tag.

In the tag-based recommender system, the ratings to the items i and i' given by user can reflect the user's preference to the items. Therefore, we denote the user's preference to different items by rating preference. By combining rating preference and the tag, which reflect the integrity of the user's preference, to assess the users' preference is useful to improve recommendation accuracy. The similarity [15], which is the one of important measures in the recommender system, always used to be the distance formulas of k-nearest neighbor (kNN) and find the k most similar neighbors for computing the predict ratings. The information retrieval is closely related with recommender system [1, 20]. In this paper, we extend the information retrieval model in [8, 18, 19] to the tag-based recommender system, and measure the preference structure relation on (user, item, tag) qualitatively. Then we assess rating preference in the tag recommender system by similarity, realize recommend and raise recommend accuracy by combining users, items and tags' kNN and reverse k-nearest neighbor ($RkNN$) to recommend.

2   The Qualitative Description for Similarity

In this section, we describe the similarity qualitatively by using the set theory. We introduce the rating preference is introduced to describe users' preference to items. We denote rating preference in item-user and tag-item space as IU and TI, respectively.

Let $D = (U, I, T)$ denote the three parts in tag recommender system. The component $U = \{u_1, u_2, ..., u_m\}$ is the set of users, $I = \{i_1, i_2, ..., i_n\}$ is the set of items, $T = \{t_1, t_2, ..., t_s\}$ is the set of tag. The (u, i, t), (u', i', t') are defined to a user-item-tag (UIT) triple. For $u, u' \in U, i, i' \in I, t, t' \in T$, let u tag i with t and u' tagged i' with t', the preference of u to i is higher than u' to i'. The rating preference defined as follows [19]:

$$(u, i, t) \succ (u', i', t') \tag{1}$$

The tag recommender system typically provides the 3-dimensional user-item-tag (UIT) space. In this paper, our direction is recommending items in which the user may be interested. So we keep tag t the same and project the 3-dimensional UIT space onto the 2-dimensional item-user (IU) space. The projected rating preference in the IU space is defined as follows:

$$(u, i) \succ_t (u', i') \tag{2}$$

Similarly, we projects the 3-dimensional UIT space onto the 2-dimensional tag-item (TI) space, the rating preference in the TI space is defined as follows:

$$(i, t) \succ_u (i', t') \tag{3}$$

In the IU space, the rating preference is $(u, i) \succ_t (u', i')$, which requires the comparison of two users with two items by tag t, may arise the difficulties. That is, we project the rating preference onto I and U from the standing point of an item and a user, respectively. For a user $u \in U$ tag items with t, we can project $\succ_t$ onto I as follows:

$$i \succ_t^u i' \Leftrightarrow (u, i) \succ_t (u, i')$$
$$\Leftrightarrow user\ u\ prefers\ item\ i\ to\ i'\ with\ tagged\ t. \tag{4}$$

According to $\succ_t^u$, we can rank items based on the given rating preferences. From the view of measure-theory, there exist a real number function to measure the $\succ_t^u$, that is $f_u : I \to \Re$:

$$i \succ_t^u i' \Leftrightarrow f_u(i) > f_u(i') \tag{5}$$

A user $u \in U$ is interested in an item $i \in I$, written $u \mathfrak{P} i$. Given a relation $\mathfrak{P}$ between U and I, one can define a mapping Γ_t which assigns a subset $\Gamma_t(i) \subseteq U$ for every $i \in I$ by [18]:

$$\Gamma_t(i) = \{u \in U | u \mathfrak{P} i,\ u\ tags\ i\ with\ a\ given\ tag\ t\} \tag{6}$$

For a single user set $\{u\}$, we have the $\underline{\omega}(\{u\})$ and $\overline{\omega}(\{u\})$ written by:

$$\underline{\omega}(\{u\}) = \{i | \Gamma_t(i) \subseteq \{u\}\} \tag{7}$$
$$\overline{\omega}(\{u\}) = \{i | \Gamma_t(i) \cap \{u\} \neq \phi\} \tag{8}$$

where $\underline{\omega}(\{u\})$ consists of these items in I, which interest in only user u. $\overline{\omega}(\{u\})$ consists of these items in I, which interest in at least one user u. Given a user u, a tag t and items i and i', let $f_u(i) = sim(\{i\}, \overline{\omega}(\{u\}))$, where sim denotes the similarity between two sets, we have

$$i \succ_t^u i' \Leftrightarrow sim(\{i\}, \overline{\omega}(\{u\})) > sim(\{i'\}, \overline{\omega}(\{u\})) \tag{9}$$

CBF tends to recommend items similar to what a given user liked in the past [1]. In Eq. (9), $sim(\{i\}, \overline{\omega}(\{u\}))$ denotes the similarity between item i and items that interest user u and tagged by t. From this point of view, we can see Eq. (9) satisfies needs of CBF and use $sim(i, i')$ to describe the rating preference.

At the same time, we also similarly let $\underline{\omega}(\{i\})$ be users in U who interest in only item i and $\overline{\omega}(\{i\})$ be users in U who interest in at least one item i. Given a user u, a tag t and items i and i', let $f_u(i) = sim(\overline{\omega}(\{i\}), \{u\})$, we have

$$i \succ_t^u i' \Leftrightarrow sim(\overline{\omega}(\{i\}), \{u\}) > sim(\overline{\omega}(\{i'\}), \{u\}) \tag{10}$$

Under the assumption that similar type of users tend to make similar decisions and to choose similar items, CF aims to recommend items to a given user based on the history of both the user and similar users [1]. In Eq. (10), $sim(\overline{\omega}(\{i\}), \{u\})$ denotes the similarity between user u and users who are interested in item i and tag t. From this point of view, we can see Eq. (10) satisfies needs of CF, and use $sim(u, u')$ to describe the rating preference.

Similarity, let $\underline{\omega}(\{i\})$ be tags in T which tagged to only item i and $\overline{\omega}(\{i\})$ be tags in T, which tagged to at least one item i. Given a tag t tagged by user u and items i and i', let $f_t(i) = sim(\overline{\omega}(\{i\}), \{t\})$, we have

$$i \succ_u^t i' \Leftrightarrow sim(\overline{\omega}(\{i\}), \{t\}) > sim(\overline{\omega}(\{i'\}), \{t\}) \tag{11}$$

In Eq.(11), $sim(\overline{\omega}(\{i\}), \{t\})$ denotes the similarity between tag t and tags that tagged to item i by user u. Obviously, we can use $sim(t, t')$ to describe the rating preference.

The qualitative description for similarity above demonstrates the possibility of studying tag recommender system methods from the view of combing the rating preference and set-oriented methods. Next we will describe similarity quantitative.

3   The Quantitative Description for Similarity

The tag recommender system typically provides a 3-dimensional relationship between users, items and tags. In this section we project this 3-dimensional space to three 2-dimensional spaces, namely IU, UT and TI, to describe similarity among items, users and tags from the quantitative view, respectively. While measure similarity, we construct the IU, UT and TI relation matrix (Figure. 1).

The rating can reflect user's preference directly and usually adopted to judge items, so we use the rating to quantitative the rating preference. For the tag recommender system $D = (U, I, T, R)$ with four parts, let the (u, i, t, r), (u', i', t', r') be the user-item-tag-rating $(UITR)$ quaternion. For $u, u' \in U, i, i' \in I, t, t' \in T$ the rating preference is defined as follows:

$$(u, i, t) \succ (u', i', t') \Leftrightarrow r > r'$$
$$\Leftrightarrow \textit{The rating of } i \textit{ tagged by } u \textit{ with } t \textit{ is higher}$$
$$\textit{than the rating } i' \textit{ tagged by } u' \textit{ with } t' \tag{12}$$

Various methods have been used for similarity computation in the tag recommender system, and the most popular method is the standard cosine (SC) similarity. In this section, we will introduce the use of this method for measure the similarity of users, items and tags.

In the IU space, the similarity of items is calculated based on the features associated with the compared items [1]. So we construct the IU relation matrix

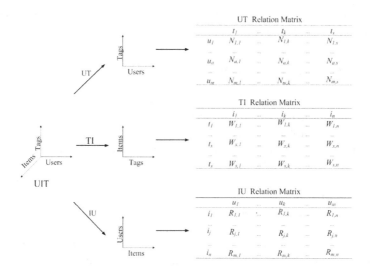

Fig. 1. Projection of the 3-dimensional UIT graph

to measure the item similarity $sim(i, i')$. Given two items $i_j, i_t \in I$, the similarity of two items $sim(i_j, i_t)$ is computed as the standard cosine of the angle [1, 3]:

$$sim(i_j, i_t) = \frac{\sum_{k=1}^{m} R_{j,k} \times R_{t,k}}{\sqrt{\sum_{k=1}^{m}(R_{j,k})^2 \times \sum_{k=1}^{m}(R_{t,k})^2}} \qquad (13)$$

where $R_j = \{R_{j,1}, R_{j,2}, ..., R_{j,m}\}$ is the weight of item i_j. $R_{j,k}$ is the rating that item i_j was rated by user u_k.

In the UT space, the similarity in taste of two users is calculated based on the similarity in the rating history of the users. Thus we construct the UT relation matrix to measure the user similarity $sim(u, u')$. Given two users $u_a, u_i \in U$, the similarity of two users $sim(u_a, u_i)$ would be computed as the standard cosine of the angle:

$$sim(u_a, u_i) = \frac{\sum_{k=1}^{s} N_{a,k} \times N_{i,k}}{\sqrt{\sum_{k=1}^{s}(N_{a,k})^2 \times \sum_{k=1}^{s}(N_{i,k})^2}} \qquad (14)$$

where $N_a = \{N_{a,1}, N_{a,2}, ..., N_{a,t}\}$ is the weight of user u_a, $N_{a,k}$ counts the number of times that user u_a used tag t_k.

In the TI space, we construct the TI relation matrix to measure the tag similarity $sim(t, t')$. Given two tags $t_x, t_y \in T$, the similarity of two tags $sim(t_x, t_y)$ would be computed as the standard cosine of the angle:

$$sim(t_x, t_y) = \frac{\sum_{k=1}^{n} W_{x,k} \times W_{y,k}}{\sqrt{\sum_{k=1}^{n}(W_{x,k})^2 \times \sum_{k=1}^{n}(W_{y,k})^2}} \qquad (15)$$

where $W_x = \{W_{x,1}, W_{x,2}, ..., W_{x,n}\}$ is the weight of tag t_x, $W_{x,k}$ counts the number of times that tag t_x was associated to item i_k.

Next we will predict ratings to measure the ability of SC similarity measure in estimating rating preference.

4 The Strategy of Recommendation

In the recommender system, the kNN (k nearest neighbors) [2] is used to find the k most similar neighbors for target user or item, and formulates a prediction by combining the preferences of these neighbors.

Chen and Yin [3] integrate the concept of influence set to compute predictions, which provides a method to find the reverse k' nearest neighbors [6] for target user, item and tag, using the user, item and tag $RkNN$ set to enhance the density of information.

In this paper, we will generate recommendation by combining the users', items' and tags' k_u, k_i, k_t nearest neighbors k_uNN, k_iNN, k_tNN, and k'_u, k'_i, k'_t reverse nearest neighbors Rk'_uNN, Rk'_iNN, Rk'_tNN to count the similarity for users, items and tags. Next we will give two predict rating formulas to assess the influence from similarity, kNN and $RkNN$. $R_{a,t}$ is the predict rating for item i_t by user u_a. $sim(u_a,u_i)$, $sim(i_j,i_t)$, $sim(t_x,t_y)$ denoted the similarity of user, item and tag.

Based on the theory above, for the item $i_t \notin I_{u_a}$, which the target user u_a haven't rating, we combine the rating preference with the user, item, tags' k-nearest neighbor and reverse k'-nearest neighbor to generate recommendation for the rating $R_{a,t}$, which rated to i_t by u_a. We denote two formulas to generate recommendation for new recommender mechanism. $div1_{a,t}$, $div2_{a,t}$, $div_{a,t}$ are denoted as follows.

$$D1\_t = \sum_{\substack{\{t_x | u_i \ tagged \ i_j \ with \ t_x\} \\ t_y \in k_t NN(t_x)}} sim(t_x, t_y)$$

$$D1\_u = \sum_{u_i \in k_u NN(u_a)} (D1\_t) \times sim(u_a, u_i)$$

$$div1_{a,t} = \sum_{i_j \in k_i NN(i_t)} (D1\_u) \times sim(i_j, i_t)$$

$$D2\_t = \sum_{\substack{\{t_{x'} | u_{i'} \ tagged \ i_{j'} \ with \ t_{x'}\} \\ t_{y'} \in Rk'_t NN(t_{x'})}} sim(t_{x'}, t_{y'})$$

$$D2\_u = \sum_{u_{i'} \in Rk'_u NN(u_a)} (D2\_t) \times sim(u_a, u_{i'})$$

$$div2_{a,t} = \sum_{i_{j'} \in Rk'_i NN(i_t)} (D2\_u) \times sim(i_{j'}, i_t)$$

$$div_{a,t} = div1_{a,t} + div2_{a,t} \tag{16}$$

where, for each user u_i who tagged i_j, $D1\_t$ quantifies how relevant the tag t_x associated by u_i to i_j are with respect to the tags t_y belonging to the $k_tNN(t_x)$

set [17], for each item i_j, $D1\_u$ quantifies how relevant the user u_a associated to i_j are with respect to the users u_i belonging to the $k_u NN(u_a)$ set. Moreover, the relevance is then magnified in a way that is proportional to item's similarity $sim(i_j, i_t)$. $k_u = k'_u$, $k_i = k'_i$, $k_t = k'_t$ are the number of user, item and tags' neighbors respectively, we set $k_u \neq k_i \neq k_t$.

Firstly, we predict rating by finding k-nearest neighbor of target user, item and tag, and detect the influence of kNN in tag recommender system:

$$R\_u = \sum_{\substack{u_i \in k_u NN(u_a)}} (\sum_{\substack{\{t_x | u_i \ tagged \ i_j \ with \ t_x\} \\ t_y \ \in \ k_t NN(t_x)}} sim(t_x, t_y)) \times (sim(u_a, u_i) + 1)$$

$$R_{a,t} = \frac{\sum_{i_j \in k_i NN(i_t)} (R\_u) \times (R_{a,j} \times sim(i_j, i_t))}{div1_{a,t}}$$

$$(17)$$

Secondly, we predict rating by combining the k-nearest neighbor and reverse k'-nearest neighbor of target user, item and tag, and detect the influence of kNN and $RkNN$ in tag recommender system:

$$R1\_u = \sum_{\substack{u_i \in k_u NN(u_a)}} (\sum_{\substack{\{t_x | u_i \ tagged \ i_j \ with \ t_x\} \\ t_y \ \in \ k_t NN(t_x)}} sim(t_x, t_y)) \times (sim(u_a, u_i) + 1)$$

$$R1_{a,t} = \frac{\sum_{i_j \in k_i NN(i_t)} (R1\_u) \times (R_{a,j} \times sim(i_j, i_t))}{div_{a,t}}$$

$$R2\_u = \sum_{\substack{u_{i'} \in Rk'_u NN(u_a)}} (\sum_{\substack{\{t_{x'} | u_{i'} \ tagged \ i_{j'} \ with \ t_{x'}\} \\ t_{y'} \ \in \ Rk'_t NN(t_{x'})}} sim(t_{x'}, t_{y'})) \times (sim(u_a, u_{i'}) + 1)$$

$$R2_{a,t} = \frac{\sum_{i_{j'} \in Rk'_i NN(i_t)} (R2\_u) \times (R_{a,j'} \times sim(i_{j'}, i_t))}{div_{a,t}}$$

$$R_{a,t} = R1_{a,t} + R2_{a,t}$$

$$(18)$$

Next we conduct some experiments to show the feasibility of our methods.

5   Results and Analysis

Our experiments make use of the publicly available $MovieLens^1$ $(Version 1.0$ $(May 2011))$ dataset. This dataset consists of users, movies, tags, ratings, movies genres information. We refine the dataset to 29,694 ratings made by 185 users on 4,238 movies. Each user rated at lease 20 movies. Ratings are expressed on an integer rating scale of 1 to 5. The users express their preference to the movies by rating, the higher the score, the more liked a movie is.

---

[1] http://movielens.umn.edu/login

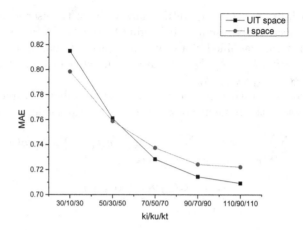

Fig. 2. The comparison of I space and UIT space with kNN only

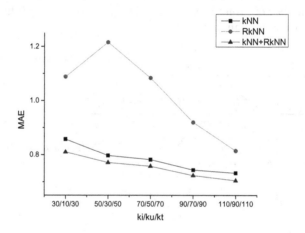

Fig. 3. The comparison of kNN and $RkNN$

In the tag-based recommender system, the mean absolute error (MAE) [9] is the most widely used measure to predict rating, the smaller the MAE is, the higher the recommendation quality is. The $R = \{r_1, r_2, ...r_N\}$ is the predict rating set for N items in the dataset, the $P = \{p_1, p_2, ...p_N\}$ is the actual rating set, so MAE is denoted as follows:

$$MAE = \frac{\sum_{i=1}^{N} |p_i - r_i|}{N} \qquad (19)$$

In this paper, we use two experiments to show the effectiveness of our proposed approach. The first is to generate recommendation in the I space and the UIT space only by kNN respectively. We use the item similarity in the I space and use the user, item and tag similarity in the UIT space to generate recommendation by Eq. (17).

In Fig. 2, the x- and y-axis show the combination of k_i, k_u, k_t, and MAE respectively. The Fig. 2 shows that the UIT space achieves lower MAE than the I space as k increased. So we conclude that take the user and tag's similarity into account can improve the accuracy of the tag recommender system.

The second is to compare the effect of kNN and $RkNN$ in recommending ratings. We use the Eq. (18) to predict ratings by kNN, $RkNN$ and $kNN + RkNN$ in UIT space. The Fig. 3 shows that the MAE of $kNN + RkNN$ is lower than kNN and $RkNN$ respectively. So we conclude that the effect of $kNN + RkNN$ is better than kNN and $RkNN$ in recommending ratings.

6   Conclusion

This paper proposed a framework of tag-based recommender systems. In this framework, the qualitative description of similarity is based on the preference structure and the quantitative description is based on the rating. Our contribution is to unify the tag-based recommender systems in the preference structure by using the similarity measure between users, items and tags, and raise recommend accuracy by combining kNN and $RkNN$ to recommend.

Acknowledgements. The China Postdoctoral Science Foundation Funded Project (2012M510298), Projected by Beijing Postdoctoral Research Foundation (2012ZZ-04), and the doctor foundation of Beijing University of Technology.

References

1. Adomavicius, G., Tuzhilin, A.: Toward the Next Generation of Recommender Systems: A Survey of the State-of-the-art and Possible Extensions. IEEE Transactions on Knowledge and Data Engineering 17(6), 734–749 (2005)
2. Cover, T., Hart, P.: Nearest Neighbor Pattern Classification. IEEE Transactions on Information Theory 13(1), 21–27 (1967)
3. Chen, J., Yin, J.: A Collaborative Filtering Recommendation Algorithm Based on Influence sets. Journal of Software 18(7), 1685–1694 (2007)
4. David, G., David, N., Brain, M.O., Douglas, T.: Using Collaborative Filtering to Weave an Information Tapestry. Communication of the ACM-Special 35(12), 61–70 (1992)
5. Deerwester, S., Dumais, S.T., Furnas, G.W., Landauer, T.K., Harshman, R.: Indexing by Latent Semantic Analysis. Journal of the American Society for Information Science 41(6), 391–407 (1990)
6. Flip, K.S.M.: Influence Sets Based on Reverse Nearest Neighbor Queries. In: Proceedings of the 2000 ACM SIGMOD International Conference on Management of Data (SIGMOD 2000), pp. 201–212 (2000)
7. Hofmann, T.: Probabilistic Latent Semantic Indexing. In: Proceedings of the 22nd Annual International ACM SIGIR conference on Research and Development in Information Retrieval (SIGIR 1999), pp. 50–57 (1999)
8. Huang, J.J.: Modeling Recommender Systems from Preference and Set-oriented Perspectives. In: Zhu, R., Ma, Y. (eds.) Information Engineering and Applications. LNEE, vol. 154, pp. 1068–1073. Springer, London (2012)

9. Jonathan, L.H., Joseph, A.K., Loren, G.T., John, T.R.: Evaluating Collaborative Filtering Recommender Systems. ACM Transactions on Information Systems (TOIS) 22(1), 5–53 (2004)
10. Kolda, T.G., Bader, B.W.: Tensor Decompositions and Applications. SIAM Rev. 51(3), 455–500 (2009)
11. Marko, B., Yoav, S.: Content-Based Collaborative Recommendation. Communications of the ACM 40(3), 66–72 (1997)
12. Rendle, S., Marinho, L.B., Nanopoulos, A., Thieme, L.S.: Learning Optimal Ranking With Tensor Factorization for Tag Recommendation. In: Proceedings of the 15th ACM SIGKDD International Conference on Knowledge Discovery and Data Mining (KDD 2009), pp. 727–736 (2009)
13. Resnick, P., Varian, H.R.: Recommender Systems. Communications of the ACM 40(3), 56–58 (1997)
14. Jäschke, R., Marinho, L., Hotho, A., Schmidt-Thieme, L., Stumme, G.: Tag Recommendations in Folk-sonomies. In: Kok, J.N., Koronacki, J., Lopez de Mantaras, R., Matwin, S., Mladenič, D., Skowron, A. (eds.) PKDD 2007. LNCS (LNAI), vol. 4702, pp. 506–514. Springer, Heidelberg (2007)
15. Spertus, E., Sahami, M., Buyukkokten, O.: Evaluating Similarity Measures: A Large-scale Study in the Orkut Social Network. In: Proceedings of the Eleventh ACM SIGKDD International Conference on Knowledge Discovery and Data Mining (KDD 2005), pp. 678–684 (2005)
16. Symeonidis, P.: User Recommendations Based on Tensor Dimensionality Reduction. In: Iliadis, Maglogiann, Tsoumakasis, Vlahavas, Bramer (eds.) Artificial Intelligence Applications and Innovations III. IFIP AICT, vol. 296, pp. 331–340. Springer, Heidelberg (2009)
17. Valentina, Z., Licia, C.: Social Ranking: Uncovering Relevant Content Using Tag-based Recommender Systems. In: Proceedings of the 2008 ACM Conference on Recommender Systems (RecSys 2008), pp. 51–58 (2008)
18. Wong, S.K.M., Yao, Y.Y.: Preference Structure, Inference and Set-oriented Retrieval. In: Proceedings of the 14th Annual International ACM SIGIR Conference on Research and Development in Information Retrieval (SIGIR 1991), pp. 211–218 (1991)
19. Wong, S.K.M., Cai, Y.J., Yao, Y.Y.: Computation of Term Associations by a Neural Network. In: Proceedings of the 16th Annual International ACM SIGIR Conference on Research and Development in Information Retrieval (SIGIR 1993), pp. 107–115 (1993)
20. Yao, Y.Y., Zhong, N., Huang, J., Ou, C., Liu, C.: Using Market Value Functions for Targeted Marketing Data Mining. International Journal of Pattern Recognition and Artificial Intelligence 16(8), 1–14 (2002)
21. Zhang, Z.K., Zhou, T., Zhang, Y.C.: Tag-Aware Recommender Systems: A State-of-the-Art Survey. Journal of Computer Science and Technology 26(5), 767–777 (2011)
22. Zhang, Y.C., Blattner, M., Yu, Y.K.: Heat Conduction Process on Community Networks as a Recommendation Mode. Physical Review Letters 99(15), 154–301 (2007)
23. Zhou, T., Ren, J., Medo, M., Zhang, Y.C.: Bipartite Network Projection and Personal Recommendation. Physical Review E 76(4), 046115 (2007)
24. Zhou, T., Kuscsik, Z., Liu, J.G.: Solving the Apparent Diversity-accuracy Dilemma of Recommender Systems. Proceedings of the National Academy of Sciences of the United States of America 107(10), 4511–4515 (2010)

Semantically Modeling Mobile Phone Data for Urban Computing

Hui Wang[1], Zhisheng Huang[1,2], Ning Zhong[1,3], and Jiajin Huang[1]

[1] International WIC Institute,
Beijing University of Technology Beijing, China
[2] Dept. of Computer Science,
Vrije University of Amsterdam Amsterdam, The Netherlands
[3] Dept. of Life Science and Informatics,
Maebashi Institute of Technology Maebashi-City, Japan
hui.wang.bjut@gmail.com, huang@cs.vu.nl, zhong@maebashi-it.ac.jp

Abstract. Urban computing aims to enhance both human life and urban environment smartly by deeply understanding human behavior occurring in urban area. Nowadays, mobile phones are often used as an attractive option for large-scale sensing of human behavior, providing a source of real and reliable data for urban computing. But analyzing the data also faces some challenges (e.g., the related data is heterogeneous and very big), and the general approaches cannot deal with them efficiently. In this paper, aiming to tackle these challenges and conduct urban computing efficiently, we propose a data integration model for the multi-source heterogeneous data related to mobile phones by using semantic technology and develop a semantic mobile data management system.

1 Introduction

Urban computing [1] is emerging as a concept where every sensor, device, person, vehicle, building, and street in the urban areas can be used as a component to probe city dynamics to further enable city-wide computing, which aims to enhance both human life and urban environment smartly. For urban computing, there are three key issues needed to be considered.

The first issue for urban computing is what we can use to probe city dynamics. Nowadays, mobile phones are often used as an attractive option for large-scale sensing of human behaviors and activities, due to the huge amount data that may be collected at the individual level, and to the possibility to obtain high levels of accuracy in time and space. These features make mobile phone data ideal candidates for a large range of applications. We will take mobile phone data as our main data source for urban computing.

The second issue for urban computing is how to integrate multi-source data. Because urban computing may involve multi-source data, and the multi-source data probably is heterogeneous, which can not be used directly without integration. The Resource Description Framework (RDF) [2] was standardized by

T. Yoshida et al. (Eds.): AMT 2013, LNCS 8210, pp. 203–210, 2013.
© Springer International Publishing Switzerland 2013

W3C as a key enabler of the Semantic Web to express web data that can be processed directly and indirectly by machines, which has been widely acknowledged in many domains, e.g., life science and information integration. In this paper, we also take RDF as our data representation model to unify the multi-source heterogeneous data related to mobile phone.

The last issue for urban computing is how to compute efficiently and effectively. This is a big challenge because the mobile phone data produced per day by CMCC in Beijing, as an example, is about 450GB, and obviously the conventional approach is not fit enough for processing such large data. The Large Knowledge Collider (LarKC) [3,4] is a platform aiming to remove the scalability barriers of currently existing reasoning systems for the Semantic Web, which is a pluggable Semantic Web framework that can be deployed on a high-performance computing cluster. Owing to its advantages in high-performance computing, we also take LarKC as a fundamental platform to tackle the grand challenge of scalability of mobile phone data for urban computing.

In this paper, our main work is to model mobile phone data based on semantic technology and provide a data management system for urban computing. The remainder is organized as follows: We briefly introduce the basic foundation about mobile phone data and the related work in Section 2. Section 3 presents the architecture of our system for urban computing based on the LarKC platform. In Section 4, we describe the proposed ontology modeling in detail. The SPARQL examples for urban computing are presented in Section 5. Finally, Section 6 gives concluding remarks.

2   Foundation and State of the Art

In this section, we give the foundation about mobile phone data obtained from the GSM (Global Systems for Mobile Communications) network, and review the related work.

2.1   Foundation

To begin with, we clarify the concepts of location area and cell briefly. In a GSM network, the service coverage area is divided into smaller areas of hexagonal shape, referred to as cells (as shown in Figure 1). In each cell, a base station is installed. And within each cell, mobile phones can communicate with a certain base station. In other words, a cell is served by a base station. A location area consists of a set of cells that are grouped together to optimize signaling, which is identified distinctively by a location area code LAC in the network. A cell is also identified uniquely by a cell identifier CI in a location area. That is to say, a cell within a GSM network is identified by a LAC and a CI. In urban areas, cells are close to each other and small in area whose diameter can be down to hundred meters, while in rural areas the diameter of a cell can reach kilometers.

And when a mobile phone corresponds with the network, the signal sent by the mobile phone contains the location information (in the form of a LAC and a

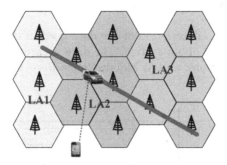

Fig. 1. Concepts of LA and Cell

CI) of the mobile phone. In order to provide service for mobile phones effectively, the location information will be stored by the network. That the mobile phone sends a signal to the network is triggered by one of the following events: 1) the mobile phone is switched on or switched off; 2) the mobile phone receives or sends a short message; 3) the mobile phone places or receives a call (both at the beginning and end of the call); 4) the mobile phone connects the Internet (for example, browsing the web); 5) the mobile phone moves into a cell belonging to a new location area, which is called Normal Location Updating; 6) the mobile phone during a call is entering into a new cell, which is called Handover; 7) the timer set by the network comes to an end when there is no any event mentioned above that happened to the mobile phone, which is called Periodical Location Updating.

2.2 State of the Art

Many studies [5,6,7,8,9,10,11] have been conducted on mobile phone data for urban computing. In [5], Caceres et al. exploit mobile phone data to drive origin-destination for traffic planning and management. In [6], Calabrese et al. develop a real-time urban monitoring system to sense city dynamics by using mobile phones. In [8] and [9], Ying et al. and Lu et al. predict the next location of the user with mobile phone data, respectively. And in [11], Liu et al. study the annotation of mobile phone data with activity purposes.

And there are also many successful applications [12,13] based on the LarKC platform. These applications mainly use the LarKC platform to tackle the challenge of large scale data.

And to the best of our knowledge, our work is the first to use semantic technology to exploit mobile phone data based on the LarKC platform.

3   Overview of Our System

In this section, we give a description of the data used in this paper and briefly illustrate the architecture of our system. The data we are manipulating consists of 1) POIs, 2) Base Stations, and 3) Mobile Phone Records.

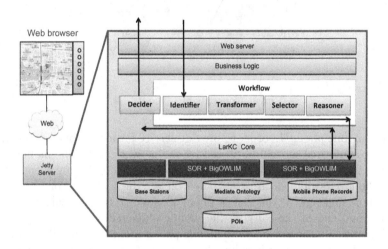

Fig. 2. System architecture

1) POIs: A POI that we use in this paper consists of PID, longitude, latitude, name, and type. We obtained the POIs within Beijing from Baidu Map and the total number of POIs is about one hundred thousand.

2) Base Stations: A base station is identified by BID, longitude, and latitude, which indicates the location of the base station. A base station is also associated with a cell. We have about twenty thousand base stations in Beijing.

3) Mobile Phone Records: A mobile phone record is made up of a phone number which is certainly anonymous, a timestamp, a LAC and a CI, the event, which means that the event happens when the user is in the cell identified by the LAC and the CI at the time. We got about ten thousand users' mobile phone records in a week from CMCC in Beijing.

The architecture of our system is depicted as Figure 2. In our system, users use the web interface to post operation requirements to the server. Those operation requirements include displacing POIs in a specified area, querying the user's trajectory in a specified time. The server sends the SPARQL queries to the SPARQL end point, which is launched by the workflow on the LarKC platform. The SOR+BIGOWLIM is located at the data layer of the LarKC platform to maintain the data sets which include the POIs, Base Stations, and Mobile Phone Records. At the same time, our system also permits users to write their own SPARQL queries and submit them by a submitting interface to the server. After processing these queries, the server will return the results in the form of files.

4 Ontology Modeling

In this section, we model our multi-source data with respect to the ontology represented in Figure 3 and briefly give some explanations about the vocabularies of the ontology. In the illustrating examples throughout this paper, we use the namespaces defined in Table 1.

Table 1. Namespace prefixes used in this paper

Prefix	Namespace URI	Comment
geo:	http://www.w3.org/2003/01/geo/wgs84/_pos	W3C Geo
urc:	http://www.wisdom.org/urc	namespace of our approach

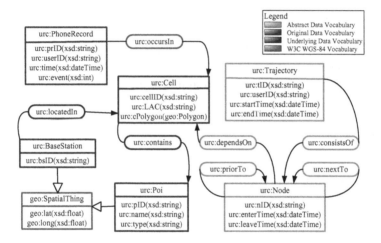

Fig. 3. Ontology modeling of multi-source data

In this section, we model our multi-source data with respect to the ontology represented in Figure 3 and briefly give some explanations about the vocabularies of the ontology. In the illustrating examples throughout this paper, we use the namespaces defined in Table 1.

As shown in Figure 3, our vocabularies are made up of four segments.

W3C WGS-84 Vocabulary: ***SpatialThing*** is a vocabulary defined by W3C, which contains the properties of longitude and latitude. In geography domain, many classes often extend this vocabulary to describe some places. In this paper, we also define some vocabularies related to place based on this vocabulary.

Underlying Data Vocabulary: we think of the data about Base Stations, Cells, and POIs as our underlying data. **BaseStation**, extending **SpatialThing**, is used to describe the location of an antenna. Cell is an irregular polygon with the type geo:polygon. Cell also has an property contains, representing the Pois inside its area. A POI is enriched with types, names.

Original Data Vocabualry: The vocabulary **PhoneRecord** is used to model the original data generated by the GSM. The property **occursIn** indicates which cell the record occurs in.

Abstract Data Vocabulary: The vocabulary Trajectory is used to model the moving behavior of a person, which consists of some consecutive Nodes and we see the trajectories as our abstract data. A Node depends on the cells. The properties **nextTo** and prior to indicates the next node and prior node, respectively. **enterTime** and **leaveTime** indicates the time the user is entering and leaving the cells.

5   Examples

In this section, we give some SPARQL query examples to demonstrate the applications on our system.

If we want to query the users who appeared in a specific area, we can use these clauses below.

```
01 select distinct ?UserID
02 where {
03          ?TrajectoryID rdf:type urc:Trajectory.
04          ?TrajectoryID urc:userID ?UserID.
05          ?TrajectoryID urc:consistsOf ?Node.
06          ?Node urc:dependsOn ?Cell.
07          ?BaseStation urc:locatedIn ?Cell.
08          ?BaseStation rdf:type urc:BaseStation.
09          ?BaseStation geo:lat ?Latitude.
10          ?BaseStation geo:long ?Longitude.
11          FILTER(?Latitude ≤ 39.87569 && ?Latitude ≥ 39.84274 &&
12                  ?Longitude ≤ 116.46142 && ?Longitude ≥ 116.37353).
13      }
14 LIMIT 300
```

Firstly, line 03-06 query the nodes of the trajectory of a user and the cells which are depended. Then, lines 07-10 query the coordinates of the base stations located in the cell. Finally, a constraint is added in lines 11-12, which indicates a specific area.

If we want to analyze the behavior of the user, we can query the POIs contained in the cell where the users stayed more than a threshold time.

```
01 select distinct ?Name ?Type ?Longitude ?Longitude
02 where {
03          ?TrajectoryID rdf:type urc:Trajectory.
04          ?TrajectoryID urc:consistsOf ?Node.
05          ?Node urc:dependsOn ?Cell.
06          ?Node urc:enterTime ?EnterTime.
07          ?Node urc:leaveTime ?LeaveTime.
08          ?Cell urc:contains ?POI
09          ?POI rdf:type urc:POI.
10          ?POI urc:name ?Name.
11          ?POI geo:lat ?Longitude.
12          ?POI geo:lat ? Longitude.
13          FILTER( ?EnterTime ≤ 30+?LeaveTIme).
14      }
15 LIMIT 300
```

As shown in these clauses above, lines 03-07 and line 13 query the cell where the user stayed more than a threshold time. Then we can use lines 08-12 to query the POIs inside the cell.

6   Conclusions

In this paper, we built the ontology to model our multi-source data and developed a semantic mobile data management system based on the LarKC platform, which is designed for urban computing. And we also presented some SPARQL query examples to demonstrate how to use our system.

Because that the trajectories describe the routes which are selected by users based on their real experiences and contain their wisdom that which routes will be suitable in a given time, we can discover the wisdom and recommend some suitable routes for users. Our future work will focus on the application research on mobile phone data and develop a routes recommendation system based on the LarKC platform.

Acknowledgements. This work is supported by the National Science Foundation of China (No. 61272345), the International Science & Technology Cooperation Program of China (2013DFA32180), and the CAS/SAFEA International Partnership Program for Creative Research Teams.

References

1. Zheng, Y., Zhou, X.: Computing with Spatial Trajectories. Springer-Verlag New York Inc. (2011)
2. Klyne, G., Carroll, J.J.: Resource Description Framework(RDF): Concepts and Abstract Syntax. Recommendation, W3C (2004)

3. Fensel, D., Harmelen, F.V., Andersson, B., Brennan, P., Cunningham, H., Valle, E.D., Fischer, F., Huang, Z., Kiryakov, A., Lee, T.K.-I., Schooler, L., Tresp, V., Wesner, S., Witbrock, M., Zhong, N.: Towards LarKC: A Platform for Web-Scale Reasoning. In: Proceedings of the 2008 IEEE International Conference on Semantic Computing, pp. 524–529 (2008)

4. Assel, M., Cheptsov, A., Gallizo, G., Celino, I., Dell'Aglio, D.: Large Knowledge Collider: A Service-Oriented Platform for Large-Scale Semantic Reasoning. In: Proceedings of the International Conference on Web Intelligence, Mining and Semantics, pp. 1–9 (2011)

5. Caceres, N., Wideberg, J.P., Benitez, F.G.: Deriving Origin Destination Data from A Mobile Phone Network. IET Intelligent Transport Systems 1(1), 15–26 (2007)

6. Calabrese, F., Colonna, M., Lovisolo, P., Parata, D., Ratti, C.: Real-Time Urban Monitoring Using Cell Phones: A Case Study in Rome. IEEE Transactions on Intelligent Transportation Systems 12(1), 141–151 (2011)

7. Calabrese, F., Lorenzo, G.D., Liu, L., Ratti, C.: Estimating Origin-Destination Flows Using Mobile Phone Location Data. IEEE Pervasive Computing 10(4), 36–44 (2011)

8. Ying, J.J.-C., Lu, E.H.-C., Lee, W.-C., Weng, T.-C., Tseng, V.S.: Mining User Similarity from Semantic Trajectories. In: Proceedings of the 2nd ACM SIGSPATIAL International Workshop on Location Based Social Networks, pp. 19–26 (2010)

9. Ying, J.J.-C., Lee, W.-C., Weng, T.-C., Tseng, V.S.: Semantic Trajectory Mining for Location Prediction. In: Proceedings of the 19th ACM SIGSPATIAL International Conference on Advances in Geographic Information Systems, pp. 34–43 (2011)

10. Steenbruggen, J., Borzacchiello, M., Nijkamp, P., Scholten, H.: Mobile Phone Data from GSM Networks for Traffic Parameter and Urban Spatial Pattern Assessment: A Review of Applications and Opportunities. GeoJournal 78(2), 223–243 (2013)

11. Liu, F., Janssens, D., Wets, G., Cools, M.: Annotating Mobile Phone Location Data with Activity Purposes Using Machine Learning Algorithms. Expert Systems with Applications 40(8), 3299–3311 (2013)

12. Balduini, M., Celino, I., DellAglio, D., Della Valle, E., Huang, Y., Lee, T., Kim, S.-H., Tresp, V.: BOTTARI: An Augmented Reality Mobile Application to Deliver Personalized and Location-Based Recommendations by Continuous Analysis of Social Media Streams. Web Semantics: Science, Services and Agents on the World Wide Web 16, 33–41 (2012)

13. Della Valle, E., Celino, I., Dell'Aglio, D., Grothmann, R., Steinke, F., Tresp, V.: Semantic Traffic-Aware Routing Using the LarKC Platform. IEEE Internet Computing 15(6), 15–23 (2011)

Action Unit-Based Linked Data for Facial
Emotion Recognition

Kosuke Kaneko and Yoshihiro Okada

Innovation Center for Educational Resource, Kyushu University Library, Fukuoka, Japan
kosukekaneko@kyudai.jp, okada@inf.kyushu-u.ac.jp

Abstract. This paper treats methodology to build linked data from the relation-ships between facial action units and their states as emotional parameters for the facial emotion recognition. In this paper, the authors are especially focusing on building action unit-based linked data because it will be possible not only to use the data for the facial emotion recognition but also to enhance the usefulness of the data by merging them with other linked data. Although in general, the repre-sentation as linked data seems to make the accuracy of the facial emotion rec-ognition lower than others, in practically the proposed method that uses action unit-based linked data has almost the same accuracy for the facial emotion rec-ognition as those of other approaches like using Artificial Neural Network and using Support Vector Machine.

Keywords: Linked Data, Semantic Data, Facial Emotion Recognition.

1   Introduction

Recently, there is a growing tendency among us to interact with intelligent computer systems. To make such interaction more meaningful, more challenging studies should be made. One of the challenges is to enable the computer to recognize the human emotion. If computers can understand our emotion, they can take their actions corres-ponding to our emotion and we will build better relationship with them.

To enable a computer to understand the human emotion, we need to build databas-es defining human emotions. One of the approaches for it is building facial action unit-based linked data. There are several representations and languages for the linked data, i.e., RDF (Resource Description Framework) [1] and OWL (Web Ontology Language) [2]. The usefulness of linked data is that the data can be merged with other linked data. For example, linked data defining emotion of a certain person can be merged with another linked data defining his/her favorite foods. This makes it possi-ble to deduce his/her favorite foods from his/her emotion states. Therefore, in this paper, we are focusing on building linked data for the facial emotion recognition, especially facial action unit-based linked data because facial expressions are different according to the corresponding emotion category and they can be defined by action unit states. Although our method is to build action unit-based linked data for the per-sonal facial emotion recognition, the same method can be applied to other person and

T. Yoshida et al. (Eds.): AMT 2013, LNCS 8210, pp. 211–220, 2013.
© Springer International Publishing Switzerland 2013

we can obtain more useful action unit-based linked data for more general facial emotion recognition by merging the built linked data of many persons with each other.

In this paper, we introduce our method that builds action unit-based linked data for the facial emotion recognition. Fig. 1 shows the overview of our method. We extract facial action unit states, which are kinds of facial feature points, from color and depth image and trace the movement of each action unit in several seconds. From the movement data of action units, our system judges the corresponding emotion category. Before that, we have to build action unit-based linked data from a training dataset.

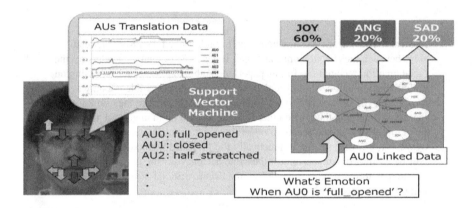

Fig. 1. Method Overview

The remainder of this paper is organized as follows: in the next section, we introduce several related works. In Section 3, we explain our method that builds facial action unit-based linked data. Section 4 shows experimental results of our approach and results of other standard approaches using Artificial Neural Network and using Support Vector Machine. Finally, we conclude the paper and present our future works in Section 5.

2   Related Works

Many researches to build linked data about the human emotion have been studied. WordNet Affect is one of the popular linked datasets defining relationships between a word and a human emotion [3]. As another approach, there is the method proposed by M. Ptaszynski, et al. [4]. Their method analyzes texts and automatically builds linked data with human emotions. The built linked data are represented as an emotion ontology using EmotionML, which is an ontology-based language to express human emotions [5]. M. Grassi defined HEO (Human Emotion Ontology) [6]. The ontology represents human emotions built based on several human emotion theories in psychology. A. García-Rojas, et al. propose an emotion ontology about the human body representation for a virtual human [7]. K. Benta, el at. propose a guide system using a human emotion ontology [8] for museums.

Several researches about relationships between facial expressions and human emotions have also been studied. N. Tsapatsoulis, et al. propose a method for facial emotion classification based on FDP (Facial Definition Parameter) defined in MPEG-4 specification [9]. The MPEG-4 specification defines feature points movement and facial expressions for several human emotions [10]. A. Azcarate, et al. propose an automatic facial emotion recognition method using Naive Bayes Classifier [11]. The system proposed by T. D. Bui, et al. generates facial expressions from human emotions using Fuzzy algorithm [12]. Another research to generate facial expression with emotion is Kozasa, et al. [13]. Their system generates facial expressions according to the emotion analyzed by Artificial Neural Network. A. García-rojas, et al. tried to build an emotion ontology for facial expression based on the MPEG-4 specification [14] for the virtual human's facial expression. They built their ontology from facial expression profiles defined by A. Raouzaiou, el at. [15]. Our approach is similar to their approach but its concept and method are different because our method builds action unit-based linked data for the human emotion recognition.

3   Proposed Method

To build action unit-based linked data for the facial emotion recognition, we use Kinect for Windows. Once, we ask one subject to make several types of facial expressions related to different emotions and record streaming data of each action unit state generated from Kinect in a few second. We analyze the streaming data to make its distribution map indicating how each action unit state changes. And then, from the distribution map, we make action unit-based linked data. We classify the states of each action unit by Support Vector Machine. The states are used as queries to search emotion categories from the linked data defining the relationships between each action unit state and its corresponding human emotion. After applying these processes to every action unit, we can find out each facial emotion category from its probability. We will explain details of the method in the following subsections.

3.1   Facial Action Unit State

We extract facial action unit states using Kinect for Windows, which is a color and depth image-based motion capture device produced from Microsoft Inc. We also use its SDK because the SDK provides a face tracking functionality and generates facial action unit states. In the SDK, the action units are defined based on Candide3 model [16]. We focus on six facial action units: Upper Lip Raiser, Jaw Lowerer, Lip Stretcher, Brow Lowerer, Lip Corner Depressor and Lip Corner Depressor because they seems important. Fig. 2 shows each action unit of them on a human face and the direction of its translation. All the action units have their translation range normalized into [-1.0 ~ 1.0]. In the followings, Upper Lip Raiser, Jaw Lowerer, Lip Stretcher, Brow Lower, Lip Corner Depressor and Lip Corner Depressor are indicated as Action Unit 0 (AU0), Action Unit 1 (AU1), … , and Action Unit 5 (AU5), respectively.

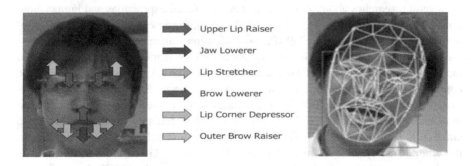

Fig. 2. Six Facial action units (Left) and facial tracking (Right)

3.2   Action Unit-Based Linked Data

This subsection explains how to build action unit-based linked data. For the linked data, we make a *Triple* as the relationship between an action unit state and its corresponding emotion category. The relationships are defined by analyzing a training dataset of each action unit state. Fig. 3 shows distribution maps built from the training dataset.

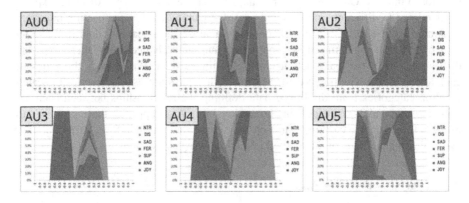

Fig. 3. Distribution maps of action unit translation values among emotion categories

The horizontal line in the charts in Fig. 3 is transition of translation values of each action unit. The vertical line is the rate of an emotion category on a translation range. The colors in the charts mean emotion categories. Each action unit has its translation range according to each of the emotion categories. We separate the distribution maps into seven intervals to be assigned to an action unit as its semantic states. It could be better to separate into more intervals if you wish more accuracy for the facial emotion recognition. For example, we assign a translation range '0.8 ~ 1.0' for Action Unit 0 as the semantic state 'full_opened'. A translation range '0.8 ~ 1.0', '0.6 ~ 0.7' and '-0.4 ~ -0.2' for Action Unit 2 are assigned as the semantic states of 'full_streatched', 'half_streatched' and 'bit_rounded', respectively. How much a certain emotion category occupies in these ranges defining semantic states can be obtained from the distribution map. We select high rate (more than 20%) emotion category and connect to a certain semantic state. For example, Action Unit 0 has high rate of JOY category on the semantic state 'full_opened' so we can make a *Triple 'Action_Unit_0 full_opened   JOY'*. An action unit doesn't have a link for a low rate category. Fig. 4 shows an example graph of the action unit-based linked data built by this method.

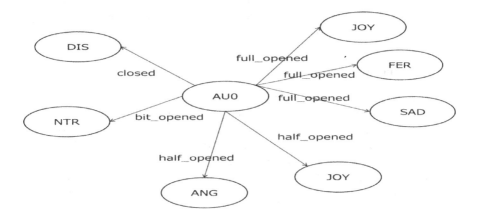

Fig. 4. Linked data defining relationships of each action unit between semantic states and their emotion categories

We use Support Vector Machine to classify action unit states of the streaming data into their semantic states. The training dataset for Support Vector Machine is the same as that of the distribution map. By using the trained Support Vector Machine, we can obtain one of the semantic states of a certain action unit, 'full_opened', 'half_streatched' or other, corresponding to the entered translation data.

We also use semantic states classified by Support Vector Machine as queries for SPARQL (Simple Protocol and RDF Query Language), which is a kind of SQL dedicated for RDF. By using SPARQL the built action unit-based linked data shows emotion categories corresponding to a certain semantic state of an action unit entered as its query. For example, if you enter 'full opened' as the semantic state of Action Unit 0, the action unit-based linked data will return emotion categories of 'JOY', 'FEAR', 'SADNESS'.

There are several ways to handle the results of each action unit. One of the simple ways is to select most common emotion category for each action unit. Another one is to use the rate of emotion category on its translation range. Because we use a semantic graph, we can deduce another action unit state from action unit results returned by SPARQL. If we combine the action unit-based linked data with other linked data, we can deduce another state. Because the search and inference need much calculation time, so it is difficult to classify a current emotion category for each frame. However, we can catch up it by treating streaming data consisting of several frames as the target.

4   Experimental Results

This section indicates several experimental results to justify the usefulness of our method. We compared our method with two other types of methods those are using Artificial Neural Network and using Support Vector Machine. Although in general, the representation as linked data seems to make the accuracy for the facial emotion recognition lower than others, in practically the proposed method uses action unit-based linked data has almost the same accuracy for the facial emotion recognition as that of the two other methods.

4.1   Experimental Environment

We used Kinect for Windows to extract facial action unit states. One subject sits in front of Kinect. The distance between the subject and Kinect was about 1 ~ 2 meters. We recorded streaming data of six action unit states for each emotion category as the training data. As a theory about the human emotion category, we referred to P. Ekman's study: JOY, ANGER, SURPRISED, FEAR, SADNESS and DISGUST [17]. In addition, we also measured the streaming data for NEUTRAL facial expression. So, we made the streaming data of six action unit state as for each of the seven emotion categories.

Actually, we asked the subject to make three different types of facial expressions for each of the seven emotion categories in 5 seconds. For example, in JOY case, a bit smile, half smile and full smile. Finally, we obtained 21 types of streaming data of six action unit states for seven emotion categories and used as the training dataset for each of the three methods.

4.2   Method Using Artificial Neural Network

As a result of comparative experiments, this subsection shows the accuracy of the method using ANN (Artificial Neural Network). The ANN was trained by 21 streaming data of six action unit states introduced in the previous subsection. The ANN has 3 layers, an input, a hidden and an output layer. The input layer has 6 nodes coincide with the number of the action units. The output layer has 7 nodes coincide with the number of emotion categories: JOY, ANGER, SURPRISED, FEAR, SADNESS, DISGUST and NEUTRAL. The hidden layer has 6 nodes. Each node is full-connected to next layer nodes in the network. We entered 21 streaming data to the ANN up to 100 times for the training. The ANN uses back-propagation algorithm for the training. Its learning rate and momentum are 0.01 and 0.99, respectively.

To make test data for the evaluation stage of the ANN method, we asked the subject to make facial expressions with an emotion in about 3 seconds and recorded the streaming data of six action unit states output from Kinect. Fig. 5 shows two types of results of the ANN method, i.e., a good case and a bad case. As the good case, the upper part of the figure indicates a result of facial expression in JOY case and as the bad case, the lower part of the figure indicates the SURPRISED case. The left line charts mean streaming data of action unit states and the right ones are classification results. The vertical line of the bar chart means the output value of ANN for each category and the horizontal line means the frame numbers. The ANN correctly classifies the streaming data with emotion 'JOY' into the corresponding emotion category. However, it classified the streaming data with emotion 'SURPRISED' into emotion category 'FEAR' and 'SAD'.

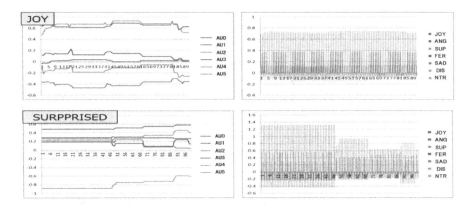

Fig. 5. Classification results of Artificial Neural Network

4.3   Method Using Support Vector Machine

As another result of comparative experiments, this subsection also shows the accuracy of the method using SVM (Support Vector Machine). The SVM was trained by the same dataset used for the ANN method and came to classify the same test data used for the ANN method into one of 7 emotion categories. We used a Radial Basis Function as the kernel function of the SVM.

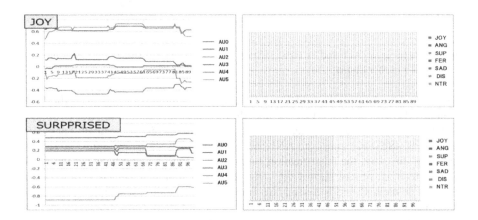

Fig. 6. Classification results of Support Vector Machine

Fig. 6 shows two classification results using the SVM of the same test data as Fig. 5. The right bar chart in Fig. 6 means streaming data of classification results for emotion categories by SVM. The horizontal line is frame number and vertical line is the result of classification which take 0 or 1 (active class has 1, otherwise 0). The SVM method obtained a better result even for the emotion category 'SURPRISED' than the ANN method. However, the SVM is still confused in the case of emotion category 'FEAR'.

4.4   Method Using Action Unit-Based Linked Data

Finally, we show two results of our method in this subsection. The test data for the results and the dataset used to build a semantic graph of action unit-based linked data are the same as those used for the ANN method and the SVM method. In each frame, once a semantic state is obtained by the SVM, furthermore it will be used as the input to obtain one of the emotion categories as an output using the semantic graph. Fig. 7 shows two tables of the probability distribution of output emotion categories for each action unit obtained as the classified result of the first frame data. The left table is the result for a JOY facial expression and the right one is SURPRISED case. A high probability cell is high-lighted. A cell with 2 values is a case that the SVM is confused to classify a semantic state. From this kind of tables for all the frames, we also obtain the two charts as the classification results of 'JOY' and 'SURPRISED' cases shown in Fig. 8. The upper part in Fig.8 is a result of JOY case and the lower part is SUPPRIZED case. The bar charts are results of classification as the total probability of each emotion category.

	JOY	ANG	SUP	FER	SAD	DIS	NTR
AU0	0.220436	0.276693	0.187141	0.120551	0.088404	0.106774	0
AU1	0.129582	0.16159	0.043366	0.07176	0.124935	0.191533	0.277233
AU2	0.439739	0.348534	0	0	0.211726	0	0
AU3	0.141487	0.131295	0.190048	0.10012	0.134293	0.016787	0.285971
AU4	0.282143	0.259524	0	0.020238	0.17381	0.157143	0.107143
AU5	0.444/ 0.113664	0.366/ 0.078597	0/ 0.076784	0.008/ 0.08948	0.122/ 0.110641	0.06/ 0.206167	0/ 0.324667

	JOY	ANG	SUP	FER	SAD	DIS	NTR
AU0	0.220436	0.276693	0.187141	0.120551	0.088404	0.106774	0
AU1	0.129582	0.16159	0.043366	0.07176	0.124935	0.191533	0.277233
AU2	0	0	0.340741	0.396298	0	0.262963	0
AU3	0.141487/ 0.258621	0.131295/ 0	0.190048/ 0	0.10012/ 0.017241	0.134293/ 0	0.016787/ 0.724139	0.285971/ 0
AU4	0/ 0.041806	0/ 0.079431	0.597647/ 0.049331	0.402393/ 0.117893	0/ 0.13796	0/ 0.199833	0/ 0.373746
AU5	0/ 0	0/ 0	0.550296/ 0.402597	0.251479/ 0.597403	0.198225/ 0	0/ 0	0/ 0

Fig. 7. Two tables in JOY and SURPRISED cases of probability distributions among emotion categories for each action unit

From the figure, it can be said that our method has similar results those of the ANN method and SVM methods. Our method shows 100% recognition accuracy for JOY category (the ANN and SVN methods show 100% and 94.5%, respectively) and 91% for SURPPRISED category (the ANN and SVN methods show 0% and 51%, respectively). In this way, the classification accuracies of the three methods for the facial emotion recognition are almost the same. However, our method has another merit because our method builds a semantic graph of action unit-based linked data. Once the semantic graph was built, it can be used as a knowledge base for other applications since the linked data can be merged with other linked data to enhance the coverage of the knowledge base.

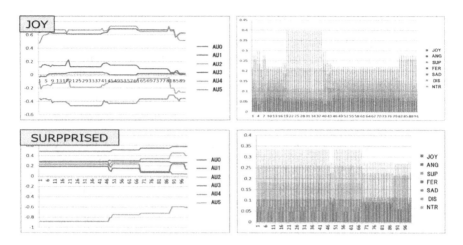

Fig. 8. Classification results of our method using action unit-based linked data

5   Conclusion and Future Works

This paper introduced the method to build a semantic graph of action unit-based linked data to be used for the facial emotion recognition. In this paper, we werc especially focusing on building action unit-based linked data because it will be possible not only to use the data for the facial emotion recognition but also to enhance the usefulness of the data by merging them with other linked data. Although in general, the representation as linked data seems to make the accuracy of the facial emotion recognition lower than others, in practically our method that uses action unit-based linked data has almost the same accuracy for the facial emotion recognition as those of other approaches using Artificial Neural Network and using Support Vector Machine. In this paper, we justified this point by showing the experimental results.

In the near future, we will try to apply the same approach of our proposed method to build more general linked data for other applications. We will also apply the proposed method to build action unit-based linked data of other persons to be used for the facial emotion recognition in more general applications.

References

1. RDF, http://www.w3.org/RDF
2. OWL, http://www.w3.org/TR/owl-features/
3. Strapparava, C., Valitutti, A.: WordNet-Affect: an Affective Extension of WordNet. In: Proceedings of the 4th International Conference on Language Resources and Evaluation, vol. 4, pp. 1083–1086 (2004)
4. Ptaszynski, M., Rzepka, R., Araki, K., Momouchi, Y.: A Robust Ontology of Emotion Objects. In: Proceedings of the Eighteenth Annual Meeting of The Association for Natural Language Processing (NLP 2012), pp. 719–722 (2012)
5. EmotionML, http://www.w3.org/TR/emotionml/

6. Grassi, M.: Developing HEO human emotions ontology. In: BioID_MultiComm 2009 Proceedings of the 2009 Joint COST 2101 and 2102 International Conference on Biometric ID Management and Multimodal Communication, pp. 244–251 (2009)

7. García-Rojas, A., Vexo, F., Thalmann, D., Raouzaiou, A., Karpouzis, K., Kollias, S.: Emotional Body Expression Parameters In Virtual Human Ontology. In: Proceedings of 1st International Workshop on Shapes and Semantics, pp. 63–70 (2006)

8. Benta, K., Rarau, A., Cremene, M.: Ontology Based Affective Context Representation. In: Proceedings of the 2007 Euro American Conference on Telematics and Information Systems (EATIS 2007), Article No. 46 (2007)

9. Tsapatsoulis, N., Karpouzis, K., Stamou, G., Piat, F., Kollias, S.A.: A Fuzzy System for Emotion Classification Based on the MPEG-4 Facial Definition Parameter Set. In: Proceedings of the 10th European Signal Processing Conference (2000)

10. Tekalp, A.M., Ostermann, J.: Face and 2-D Mesh Animation in MPEG-4. Signal Processing: Image Communication 15(4), 387–421 (2000)

11. Azcarate, A., Hageloh, F., van de Sande, K., Valenti, R.: Automatic Facial Emotion Recognition, Univerity of Amsterdam (2005)

12. Bui, T.D., Heylen, D., Poe, M., Nijholt, A.: Generation of facial expressions from Emotion using a Fuzzy Rule Based System. In: Proceedings of the 14th Australian Joint Conference on Artificial Intelligence: Advances in Artificial Intelligence (AI 2001), pp. 83–94 (2001)

13. Kozasa, C., Fukutake, H., Notsu, H., Okada, Y., Niijima, K.: Facial Animation Using Emotional Model. In: International Conference on Computer Graphics, Imaging and Visualisation (CGIV 2006), pp. 428–433 (2006)

14. García-Rojas, A., Vexo, F., Thalmann, D., Raouzaiou, A., Karpouzis, K., Kollias, S., Moccozet, L., Thalmann, N.M.: Emotional face expression profiles supported by virtual human ontology. In: Computer Animation and Virtual Worlds (CASA 2006), vol. 17(3-4), pp. 259–269 (2006)

15. Raouzaiou, A., Tsapatsoulis, N., Karpouzis, K., Kollias, S.: Parameterized facial expression synthesis based on MPEG-4. EURASIP Journal on Applied Signal Processing archive 2002(1), 1021–1038 (2002)

16. Candide3 model, http://www.icg.isy.liu.se/candide/

17. Ekman, P., Friesen, W.V., Ellsworth, P.: What emotion categories or dimensions can observers judge from facial behavior? In: Ekman, P. (ed.) Emotion in the Human Face, pp. 39–55 (1982)

Online Visualisation of Google Images Results

Gerald Schaefer[1], David Edmundson[1], and Shao Ying Zhu[2]

[1] Department of Computer Science, Loughborough University, Loughborough, U.K.
`gerald.schaefer@ieee.org, d.edmundson@lboro.ac.uk`
[2] School of Computing and Mathematics, University of Derby, Derby, U.K.
`s.y.zhu@derby.ac.uk`

Abstract. Visual information, especially in the form of images, is becoming increasingly important, and consequently there is a rising demand for effective tools to perform online image search. However, image search engines such as Google Images, are based on the text surrounding the images rather than the images themselves. At the same time, while the employed keyword-based search provides a basic level of filtering, it is not sufficient to handle large search results. Image database visualisation, which provides a visual overview of an image collection, could be applied to the retrieved images, but the associated overheads, both in terms of bandwidth and computational complexity, are prohibitive.

In this paper, we introduce an image browsing system that does not suffer from these drawbacks. In particular, we construct an interactive image database navigation application that uses the Huffman tables available in the JPEG headers of Google Images thumbnails directly as image features, and projects images onto a 2-dimensional visualisation space based on principal component analysis derived from the Huffman entries. Images are dynamically placed into a grid structure and organised in a tree-like hierarchy for visual browsing. Since we utilise information only from the JPEG header, the requirement in terms of bandwidth is very low, while no explicit feature calculation needs to be performed, thus allowing for interactive browsing of online image search results.

Keywords: image databases, content-based image retrieval, image browsing, Google Images.

1 Introduction

Visual information, especially in the form of images, is becoming increasingly important. Consequently, while the majority of information that is searched on the web is still of textual nature, there is a rising demand for online tools that perform image search, with Google Images[1] being one of the most popular platforms. However, the way Google returns image search results is similar to how text results are processed in that the results are primarily based on text in the webpages where the images reside rather than the content of the image files themselves.

---

[1] `http://images.google.com`

T. Yoshida et al. (Eds.): AMT 2013, LNCS 8210, pp. 221–230, 2013.
© Springer International Publishing Switzerland 2013

Furthermore, results are presented essentially in a linear fashion, with images arranged on a grid and pages of results to look through. As many queries will produce ambiguous or non-relevant results, this may make the search task difficult and time consuming.

An alternative form of presenting the results would to apply an image database navigation approach to give a visual overview of the retrieved image set. In recent years, various approaches for intuitive image browsing based on concepts from content-based image retrieval (CBIR) have been proposed [7,15,20]. The main idea behind most of these techniques is to extract CBIR features [23,2] and visualise an image set so that visually similar images are located close to each other in the visualisation space. This is often achieved through application of a dimensionality reduction technique such as principal component analysis (PCA) or multi-dimensional scaling (MDS).

In fact, such an approach has been suggested in [10]. Here, the authors propose the use of features based on an attention model that involves segmentation and extraction of region features which are matched at global and local scales. MDS is then employed for visualising the image set, and images are fitted to a 2-dimensional grid. Unfortunately, this approach is not actually feasible as it would not allow for an interactive operation. The reason for this is twofold: first, in order to extract the image features, the images need to be downloaded and decompressed which for larger results is too time consuming, while second, feature calculation is performed which puts a computational burden on the system and hence necessitates further time.

In this paper, we present an effective approach to visualise and browse image search results from Google Images that is efficient enough to allow for interactive operation. To do so, we exploit our earlier work [5] which allows for very fast CBIR feature extraction of JPEG images. Since this approach, which uses tuned JPEG Huffman tables as image features, requires only the header information of the image files, this significantly reduces the required bandwidth. At the same time, no explicit feature calculation is necessary, maintaining a feasible level in terms of computational complexity. The retrieved images are placed, by mutual visual similarity, onto a 2-dimensional grid structure based on the projection of the Huffman tables by principal component analysis. By employing a tree-like browsing hierarchy, larger retrieval results can be effectively navigated in an intuitive manner, as our developed application demonstrates.

2   JPEG Image Compression

JPEG [24] is the current de-facto standard for still picture coding. It is based on the discrete cosine transform (DCT), a derivative of the discrete Fourier transform. First, an (RGB) image is usually converted into the YCbCr colour space. The reason for this is that the human visual system is less sensitive to changes in the chrominance (Cb and Cr) channels than in the luminance (Y) channel. Consequently, the chrominance channels can be downsampled by a factor of 2 without significantly reducing image quality, resulting in a full resolution Y and downsampled Cb and Cr components.

The image is then divided (each colour channel separately) into 8×8 pixel sub-blocks and DCT applied to each such block. The 2-d DCT for an 8×8 block $f_{xy}, x, y = 0 \ldots 7$ is defined as

$$F_{uv} = \frac{C_u C_v}{4} \sum_{x=0}^{7} \sum_{y=0}^{7} f_{xy} \cos\left(\frac{(2x+1)u\pi}{16}\right) \cos\left(\frac{(2y+1)v\pi}{16}\right) \qquad (1)$$

with $C_u, C_v = 1/\sqrt{2}$ for $u, v = 0$, $C_u, C_v = 1$ otherwise. Of the 64 coefficients, the one with zero frequency (i.e., F_{00}) is termed "DC coefficient" and the other 63 "AC coefficients". The DC term describes the mean of the image block, while the AC coefficients account for the higher frequencies. As the lower frequencies are more important for the image content, higher frequencies can be neglected which is performed through a (lossy) quantisation step that crudely quantises higher frequencies while preserving lower frequencies more accurately.

The DC and AC components of the image are stored in separate streams. The DC stream (essentially the equivalent of a downsampled version of the image) is differentially encoded; i.e. rather than storing the actual DC values, differences between DC values are saved. As DC values range in $[-1024; 1024]$, the maximum possible difference between the DC components is in $[-2048; 2048]$. This difference value is stored as two components: the first component known as the "DC code", while the second component stores the actual difference between the DC blocks. The DC code data is then entropy coded for which Huffman coding is employed.

While standard Huffman tables (one each for luminance and chrominance DC data) are defined, these tables can also be optimised to match the image data and consequently lead to slightly better compression with no loss of image data. This process is applied by various image websites including Flickr[2] (during photo upload) and Google Images (for the thumbnails stored on Google's servers).

3   JPEG Huffman Table CBIR Features

As the table optimisation process assigns the smallest codes to the most commonly occurring DC codes, we can use the thus adapted Huffman tables as an indication of the frequencies of each code. Similar images should thus lead to similar Huffman tables which is demonstrated in Fig. 1.

In [5], we consequently introduced a method to compare two images based solely on comparing their DC Huffman tables. In order to compare the tables from two images, and hence to arrive at a measure of (dis)similarity between the images, we use the bitlengths of the DC codes directly as a feature. As dissimilarity measure, we utilise the L_1 norm between the feature vectors. That is, if we have two images I_1 and I_2 with bitlength vectors f_{I_1} and f_{I_2}, the dissimilarity between the images is calculated as

$$d(I_1, I_2) = \sum_{i=1}^{12} |f_{I_1}(i) - f_{I_2}(i)|. \qquad (2)$$

---

[2] http://flickr.com

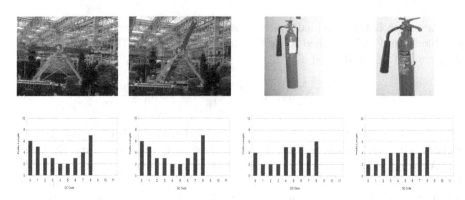

Fig. 1. Sample UCID images together with their prefix code sizes of the luminance DC tables

In cases where an entry does not exist in the Huffman table, the corresponding bitlength in the feature vector is set to the maximum over all other bitlengths plus 1 to indicate that the corresponding DC code appears even less frequently than all the other ones. To incorporate both intensity and colour information, we use both the luminance and chrominance DC tables, calculate the distances between the corresponding tables and sum the two distances to arrive at a combined measure of dissimilarity.

The approach was shown to be very efficient as no decompression at all is required and only a small part of the image file, the JPEG header, needs to be read. This is in particular useful for online image retrieval [4] where features are not pre-computed but are derived during the retrieval process.

4   Image Database Browsing

Image database navigation systems have been shown to provide an interesting and useful alternative to image retrieval systems [7,15,20]. The idea here is to provide a visualisation of a complete image collection together with browsing tools for an interactive exploration of the database.

Visualisation methods for image repositories can be grouped into three main categories: mapping-based, clustering-based, and graph-based approaches [15,14]. Mapping-based techniques employ dimensionality reduction to map high-dimensional image feature vectors to a low-dimensional space for visualisation. Typical examples examples use principal component analysis (PCA) [11,9,16], multi-dimensional scaling (MDS) [18,21], or non-linear embedding techniques [12] to define a visualisation space onto which to place images. Clustering-based visualisations group visually similar images together, often in a hierarchical manner [1,6,9,19]. In graph-based navigation systems, images are the nodes of a graph structure, while the edges of the graph show relationships between images [8,3,25].

Once a database has been visualised, it should then be possible to browse through the collection in an interactive, intuitive and efficient way [15,13]. We can distinguish between horizontal browsing which works on images of the same visualisation plane, and includes operations such as panning, zooming or magnification, and vertical browsing which allows navigation to a different level of a hierarchically organised visualisation.

5   Browsing of Google Images Results

Google Images provides a search engine in the same fashion as Google's main search site but one that returns images rather than webpages. After searching for a keyword, navigation proceeds in a linear fashion with the system showing pages of image thumbnails arranged in a grid array.

Fig. 2 gives an example for the query "Eiffel Towel". As we can see from there, despite the deficiencies of the text based search, the query does provide a good set of results, effectively all showing the tower, while later results also reveal different kinds of images including a variety of photos that are not actually related to the query. At the same time, we notice that there is relatively little variation between the images with all results showing the monument from a far perspective (at different times of the day). If we were interested in close up views, we would need to browse down or through the next pages, making exploration of the collection a slow and time consuming process.

An image database navigation approach would hence provide a more intuitive interface and it is this approach that we pursue in this paper. Image database navigation systems, or image database browsers, are typically based on CBIR features and arrange image thumbnails based on their mutual similarity derived from these features. However, such an approach would require feature extraction on the client side and hence download of the image files resulting in infeasibly high bandwidth and computational complexity requirements.

In this paper, we show that by exploiting the fact that the image thumbnails that Google Images returns have undergone JPEG Huffman table optimisation coupled with our Huffman table based CBIR method from [5], these requirements can be lowered sufficiently to allow for interactive visualisation and browsing of image search results.

As in [5], we use the bitlength of each DC code; combining the entries from both luminance and chrominance tables we hence have feature vectors of length 24. Similar to other image browsing methods [11,16], we perform principle component analysis (PCA) to project these 24 features into a 2-dimensional space where the first two principal components define the visualisation space. Fig. 3 shows the UCID dataset [22] projected onto that space and confirms that visually images appear clustered together.

Our browsing application works as follows. Based on a keyword query, we initiate Google Images to return images matching the query. The application then starts downloading and analysing the header for each image thumbnail in the results list. The extracted Huffman tables of these images are then projected

Fig. 2. Google Images retrieval results based on the query "Eiffel Tower"

onto the PCA space which only requires a simple matrix mutliplication since we employ the projection matrix derived from the UCID dataset to define the visualisation space. In order to avoid image overlap and hence improve the browsing experience [17], images are displayed on a regular grid of $N \times N$ cells (with $N = 7$ in our current implementation). If an image falls into an empty grid cell, we download the thumbnail and display it in the cell. If the cell already contains an image, the image ID and co-ordinates are simply stored, in a hierarchical tree structure, for future use.

Once an image is selected by the user, the applications "zooms in" on the respective area of the visualisation space, hence performing a vertical browsing operation. This will show more images that also fall into the same area of visualisation space as the selected thumbnail. These images are again shown in a grid layout which now corresponds to a partitioning of the visualisation space of the original cell in the same manner as before. To deal with images having the same or very similar co-ordinates in PCA space (e.g. duplicates) and maximise the use of visualisation space, a spreading algorithm is applied: if a cell is already occupied in the zoomed in view, its neighbouring cells are searched and empty cells filled. Once on a lower browsing layer, the user can also directly navigate (up/down/right/left) to view the contents of neighbouring parent image cells.

At any stage, only images that are shown to the user are downloaded in full for display purposes while for all other images only the headers are retrieved, thus reducing the amount of bandwidth required.

Fig. 3. PCA-visualised UCID dataset

Fig. 4. Root view of retrieved results for the query "Eiffel Tower"

Fig. 5 shows the initial view for the query "Eiffel Tower". As can be seen, there are several distinct clusters of images: while the left hand side shows images with the tower far away and with large regions of sky, towards the right we see pictures

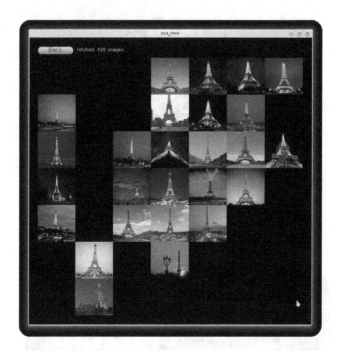

Fig. 5. Zoomed in view of retrieved "Eiffel Tower" images

with high activity changes such as those found when zooming in closer on the metal structure of the tower[3].

Fig. 5 shows the result of zooming in on a selected image from the root view. As we can see, the majority of images there are visually quite similar to the selected sample image on the root layer.

As our approach requires no feature calculation, the images are efficiently visualised "on-the-fly", i.e. while they are being retrieved, and consequently most of the root layer is filled very quickly. Overall, for the 420 images retrieved, the required bandwidth was measured to be 1,175,722 bytes (which included all associated network overheads) which represents a quite reasonable load for today's home broadband connections

6   Conclusions

In this paper, we have presented an approach to visualise and browse retrieval results from Google Images. For this we exploit the fact that Google stores image

---

[3] We also notice, towards the bottom left, an image that does not match the query term. Since our approach merely deals with visualisation and browsing of the re-trieved images, any inaccuracies present in Google Images search will still be present in our application.

thumbnails with optimised Huffman tables and employ the Huffman tables directly as image features. Images are projected onto a 2-dimensional visualisation space using principal component analysis and our browsing application places image thumbnails on a regular grid structure while providing hierarchical access to larger retrieval results in an efficient and intuitive manner.

References

1. Bartolini, I., Ciaccia, P., Patella, M.: Adaptively browsing image databases with PIBE. Multimedia Tools and Applications 31(3), 269–286 (2006)
2. Datta, R., Joshi, D., Li, J., Wang, J.Z.: Image retrieval: Ideas, influences, and trends of the new age. ACM Computing Surveys 40(2), 1–60 (2008)
3. Dontcheva, M., Agrawala, M., Cohen, M.: Metadata visualization for image browsing. In: 18th Annual ACM Symposium on User Interface Software and Technology (2005)
4. Edmundson, D., Schaefer, G.: Efficient and effective online image retrieval. In: IEEE Int. Conference on Systems, Man, and Cybernetics (2012)
5. Edmundson, D., Schaefer, G.: Fast JPEG image retrieval using optimised Huffman tables. In: 21st Int. Conference on Pattern Recognition, pp. 3188–3191 (2012)
6. Gomi, A., Miyazaki, R., Itoh, T., Li, J.: CAT: A hierarchical image browser using a rectangle packing technique. In: 12th Int. Conference on Information Visualization, pp. 82–87 (2008)
7. Heesch, D.: A survey of browsing models for content based image retrieval. Multimedia Tools and Applications 40(2), 261–284 (2008)
8. Heesch, D., Rüger, S.: NNk networks for content-based image retrieval. In: European Conference on Information Retrieval, pp. 253–266 (2004)
9. Keller, I., Meiers, T., Ellerbrock, T., Sikora, T.: Image browsing with PCA-assisted user-interaction. In: IEEE Workshop on Content-Based Access of Image and Video Libraries, pp. 102–108 (2001)
10. Liu, H., Xie, X., Tang, X., Li, Z.W., Ma, W.Y.: Effective browsing of web image search results. In: ACM Int.l Workshop on Multimedia Information Retrieval, pp. 84–90 (2004)
11. Moghaddam, B., Tian, Q., Lesh, N., Shen, C., Huang, T.S.: Visualization and user-modeling for browsing personal photo libraries. Int. Journal of Computer Vision 56(1-2), 109–130 (2004)
12. Nguyen, G.P., Worring, M.: Interactive access to large image collections using similarity-based visualization. Journal of Visual Languages and Computing 19(2), 203–224 (2008)
13. Plant, W., Schaefer, G.: Navigation and browsing of image databases. In: Int. Conference on Soft Computing and Pattern Recognition, pp. 750–755 (2009)
14. Plant, W., Schaefer, G.: Visualising image databases. In: IEEE Int. Workshop on Multimedia Signal Processing, pp. 1–6 (2009)
15. Plant, W., Schaefer, G.: Visualisation and browsing of image databases. In: Lin, W., Tao, D., Kacprzyk, J., Li, Z., Izquierdo, E., Wang, H. (eds.) Multimedia Analysis, Processing and Communications. SCI, vol. 346, pp. 3–57. Springer, Heidelberg (2011)
16. Plant, W., Schaefer, G.: Interactive exmploration of large remote image databases. In: 20th ACM Int. Conference on Multimedia (2012)

17. Rodden, K.: Evaluating Similarity-Based Visualisations as Interfaces for Image Browsing. PhD thesis, University of Cambridge Computer Laboratory (2001)
18. Rubner, Y., Guibas, L., Tomasi, C.: The earth mover's distance, multi-dimensional scaling, and color-based image retrieval. In: Image Understanding Workshop, pp. 661–668 (1997)
19. Schaefer, G.: A next generation browsing environment for large image repositories. Multimedia Tools and Applications 47(1), 105–120 (2010)
20. Schaefer, G.: Interacting with image collections – Visualisation and browsing of image repositories. In: 20th ACM Int. Conference on Multimedia (2012)
21. Schaefer, G., Ruszala, S.: Image database navigation on a hierarchical MDS grid. In: Franke, K., Müller, K.-R., Nickolay, B., Schäfer, R. (eds.) DAGM 2006. LNCS, vol. 4174, pp. 304–313. Springer, Heidelberg (2006)
22. Schaefer, G., Stich, M.: UCID - An Uncompressed Colour Image Database. In: Storage and Retrieval Methods and Applications for Multimedia. Proceedings of SPIE, vol. 5307, pp. 472–480 (2004)
23. Smeulders, A.W.M., Worring, M., Santini, S., Gupta, A., Jain, R.: Content-based image retrieval at the end of the early years. IEEE Trans. Pattern Analysis and Machine Intelligence 22(12), 1249–1380 (2000)
24. Wallace, G.K.: The JPEG still picture compression standard. Communications of the ACM 34(4), 30–44 (1991)
25. Worring, M., de Rooij, O., van Rijn, T.: Browsing visual collections using graphs. In: Int. Workshop on Workshop on Multimedia Information Retrieval, pp. 307–312 (2007)

The Roles of Environmental Noises and Opinion Leaders in Emergency

Yiyi Zhao and Yi Peng[*]

School of Management and Economics, University of Electronic Science and Technology of
China, Chengdu 610054, China
pengyicd@gmail.com

Abstract. This paper proposes a dominant-submissive agent model on bounded
confidence opinion dynamics under an emergency environment. In the proposed model, environmental noises and opinion leaders are involved in the collective opinion formation. A series of computer simulations demonstrate that
environmental noises have a great impact on the collective opinion evolution.
The interactions among individuals are strengthened as the variances of the environmental noises increase, and then a global group behavior emerge with a
higher probability. On the other hand, the influence of opinion leaders on the
collective opinion dynamics is limited. Firstly, when the fraction of opinion
leaders is fixed in the social network, the number of agents following the opinion leaders decreases as the variance of the environmental noise exceeds a certain threshold. Secondly, the number of agents following the opinion leaders
does not change obviously as the fraction of opinion leaders increases under a
constant noisy environment.

Keywords: Environmental noises, Opinion leaders, Emergency, Opinion
propagation, Bounded confidence.

1 Introduction

In a social system, individuals are not isolated and their opinions on a commonly
focused incident or person are often influenced by the neighbors and various public
media. In most normal cases, individual decisions are diverse due to distinct differences of inherent characters, such as age, characteristic, educational level, professions
and so on, which generally causes a fragmentation of the collective opinions of a
group of individuals. However, in many emergency cases, individuals may lack rational judgment and easily believe other individuals' opinions, thereby, a collective
behavior emerges in large scale. For example, people usually withdraw cash from a
bank according to their own actual demands, but if a rumor says that the bank will go
bankrupt, then people may draw the total money from the banks and a financial con-
fusion may occur. For another example, iodized salt is just a kind of ordinary daily
necessities. But in March 2011, a panic buying of iodized salt was triggered by some

---

[*] Corresponding author.

T. Yoshida et al. (Eds.): AMT 2013, LNCS 8210, pp. 231–240, 2013.
© Springer International Publishing Switzerland 2013

opinion leaders in many regions in China after the Fukushima Daiichi nuclear accident. These examples indicate that environmental influence and opinion leaders play an important role in the opinion propagation.

Recently, the topic of opinion dynamics has been paid much attention by more and more interdisciplinary researchers. A wide variety of models have been developed in order to deal with the collective opinion evolution phenomena observed in real societies. Some studies focused on the influence of different network structures on the mechanism of opinion formation [1] - [6]. Some other references considered the role of various individuals in opinion propagation [7] - [11]. In addition, a large number of works investigated, under the framework of bounded confidence, the opinion evolution in a certain social network, where agents have different bounded confidence levels [12] -[15]. However, few results can be found to investigate the relation between opinion propagation and environmental uncertainties. The aim of this paper is to consider the collective opinion dynamics with opinion leaders and environmental noises, and provide some helpful suggestions and scenarios for public emergency solutions.

The rest of this paper is organized as follows. Section 2 gives a detailed description on bounded confidence models, as well as a brief survey of some results available in recent years with regard to the bounded confidence model and its extensions. An extended opinion dynamics model with heterogeneous confidence levels is proposed based on Hegselmann and Krause model in Section 3. Section 4 presents simulation results, which studies the impacts of opinion leaders and environmental uncertainties on the opinion propagation of the proposed heterogeneous opinion dynamics. Section 5 concludes the paper.

2   Bounded Confidence Models

In most real cases, agents interact only if their opinions are sufficiently close to each other, a situation referred to as bounded confidence (BC). Bounded confidence plays an important role in many social dynamics models.

Mathematical models of opinion dynamics under bounded confidence have been presented independently by Deffuant and Weisbuch (DW) model [12] and by Hegselmann and Krause (HK) model [13] – [14]. The above mentioned two BC models assume that each agent is homogeneous, namely, all agents have the identical confidence level and influence on other agents within the threshold of confidence level.However, due to complex physiological or psychological factors, each social agent may have diverse confidence level thresholds. Thus a heterogeneous bounded confidence model is more appropriate for the opinion evolution with agent-dependent confidence levels. A heterogeneous HK model was reformulated as an interactive Markov chain in [16]. Both heterogeneous DW model and HK model were respectively proposed for opinion dynamics with agent-based version and density-based version in [15] and [17]. The multi-level heterogeneous opinion formation model was formulated to study the influences of the fractions of heterogeneous agents, the initial opinions, and the group size on the collective opinion evolution in [18].The effects of heterogeneous confidence bounds were analyzed by a series of computer experiments.

The agents were classified into two classes, essential and inessential, or close-minded and open-minded, according to the diverse confidence levels.

3   Dominant-Submissive Opinion Dynamics Model

In a real social group, agents may have heterogeneous confidence levels and different interaction influence [19]. Some have higher confidence levels and lower influence while others are just the opposite for a certain issue. Thus, it is reasonable and important to investigate how collective opinions evolve when agents have heterogeneous confidence thresholds and influence hierarchy. Some interesting problems need to be addressed. How do agents update their opinions when they communicate with opinion leaders under different noisy environments, especially, in public crisis and emergence? What is the evolution of the final collective opinion? Do opinion leaders play important role in the evolution of the collective opinion?

Motivated by these considerations, a dominant-submissive opinion dynamics model is proposed. In the model, each agent has a different confidence level, and the whole group is divided into two subgroups according to the influences of agents. The first subgroup is called opinion leaders who have comparatively complete information about a certain object and are hardly impacted by other agents, besides the agents belonging to the leader subgroup. The second subgroup is called submissive agents who have incomplete information on the same object and can easily adopt others' opinions, especially, the opinions of leaders. For simplicity, we assume that the opinion space is one dimensional, that is, there exists one bounded opinion interval such that every agent is initially assigned an opinion, which is denoted by a real number. Without loss of generality, the opinion space is the interval $[0,1]$. Consider a social group having N agents. Each agent i at time t $(=0,1,\cdots)$ has a continuously varying opinion state $x_i(t) \in [0,1]$. The initial opinion $x_i(0)$ obeys the uniform distribution. Each has its own opinion showing what degree of adopting or rejecting a certain object. The state 1 represents "agreement", 0 represents "refusal" and the numbers in $(0,1)$ represent the fuzzy levels of opinion. The N-dimensional vector $x(t) = (x_1(t), \cdots, x_N(t))^T \in R^N$ denotes an opinion profile, which is a group opinion and aggregates all the private opinions.

Suppose that the confidence levels of agents obey an uniform distribution in $[0,1]$. The whole group is divided into two subgroups: N_1 submissive agents and $N - N_1$ dominant agents. During the evolution of the collective opinions, at each time step, each submissive agent i firstly searches his neighbors according to his own confidence level. If the opinion of agent j satisfies $|x_i(t) - x_j(t)| \le \varepsilon^i$ for $j = 1, \cdots, N$, agent j is called a neighbor of agent i. Then, the submissive agent i updates his opinion according to the following model.

$$x_i(t+1) = (1-\alpha_i)\frac{1}{N_i^S(t)}\sum_{j=1}^{N_1} a_{ij}(t)x_j(t) + \alpha_i \frac{1}{N_i^L(t)}\sum_{j=N_1+1}^{N} a_{ij}(t)x_j(t) + \xi_i(t) \qquad (1)$$

where $i=1,\cdots,N_1, j=1,\cdots,N$, $a_{ij}(t)=1$ if $\parallel x_i(t)-x_j(t)\parallel \leq \varepsilon_i$ and 0 other-

wise, $N_i^S(t) = \sum_{j=1}^{N_1} a_{ij}(t)$ and $N_i^L(t) = \sum_{j=N_1+1}^{N} a_{ij}(t)$ are the total numbers of neighbors of

agent i in submissive agent subgroup and opinion leader subgroup, respectively. The parameter $0 \leq \alpha_i <1$ is the degree of dependence of submissive agent i on opinion leader. The random variable $\xi_i(t)$ is environmental noise factor at time t, and follows a Gauss distribution with mean 0 and standard variance $\sqrt{\sigma_1}$.

Furthermore, consider the opinion update of dominant agents, i.e., opinion leaders, who is generally influenced by two aspects. The one is other opinion leaders within its confidence level and the other is the global goal. The opinion leader i updates his opinion according to the following model.

$$x_i(t+1) = (1-w_g)\frac{1}{N_i^L(t)}\sum_{j=N_1+1}^{N} a_{ij}(t)x_j(t) + w_g d + \xi_i(t), i, j = N_1+1,\cdots,N \qquad (2)$$

where $N_i^L(t) = \sum_{j=N_1+1}^{N} a_{ij}(t)$ is the neighbor number of agent i at time t. The constant

parameter $0 \leq d \leq 1$ represents the goal value expected by opinion leaders. W_g is denoted an influence weight of opinion leaders on the global goal. $\xi_i(t)$ is an environmental noise factor and follows a Gauss distribution with mean 0 and standard variance $\sqrt{\sigma_2}$, which usually satisfies $0 \leq \sigma_2 < \sigma_1 \ll 1$.

4   Simulation Results

In this section, simulation studies are given to the proposed dominant-submissive opinion formation model (1) - (2). We focus mainly on the roles of environmental noise factor and opinion leader in the emergency management. For giving better explanations in the two aspects mentioned above, we investigate the collective opinion dynamics according to three cases: no opinion leader, single opinion leader and multiple opinion leaders.

The size of the considered social network is selected as $N =1000$, including two subgroups, opinion leaders and submissive agents. In the subsequent experiments, the initial opinion of each agent obeys a uniform distribution between 0 and 1. The confidence levels of submissive agents follow uniform distribution between 0 and 1 while those of opinion leaders are suppose to be larger than 0.25. Each experiment runs 100 times Monte Caro simulations and 1000 time steps. The evolution of the collective opinion will be studied in comparisons with no opinion leader, single opinion leader, multiple opinion leaders under different environmental noises.

4.1   No Opinion Leader

In this case, there is no opinion leader and all agents are submissive ones, thus each one updates his opinion according to the model (1). The initial opinions and the confidence levels of submissive agents follow uniform distribution between 0 and 1. The evolution of the collective opinions is investigated under different environmental noises, i.e., the variance of the environmental noise changing from 0 to 0.15. To eliminate the effect of the initial opinions of agents on the opinion formation, it is assumed that agents have the same initial opinions in all experiments.

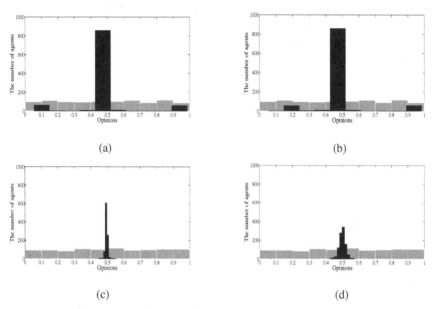

(a)   (b)

(c)   (d)

Fig. 1. Opinion distribution of model (1) without opinion leader. The green histogram denotes the distribution of the initial profile while the blue histogram denotes the final profile. (a) The standard variance $\sqrt{\sigma_1}$ of environment noise is 0; (b) The standard variance $\sqrt{\sigma_1}$ of environment noise is 0.0001; (c) The standard variance $\sqrt{\sigma_1}$ of environment noise is 0.05; (d) The standard variance $\sqrt{\sigma_1}$ of environment noise is 0.15.

When there is not environmental noise or the noise is very small, namely, no emergency case happens, the final collective opinion is fragmental due to the difference of the initial opinion and the diversity of the confidence levels of agents. As shown in Fig. 1(a), the final opinions of most of agents (over 85% of the whole group) converge to the mean value 0.5 of the initial opinions and form the largest final opinion cluster. Simultaneously, the rest agents in the other distributions have small size. The result is identical to the heterogeneous HK model (see [17]). In Fig. 1(b), when the environmental noise has a slight increase from 0 to 0.0001, the change of the distribution of the final profile is not obvious. However, the opinions closing to zero approach to the mean values. As the environmental noises increase further, it is noted that all opinions approach to the mean value, however, the number of agents having opinions around the mean value decreases. Thus, the opinions under small environmental noises are close to the center of the opinion interval; on the other hand, the opinions located around the mean value are splitting. As a result, multiple opinion clusters emerge around the mean value of the opinion interval, as shown in Fig. 1 (c) and Fig. 1 (d).

The results in Fig. 1 show that the environmental uncertainty brings a high probability of group behavior, such as panic buying or crowd. Since the environmental noises are turning large, an agent will be impacted more easily by the other agents, therefore, the collective opinions seem fragmental in small noises and become intensive in large noises. The extreme agents, who have lower confidence levels and extreme opinions, also follow the majority of the group and change their opinions to a more neutral state.

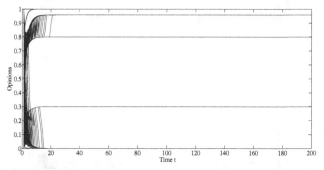

Fig. 2. The collective opinion evolution of submissive agents and the single opinion leader. $\varepsilon_f \in [0,1]$, $\varepsilon_f \in [0,1]$, $\alpha_i = 0.5$, $w_g = 0.5$, $d = 1$, $\sqrt{\sigma_1} = 0, \sqrt{\sigma_2} = 0$. The red line denotes the opinion evolution of the opinion leader. The blue lines denote the opinion evolutions of submissive agents.

4.2   Single Opinion Leader

Now let us consider the second case, where there is only one opinion leader and the others are submissive agents. The dependence degree on the opinion leader, α_i, for submissive agent i is 0.5. Since the opinion leader is the only one and he will not communicate with other agents, so the opinion of the leader is completely decided by the global goal during its opinion evolution. Here, it is assumed that the final goal value d is 1 and the influence weight W_g on the opinion leader equals 0.5. In the subsequent simulations, the variance values of environmental noises are respectively set as 0, 0.0001, 0.0005, 0.001, 0.005, 0.01, 0.02, 0.03, 0.04, 0.05, 0.06, 0.07, 0.08, 0.09 and 0.1 for submissive agents.

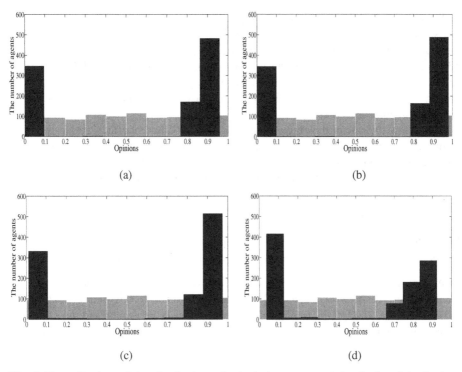

Fig. 3. The collective opinion distributions of submissive agents and the single opinion leader. $\varepsilon_f \in [0,1]$, $\varepsilon_f \in [0,1]$, $\alpha_i = 0.5$, $w_g = 0.5$, $d = 1$, $\sqrt{\sigma_1} = 0, \sqrt{\sigma_2} = 0$. The green histogram denotes the initial distribution of the submissive agents. The blue histogram denotes the final distribution of the submissive agents. (a) $\sqrt{\sigma_1} = 0, \sqrt{\sigma_2} = 0$; (b) $\sqrt{\sigma_1} = 0.001, \sqrt{\sigma_2} = 0.0001$; (c) $\sqrt{\sigma_1} = 0.03, \sqrt{\sigma_2} = 0.003$; (d) $\sqrt{\sigma_1} = 0.1, \sqrt{\sigma_2} = 0.01$.

Comparing with case without opinion leader, the evolution of the collective opinion has an obvious difference when there is one opinion leader in the social network. The phenomenon that the final opinion cluster locates around the mean value of the opinion interval in the no leader case disappears. Now three final opinion clusters for most agents form. The cluster closing to 1 has the largest size, which is influenced by the opinion leader. Another final opinion cluster with a larger size approaches to 0. There are some clusters with different size distributed between the two large final clusters and most of them are also close to the opinion leader. When there is no environmental uncertainty, 482 submissive agents are influenced by the opinion leader and 170 submissive ones have opinions ranging from 0.8 to 0.9. Additionally, there are nearly 400 submissive agents having opinions closing to 0. As the variance of the environmental noise increases from 0 to 0.03, as shown in Fig. 3 (a), (b) and (c), the opinion leader has a stronger influence on the submissive agents. On the one hand, some agents having opinions between 0.8 and 0.9 under zero noise are attracted to 1 by the opinion leader as the noises increase. On the other hand, those individuals holding opinions around 0 are gradually diverging in the opinion interval and inclining towards 1. However, it is noticed that the influence of the opinion leader starts to decline when the variance of the environmental noise increase after 0.03. The number of the submissive agents approaching to the opinion leader decreases from 530 to 286 while the variance of the environmental noise increase from 0.03 to 0.1, as shown in Fig. 3 (c) and (d). On the contrary, the number of the submissive agents having opinions near 0 decreases firstly and increase lately as the variance of the environmental noise increases.

As a result, the role of the single opinion leader has different effects under different environment noises. When the environmental noise is large enough so that the submissive agents cannot identify the single opinion leader, they are difficult to make a rational judgment, e.g., right and wrong and more opinion clusters emerge.

4.3   Multiple Opinion Leaders

In the third case, assume that there are 50 opinion leaders in the social network with size $N = 1000$. The final goal value d equals 1, and its influence weight w_g on the opinion leader equals 0.5. In the subsequent simulations, the variance of the environmental noise is respectively set as 0, 0.0001, 0.0005, 0.001, 0.005, 0.01, 0.02, 0.03, 0.04, 0.05, 0.06, 0.07, 0.08, 0.09 and 0.1 for the submissive agents, and the variance of the environmental noise is one-tenth of that of the submissive agents for the opinion leaders.

From Fig. 4, it is interesting to find that, as the number of opinion leader changes from 1 to 50, the influence on the submissive agents becomes stronger under a constant noisy environment. For example, under a zero environment noise, the number of agents following the opinion leaders increases from 482 to 588. The whole evolution of the collective opinions is similar with that of the single opinion leader as the environmental noise changes. The influence of the opinion leaders on the submissive agents is strong at the initial period and then becomes weak after a noise threshold. Similarly, the number of agents having opinions close to 0 decrease firstly and increase later. At the same time, the final collective opinions become more fragmental.

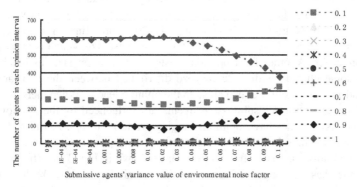

Fig. 4. The number distribution of the submissive agents under different noisy environments. Each line denotes the evolution of the opinions near the state given on the right-side.

5   Conclusion

In this paper, a dominant-submissive agent-based model was proposed to investigate the opinion dynamics under the bounded confidence. Two factors, the environment noise and the opinion leaders, were considered in the collective opinion formation. Simulation results were provided to analyze the impacts of the two factors on the evolution of the collective opinions. On the one hand, as the variance of the environmental noise increases, though the agents become more open-minded, but they are changing opinions with a larger randomness. Thus, it is difficult to form a global consensus or polarization. On the other hand, the opinion leaders can make a nontrivial contribution to guild the submissive agents below a certain noise threshold, but once the environment noise turns large enough, the opinion leaders have weaker influence on the submissive agents. Our results show that there are close relations between the environment noise and the opinion leaders in the collective opinion dynamics. The two factors should be treated prudently in public crisis to avoid the emergence of a harmful global behavior.

Acknowledgements. This research has been partially supported by grants from the National Natural Science Foundation of China (#91224001 and #71173028).

References

1. Boccalettl, S., Latora, V., Moreno, Y., Chavez, M., Hwang, D.-U.: Complex networks: structure and dynamics. Physics Reports 424, 175–308 (2006)
2. Castellano, C., Vilone, D., Vespignani, A.: Incomplete ordering of the voter model on small-world networks. Europhysics Letters 63(1), 153–158 (2003)
3. Bartolozzi, M., Leinweber, B., Thomas, A.W.: Stochastic Opinion Formation in Scale-Free Networks. Physical Review E 72, 046113 (2005)
4. Sood, V., Redner, S.: Voter Model on Heterogeneous Graphs. Phys. Rev. Lett. 94, 178701 (2005)
5. Lambiotte, R., Ausloos, M., Holyst, J.: Majority Model on a network with communities. Phys. Rev. E 75, 030101 (2007)
6. Gabbay, M.: The effects of nonlinear interactions and network structure in small group opinion dynamics. Physica A: Statistical Mechanics and its Applications 378(1), 118–126 (2007)
7. Galam, S.: Heterogeneous beliefs, segregation, and extremism in the making of public opinions. Phys. Rev. E 71, 046123 (2005)
8. Deffuant, G., Amblard, F., Weisbuch, G., Faure, T.: How can extremism prevail? A study based on the relative agreement interaction model. Journal of Artificial Societies and Social Simulation 5(4), 1–10 (2002)
9. Couzin, I.D., Krause, J., Franks, N.R., Levin, S.A.: Effective leadership and decision-making in animal groups on the move. Nature 433(2), 513–516 (2005)
10. Valente, T.W., Pumpuang, P.: Identifying Opinion Leaders to Promote Behavior Change. Health Education & Behavior 34(6), 881–896 (2006)
11. Afshar, M., Asadpour, M.: Opinion Formation by Informed Agents. Journal of Artificial Societies and Social Simulation 13, 5 (2010)

12. Deffuant, G., Neau, D., Amblard, F., Weisbuch, G.: Mixing beliefs among interacting agents. Adv. Complex Sys. 3, 87–98 (2000)
13. Krause, U.: A discrete nonlinear and non-autonomous model of consensus formation. In: Proc. Commun. Difference Equations, pp. 227–236 (2000)
14. Hegselmann, R., Krause, U.: Opinion dynamics and bounded confidence: models, analysis and simulation. Jr. of Art. Soc. and Social Simulation 5, 1–33 (2002)
15. Lorenz, J.: Continuous opinion dynamics under bounded confidence: a survey. Int. Journal of Modern Physics C 18, 1819–1838 (2007)
16. Lorenz, J.: Consensus strikes back in the Hegselmann-Krause model of continuous opinion dynamics under bounded confidence. Journal of Artificial Societies and Social Simulation 9, 8 (2006)
17. Lorenz, J.: Heterogeneous bounds of confidence: Meet, discuss and find consensus! Complexity 4, 43–52 (2010)
18. Kou, G., Zhao, Y.Y., Peng, Y., Shi, Y.: Multi-level opinion dynamics under bounded confidence. PLoS One 7(9), e43507 (2012), doi:10.1371/journal.pone.0043507
19. Peng, Y., Kou, G., Shi, Y., Chen, Z.: A Descriptive Framework for the Field of Data Mining and Knowledge Discovery. International Journal of Information Technology & Decision Making 7(4), 639–682 (2008)

Lexical-Syntactical Patterns for Subjectivity Analysis of Social Issues

Mostafa Karamibekr and Ali Akbar Ghorbani

Faculty of Computer Science
University of New Brunswick
Fredericton, New Brunswick, Canada
{m.karami,ghorbani}@unb.ca
http://ias.cs.unb.ca/

Abstract. Subjectivity analysis investigates attitudes, feelings, and expressed opinions about products, services, topics, or issues. As the basic task, it classifies a text as subjective or objective. While subjective text expresses opinions about an object or issue using sentiment expressions, objective text describes an object or issue considering their facts. The presence of sentiment terms such as adjectives, nouns and adverbs in products reviews usually implicates their subjectivity, but for comments about social issues, it is more complicated and sentiment phrases and patterns are more common and descriptive. This paper proposes a lexical-syntactical structure for subjective patterns for subjectivity analysis in social domains. It is employed and evaluated for subjectivity and sentiment classification at the sentence level. The proposed method outperforms some similar works. Moreover, its reasonable F-measure implicates its usability in applications like sentiment summarization and opinion question answering.

Keywords: Sentiment analysis, Subjectivity analysis, Subjectivity classification, Subjective pattern.

1 Introduction

Nowadays government, organizations, and research centers who are responsible for providing solutions for social, political, economical, and cultural problems, are looking for the attitudes and feelings of the public. It helps them to make right decisions that have less side affects on the social life as well as satisfy more people.

From sentiment perspective, facts and opinions are two kinds of textual information. Facts are objective statements about products and services as well as their features. Opinions are subjective expressions that describe feelings, attitudes, appraisals, ideas, judges, and beliefs expressed about products, services, their features, and also topics and issues. Opinions are expressed in various textual formats such as articles, blog posts, reviews, comments, forums, and tweets as well as web pages designed specifically for voting and opinions such as Debate.org and ProCon.org.

T. Yoshida et al. (Eds.): AMT 2013, LNCS 8210, pp. 241–250, 2013.
© Springer International Publishing Switzerland 2013

Sentiment analysis or opinion mining is an interdisciplinary field that crosses natural language processing, artificial intelligence, and text mining. It has emerged as a subfield of text mining because it analyzes opinions and most opinions are expressed in the text format[2]. Classifying the text as subjective or objective and the text orientation as positive or negative are two basic tasks of sentiment analysis. It is used in various applications from the domains of shopping, marketing, entertainment, education, politics, and social[2,10,6,4].

People express their opinions not only about the products and services, but also about various topics and issues especially those that have influences on their social lifes. A social issue is an issue that relates to people's personal lives and interactions. Public opinions about social issues are used to make decisions that satisfy people's right as well as to bias the public attitudes and opinions. It is statistically proven that products and social issues are different from sentiment perspective[7].

Sentiment analysis is highly domain and context dependent because it benefits from the sentiments of words that depend on not only the domain, but also the context. Context means the combination of various terms as well as the lexical and syntactical structure of sentence. For example, while we expect positive sentiment for "lenient" and negative sentiment for its antonym "strict", but both make negative sentiments in "too lenient" and "too strict" about punishment. Therefore, the roles of domain and context have to be considered in applied techniques for sentiment analysis. This paper focuses on the lexical-syntactical patterns for subjectivity analysis at the sentence level.

The rest of paper is organized as follows. Section 2 explains the subjectivity analysis as well as some previous works. Section 3 introduces a method based on the lexical-syntactical patterns for subjectivity analysis at the sentence level in social domains. In Section 4, we present the evaluation results of our proposed method. In Section 5, we conclude this paper by summarizing the results and provide some future directions.

2   Subjectivity Analysis

From philosophical perspective, subjectivity refers to the subject's perspective, feelings, beliefs, and desires. It usually contrasts with objectivity[14]. In sentiment analysis, private state is described as a general covering term for subjectivity[17].

Subjectivity analysis investigates the presence of opinions in a given text.It consists three primitive tasks: recognizing the subjective clues, finding their relatedness to the opinion target, and classifying the text as subjective or objective. The third task is specifically called subjectivity classification that is distinct than sentiment classification[16,11,19].

Subjectivity analysis is performed at the levels of document, sentence, phrase, and word. Word sentiment extraction can be interpreted as subjectivity classification at the word level that determines the sentiment of word[5,18]. Subjectivity analysis at the sentence level is useful for subjectivity and sentiment classification at the sentence and document level[9], opinion question answering[21], document opinion summarization[11], and information extraction[13].

Most approaches for subjectivity analysis are based on machine learning techniques, which employ lexical features such as adjective, adverbs, nouns, and verbs that have sentiment[19,12,21,1,10]. The combination of opinion terms with other lexical or syntactical information makes subjectivity patterns. While they are domain dependent, but there are common patterns for each domain that people employ to express their opinions. We propose a method for subjectivity classification at the sentence level using subjective patterns.

A sentence may contain some sentiment words but it necessarily does not even express any opinion or its opinions are about another issue rather than the particular opinion target. It means that a sentence may be generally subjective but be objective about a particular opinion target. We are looking for the opinions regarding the particular opinion target. Therefore, we categorize a subjective sentence that is not subjective regarding that target as objective(See Figure 1) From sentiment perspective, a related subjective sentence to social issue is categorized as *for*, *neutral*, or *against*.

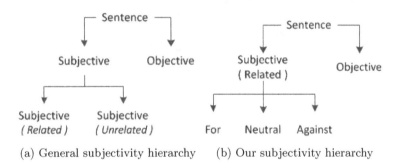

(a) General subjectivity hierarchy   (b) Our subjectivity hierarchy

Fig. 1. Sentence subjectivity hierarchies (general and topic-related)

2.1 Previous Works

Sentence similarity and Naive Bayes classifier have been used to separate the facts from opinions in an opinion question answering application[21]. Pang and Lee[11] benefit from the minimum cut algorithm to select subjective sentences from text. Riloff et al.[13] employ a rule-based classifier to extract the subjectivity clues for train the pattern extraction. Then, a Naive Bayes classifier categorizes the sentence as subjective or objective in the sake of performance improvement of information extraction.

Wilson et al.[20] have used a supervised machine learning technique for sentence subjectivity classification considering four strength levels: *neutral*, *low*, *medium*, and *high*. Barbosa and Feng[1] employ sentiment features as well as tweet specific features such as positive polarity, link, strong subjective, upper case and verbs for subjectivity classification.

Wiebe and Mihalcea[18] have studied the associations between sentence subjectivity and word senses through some empirical evidences. Probabilistic classifier has been employed for sentence subjectivity classification[17]. Conrad et al.[3] have focused on detecting and labeling arguing subjectivity at the sentence level.

3   Lexical-Syntactical Subjective Patterns

People usually express their opinions using common patterns consisting a set of opinion and non opinion words(See Table 1). For example, "please" is very common to ask for something politely. A pattern is subjective when it has sentiment or is used to express opinion.

Table 1. Examples of opinion words

Type		Examples
Opinion Term		good, excellent, bad, wrong, true,...
Verb	Opinion Verb	like, agree, hate, believe, think,...
	To-Be	am, is, are, was, were, ...
	Modal	can, must, should, shall, have to, has to, ...
Negation Clue		not, none, no, never,...

Subjective patterns usually have common terms. We employ the Part Of Speech (POS) tags to generalize them. For example, *"am nervous"*, *"is wrong"*, and *"are great"* are generalized to the lexical-syntactical pattern, *"VB{ To-Be verbs} + JJ{opinion adjective}"*. Table 2 shows some subjective sentences in which opinion expressions are underlined. Almost all of them follow some common subjective patterns.

3.1   Subjective Pattern Structure

Subjective pattern is a sequence of words which either have sentiments or impact the sentiment of the sentence. We define a structure using POS tags as well as lexical terms for subjective pattern. Suppose PT = [POS]{term} shows a pattern term, that means a term with the specified POS tag. In most cases, the term is selected from a list of words. For example, JJ{legal, illegal, wrong, murder} is a pattern that represents an adjective from the provided list. Subjective pattern is defined as a sequence of pattern terms(Eq. 1). For example, *VBZ{is} JJ{legal, illegal, wrong, murder}* is a subjective pattern.

$$SP(\ Subjective\ Pattern\ ) := \{PT_1\ PT_2\ PT_3\ ...\} \tag{1}$$

Table 2. Samples from three data-sets

#	Review/Comment
1	Abortion <u>IS Murder</u>!
2	I hope I dont offend anyone, but anyone who aborts a child (even the doctor who did it) <u>should be arrested</u>!
3	A woman <u>has the right</u> to decide what to do with her body.
4	Also, <u>I don't believe</u> life starts at conception.
5	I'm <u>in agreement</u> with you.
6	If you believe abortion is wrong, <u>don't have one</u>.
7	It <u>is not murder</u> <u>I don't believe</u> because a <u>fetus IS NOT a baby</u>.

PT_i represents a pattern term. Table 3 shows some samples of subjective patterns. The list of subjective patterns is manually extracted from a portion of comments from the data-set. For example, "I agree", "I hate", "I like", "I believe", "I think", and "I guess" result the pattern "I VB{opinion verbs}".

Table 3. Some samples of subjective patterns

#	Pattern	Examples
1	VBZ{is} JJ{legal,illegal,wrong,murder}	is wrong, is illegal
2	NN{baby,fetus,woman,child} has right	baby has right, woman has right
3	PRP{I,she,we,he,it} MD{can,could,must,should}	We must, she should
4	PRP{I,we} VB{hate,agree,disagree,love}	I hate , we love

Existence of a subjective pattern does not necessarily implicate any opinion regarding the social issue. We classify a sentence as subjective when at least one of its subjective patterns is related to the social issue. For example, Sentence #5 from Table 2 is classified as objective regarding "abortion" because there is no evidence of "abortion", but it is generally subjective.

We define a list of representative terms for the social issue (as opinion target). For example, "abortion", "pro-choice", "pro-life", "pro-abortion", and "abort" are defined as representative terms for "abortion". A sentence is related to the social issue when it refers to at least one of its representative terms.

3.2   Implementation

We have implemented a system consists of five components: preprocess, pattern extraction, relatedness analysis, subjectivity classification, and sentiment classification(Fig. 2). The preprocess component removes useless characters such as additional spaces and replaces the abbreviations with the corresponding phrases.

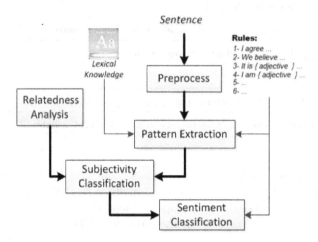

Fig. 2. A general architecture for proposed architecture

The pattern extraction component recognizes and extracts subjective patterns. Relatedness analysis searches the relatedness between the sentence and the social issue. Subjectivity classification component uses the extracted subjective patterns and the relatedness to classify a sentence as subjective or objective. Sentiment classification computes the sentiment of related subjective patterns regarding the social issue and categorizes the sentence as *for, neutral,* or *against.* We define sentiment polarity and strength for each subjective pattern.

Let $S = \{SP_1,\ SP_2,\ ...,\ SP_n\}$ represents a subjective sentence. SP_i represents a related subjective pattern. The sum of sentiment values of subjective patterns is assigned to the sentiment of the sentence (S_S in Eq. 2).

$$S_S := \sum_i SSP_i, \tag{2}$$

SSP_i represents the sentiment of ith subjective pattern. If the calculated sentiment of a subjective sentence is positive, it is classified as *for,* if equals to zero, it is classified as *neutral,* otherwise it is classified as *against.* The sentiment of a document is the cumulative of sentiments of its sentences.

Opinion terms and verbs are the lexical inputs of subjective patterns. For example, all adjectives in *NP is ADJ{wrong, right, legal}* have sentiments. Therefore, opinion adjectives, verbs, adverbs, and nouns are needed. We use two dictionaries of 14320 opinion terms and 3231 opinion verbs, which are enriched version of our previous opinion dictionaries[8]. The sentiment polarity and strength of each opinion term are defined in the dictionary. The sentiment strength can be -2, -1, 1, or 2. {murder, negative, -2} is an example of opinion terms.

The dictionary of opinion verbs consists 3231 verbs gathered from various online dictionaries includes Dictionary.com and TheFreeDictionary.com. The root, 3th person, gerund, past simple, past participle, transitivity, sentiment polarity, and sentiment strength of opinion verbs are defined in the dictionary. {terrify, terrifies, terrifying, terrified, terrified, transitive, negative, -2} is a record in dictionary of opinion verbs.

4   Experimental Evaluation

For this study, we have collected 1125 comments about abortion under the controversial news, questions, and votes from CNN, NY-Times, Debate.org, Pro-Con.org, SodaHead.com, Yahoo Answers, and Ask.com. We have split the comments into their sentences and then manually labeled each sentence as *subjective* or *objective* considering "abortion" as the social issue(See Table 4 for statistics). At the next level, we have labeled each subjective sentence as *for*, *neutral*, or *against* considering its sentiment polarity.

Table 4. Dataset summary statistics

Subjectivity	polarity	Number of Sentences
Objective		697
Subjective	Unrelated	406
Strong subjective	(for , against)	(427 , 656)
Subjective	neutral	140
Weak subjective	(for , against)	(484 , 585)
Total		3395

We have considered the explicit subjective sentence as strong and the implicit subjective sentence as weak. More than 46% of subjective sentences are weak that follows the argument that the rate of implicit opinions compare to explicit ones in social domains is higher than the products[7]. We manually extracted 668 subjective expressions from 100 comments and then defined 228 covering subjective patterns.

We have implemented our pattern-based method as well as the Bag Of Word (BOW) approach for subjectivity and sentiment classification. In the BOW approach, the presence of only one opinion term or subjective expression in a sentence is considered as a clue for subjectivity.

We have measured the performance of our method for subjectivity classification on the whole data-set using the standard precision, recall, F-measure, and accuracy(See Table 5). The F-measure scores of 83.86% and 92.43% indicate a good overall performance. The accuracy of our method on strong opinions is higher than the reported accuracy by [17], while for the weak subjective sentences the results are close.

Table 5. Performance measures of our subjectivity classification and Others

Approach	Data	Strength	Precision	Recall	F-Measure	Accuracy
BOW	abortion	general	71.36%	93.28%	80.86%	70.19%
Our	abortion	general	81.56%	86.30%	83.86%	**80.15%**
Our	abortion	strong	91.25%	93.64%	92.43%	**87.46%**
Our	abortion	weak	77.10%	83.47%	80.16%	71.81%
Wiebe et al.[17]	Treebank	strong	–	–	–	81.5%
Wiebe et al.[17]	Treebank	weak	–	–	–	72.17%

Table 6 shows the comparison of our proposed method with the Naive Bayes classifier employed by Yu et al.[21] as well as high precision classifier employed by Wiebe et al.[19]. The overall F-measure score of our method is slightly higher than others while the precision of [19] and recall of [21] are higher than ours.

Table 6. Performance comparison of our method vs others

Method	Sentiment	Precision	Recall	F-Measure
Our Method	Subjective	81.56%	**86.30%**	**80.15%**
	Objective	63.40%	**54.26%**	**58.47%**
Wiebe et al.[19]	Subjective	91.7%	30.9%	46.2%
	Objective	83%	32.8%	47.1%
Yu et al.[21]	Opinion	69%	91%	78.48%
	Fact	43%	15%	22.24%

We have measured the performance of our method on sentiment classification at the sentence and document levels (See Table 7). Our method's performance at the document level is higher than the sentence level because each document usually contains subjective sentences with similar subjectivity and sentiment polarity. Our method's accuracy at the document level is slightly higher than the accuracy of a supervised technique used by [15].

We have to notice that the accuracy of our method in social domain is even higher than some previous works on products such as 62.5% on car models[6] and 66% on movies[16], while it has been statistically proven that the sentiment analysis of social issues is harder than products[7].

Table 7. Performance measures of our sentiment polarity classification

Approach	Level	Precision	Recall	F-Measure	Accuracy
BOW	sentence	48.62%	53.74%	51.05%	42.03%
Our approach	sentence	74.63%	80.15%	77.29%	**68.45%**
Our approach	strong sentences	76.32%	82.96%	79.50%	**73.34%**
BOW	document	49.85%	56.02%	52.76%	47.65%
Our approach	document	74.97%	81.30%	78.01%	**69.23%**
Somasundaran[15]	document	-	-	-	62.5%

5   Conclusion

Subjectivity analysis investigates the presence of opinions in the text. In this paper, we focused on subjectivity analysis of social issues at the sentence level based on the idea that in social domains, people use common patterns to express their opinions. We defined a lexical-syntactical structure for subjective patterns using the lexical knowledge as well as POS tags.

The experiments shows that our method has a good performance on subjectivity and sentiment classification at the sentence level considering the fact that a general subjective sentence may not be subjective regarding the particular social issue (as opinion target). The overall recall and F-measure of our methods shows its usability in some applications of sentiment analysis such as opinion question answering and opinion-oriented text summarization. Some of our performances' values are slightly higher the similar works.

In the future works, we plan to work on how to define subjective patterns that cover more opinions. We will also work on the implicit opinions because the extraction of subjective patterns and their relatedness to the opinion target are more complicated than the explicit ones. Finally, we will evaluate our method in a sentiment-based application such as document summarization.

References

1. Barbosa, L., Feng, J.: Robust sentiment detection on twitter from biased and noisy data. In: Proceedings of the 23rd International Conference on Computational Linguistics: Posters, pp. 36–44. Association for Computational Linguistics (2010)
2. Binali, H., Potdar, V., Wu, C.: A state of the art opinion mining and its application domains. In: IEEE International Conference on Industrial Technology, pp. 1–6 (February 2009)
3. Conrad, A., Wiebe, J., Hwa, R.: Recognizing arguing subjectivity and argument tags. In: Proceedings of the Workshop on Extra-Propositional Aspects of Meaning in Computational Linguistics, pp. 80–88. ACM (2012)
4. Dave, K., Lawrence, S., Pennock, D.M.: Mining the peanut gallery: opinion extraction and semantic classification of product reviews. In: Proceedings of the 12th International Conference on World Wide Web, pp. 519–528. ACM (2003)

5. Esuli, A., Sebastiani, F.: Sentiwordnet: A publicly available lexical resource for opinion mining. In: Proceedings of the 5th Conference on Language Resources and Evaluation, pp. 417–422 (2006)
6. Gamon, M., Aue, A., Corston-Oliver, S., Ringger, E.: Pulse: Mining customer opinions from free text. In: Famili, A.F., Kok, J.N., Peña, J.M., Siebes, A., Feelders, A. (eds.) IDA 2005. LNCS, vol. 3646, pp. 121–132. Springer, Heidelberg (2005)
7. Karamibekr, M., Ghorbani, A.A.: Sentiment analysis of social issues. In: Proceedings of ASE International Conference on Social Informatics (SocialInformatics 2012), pp. 215–221. ASE, Washington, D.C. (2012)
8. Karamibekr, M., Ghorbani, A.A.: Verb oriented sentiment classification. In: Proceedings of IEEE/WIC/ACM International Conferences on Web Intelligence and Intelligent Agent Technology, pp. 327–331. IEEE/WIC/ACM, Macau, China (2012)
9. Karamibekr, M., Ghorbani, A.A.: Sentence subjectivity analysis in social domains. In: International Conference on Web Intelligence (WI 2013). IEEE/WIC/ACM (November 2013)
10. Liu, B.: Sentiment Analysis and Opinion Mining. Morgan and Claypool (2012)
11. Pang, B., Lee, L.: A sentimental education: sentiment analysis using subjectivity summarization based on minimum cuts. In: Proceedings of the 42nd Annual Meeting on Association for Computational Linguistics, pp. 271–278. ACL (2004)
12. Riloff, E., Wiebe, J.: Learning extraction patterns for subjective expressions. In: Proceedings of the Conference on Empirical Methods in Natural Language Processing, pp. 105–112. ACM (2003)
13. Riloff, E., Wiebe, J., Phillips, W.: Exploiting subjectivity classification to improve information extraction. In: Proceedings of the 20th National Conference on Artificial Intelligence, pp. 1106–1111. AAAI (2005)
14. Solomon, R.C.: Subjectivity in Henderich Ted. Oxford University Press (2005)
15. Somasundaran, S., Wiebe, J.: Recognizing stances in ideological on-line debates. In: Workshop on Computational Approaches to Analysis and Generation of Emotion in Text, pp. 116–124. ACM (2010)
16. Turney, P.D.: Thumbs up or thumbs down? semantic orientation applied to unsupervised classification of reviews. In: Proceedings of the 40th Annual Meeting, pp. 417–424. Association for Computational Linguistics (ACL) (2002)
17. Wiebe, J., Bruce, R., O'Hara, T.: Development and use of a gold-standard data set for subjectivity classifications. In: Proceedings of the 37th Annual Meeting of the Association for Computational Linguistics, pp. 246–253. ACL (1999)
18. Wiebe, J., Mihalcea, R.: Word sense and subjectivity. In: Proceedings of the 21st International Conference on Computational Linguistics and the 44th Annual Meeting of the Association for Computational Linguistics, pp. 1065–1072. ACM (2006)
19. Wiebe, J., Riloff, E.: Finding mutual benefit between subjectivity analysis and information extraction. IEEE Transactions on Affective Computing 2(4), 175 191 (2011)
20. Wilson, T., Wiebe, J., Hwa, R.: Just how mad are you? finding strong and weak opinion clauses. In: Proceedings of AAAI, pp. 761–769 (2004)
21. Yu, H., Hatzivassiloglou, V.: Towards answering opinion questions: separating facts from opinions and identifying the polarity of opinion sentences. In: Proceedings of the Conference on Empirical Methods in Natural Language Processing, pp. 129–136. ACL (2003)

Technology and Cognition: Does the Device We Use Constrain the Way We Retrieve Word Meanings?

Tania Cerni and Remo Job

Department of Psychology and Cognitive Science
University of Trento, Italy

Abstract. We examined the possible implication of two different technological tools, the touch screen and the keyboard, on cross-modal interaction in writing. To do this, we revisit experiments (e.g. [1]) that investigated the recovery of spatial iconicity in semantic judgment and applied them in writing to dictation. In the present experiment participants had to type or to handwrite on a touchscreen, in the upper part or in the lower part of the screen, words whose referents are typically associated with the top or the bottom part of space. In this way congruent (e.g. cloud at the top of the screen) or incongruent (e.g. cloud at the bottom of the screen) conditions were created. The hypothesis was that incongruent conditions give rise to a delay in starting to write more pronounced for touch screen session than for the keyboard one. Results are discussed in terms of embodied cognition theory.

1 Introduction

Do technologies shape cognition, and/or do the way people use and interact with technologies affect their performance in cognitive tasks? The response to both questions are clearly positive if we consider broad levels of analyses (e.g. [2]), but empirical evidence is still lacking for more focused, fine-grained analyses of specific tasks. In this paper we report a study aimed at investigating whether technologically mediated linguistic performance reflects cross-modal interaction and whether it is modulated by the technology used. To do so we will focus on some effects that have been motivated by the embodied approach to cognition.

Accounts of embodied cognition propose that perceptual and motor aspects of a word meaning are intimately related, and this may have implication for language processing as well as for language representation. Specifically, the view that language is closely linked to perceptual and motor representations allows postulating that perceptual and motor aspects of meaning are activated during language processing and may thus interact with the actual processing of the linguistic message. To illustrate, recognizing a word or a sentence would require the re-enactment of the perceptual and motor-related processing performed during actual learning and everyday experiences with the concept the word refers to (see, among others, [1, 3–5]).

T. Yoshida et al. (Eds.): AMT 2013, LNCS 8210, pp. 251–257, 2013.
© Springer International Publishing Switzerland 2013

Several studies have been performed to test the embodied view of language processing. In particular, several experiments have shown systematic effects due to (a) perceptual factors such as concepts typical position and the actual position of the stimulus word on the screen such that congruent trials (e.g. the word cloud presented on the top of the screen) are responded to faster that incongruent trials (e.g. the word cloud presented at the bottom of the screen) (see, e.g. [1]); (b) response direction, such that the direction of the response with respect to a persons body interacts with the sentence the person is exposed to. This is best exemplified by the actionsentence compatibility effect (ACE) [6]: when a sentence implies action in one direction (e.g., Close the window) and participants move their hand in the same direction (i.e. away from the body) to respond responses are faster than in the condition in which the sentence meaning and the direction of the response are incongruent; (c) response position, such that congruency between the directionality of an action implied by a verb and the location of the response key with respect to the body speed up response time [7].

While accounts of these results have been offered that did not make reference to the embodied approach (e.g. [7, 8]) the pattern show a congruency effect between the meaning of the words and the locations of the stimuli on the screen. Most of the studies have focused on reading (and comprehension) while only few have addressed writing (and production). Among the latter, Jasmine and Casasanto (2012) have reported the QWERTY effect, named after the disposition of the keys on the keyboard [9]. Some words are spelled with more letters on the right side of the keyboard and others with more letters on the left side, and the authors found a relationship between the emotional valence of the words and key position so that words with more letters on the right were rated as more positive than words with more letters on the left. Interestingly, the effect was stronger for more recently coined words.

In this study we report an experiment in which we investigated whether the interaction between perception and language reported by [1] would emerge in writing. Furthermore, we investigated whether the technical device used for writing modulates the participants performance. Specifically, we used both a touch screen and a keyboard, on the assumption that the use of the pen on the touch screen would favor the asymmetry between congruent and incongruent location effect while the keyboard would be less prone to let the effect to emerge.

2   The Study

2.1   Method

Participants. Twenty-four Italian native speakers (mean age = 23), students and researchers of the University of Trento, Italy, were recruited for free participations or for course credits. In a brief pre-experimental sessions each participant completed a short questionnaire about his/her habits of writing with pen, keyboard and touch screen in order to expunge participants unfamiliar with any of these devices.

Procedure. A set of 40 Italian words referring to animals and 40 referring to objects were selected. In each category, half of the concepts had a typical location at the top (i.e. eagle, sky) and a half at the bottom (i.e. mouse, carpet) of space. All 80 words were vocally registered with Cool Edit Pro 2.0 reducing rumor and hiss and controlling the length of the silence at the beginning and at the end of each word (120 ms). The experimental procedure consisted of two writing conditions performed in succession by each participant. In the first condition, participants had to manually write with a pen on a touch screen (LG FLATRON T1910); in the second condition, participants had to write with the keyboard with the support of the same screen used as a normal computer screen.

All of the 80 words were automatically dictated by the computer in each condition. The procedure of stimulus presentation and the collection of data (reaction time) in both conditions were done using E-Prime 1.1.4.1, with a similar configuration of the visual cues presented on the screen to make them comparable as much as possible. Specifically, in each writing condition, at the end of each word dictated by the software, a visual cue, precisely a line, was presented at a fixed point at the top or at the bottom of the touch screen where participants have to directly handwrite the stimulus, or to type the word they heard. The line where the words have to be written, in both conditions, appeared at the middle of the audio file of the presented word. (See Fig. 1 for a schematic pattern of the experiment).

Each word was presented twice so that one time it was written at the top and one time at the bottom of the screen for each writing condition (pen vs. keyboard) for a total of 160 stimuli divided into four blocks per condition. In this way, a congruent (i.e. eagle at the top of the screen) and an incongruent (i.e. eagle at the bottom of the screen) condition was devised. Words presentation was random for each block and for each participant with the only constraint that the same word did not appear in the same block or in consecutive blocks.

The difference between the procedures of the two conditions consisted in the input system of writing. For the handwriting condition, the experimenter fixed the screen in a sloping position in order to facilitate writing on the screen and to prevent fatigue of the arm. During the experiment, the participant had to position the pen at the center of the screen, on a specified fixed point. When the software recorded the position of the pen, the audio of the word started. In this way we are able to control the position of the participant for every trial. Then, when the participant finished writing, he/she had to return to the central point to stop the current trial and to begin the next.

During the keyboard session, the screen was positioned vertically and the keyboard in front of the participant. To build a graphical design similar to the handwriting condition, in the middle of the screen were put a fixation point that the participant had to fix before starting to write. Furthermore, to control the position of the hands and to avoid that they were put on the letters of the keyboard, the participants were instructed to put the hands on the space bar and to keep it pressed. This input activated the audio of the word and, once the line appeared on the screen, the participants had to release the bar and write the corresponding word. The end of the trial coincided with the press of the enter key and then the return to the bar space to hear the next word.

Participants were instructed to be fast an accurate. Reaction time from the presentation of the to-be-written-on line to the beginning of writing was registered and collected. At the beginning of each condition, a brief training was administered that contained, unlike the experiment, some comprehension questions to verify access to the meaning of the dictated words. The stimuli were seven words, without any specific reference to the top or to the bottom of the space. Associated with three of these words we used three simple yes-no questions (is it an object?, is it an animal?, has it the wings?) that the participant answered orally, for handwriting condition, and pressing the corresponding keys, for the keyboard condition. Erroneous responses were very few (. 17), confirming access to semantics in this condition.

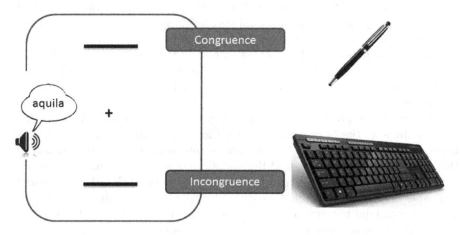

Fig. 1. A schematic pattern of the experimental design

Results. Response times (RTs) shorter than 250 ms and longer than 2000 ms (3and errors (0,13RTs for correct responses are plotted in Figure 2, separately for touch screen and keyboard, as a function of the condition (congruent vs. incongruent) and block (1 to 4). A repeated measure ANOVA with device (touch screen and keyboard) and location congruency (congruent and incongruent) was run on the RTs. The effect of the device is significant ($F1 (1,79) = 6.91$, $F2 (1,23) = 4.86$), with responses via keyboard being faster than response via touch screen. The effect of the location congruency is not significant (Fs ¡ 1): independently from the writing device, words written in a congruent or in an incongruent location were equally fast.

A piecewise linear model shows that there is an effect of block, indicating a learning curve which is modulated by the writing device: for the touch screen performance is increasing continuously across blocks while for the keyboard, after an increased from block 1 to block 2 there is no further increase from block 2 to block 4 (paired ttest: $t (23) = 3.44$, p¡.002).

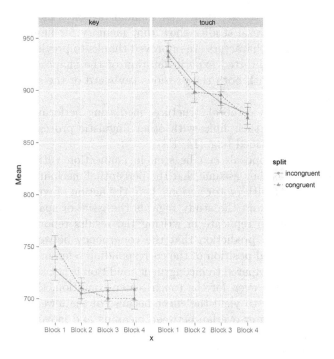

Fig. 2. Plot of the response times means along the four experimental blocks

3   Discussion

Two main findings emerge from the study. First, the expected difference between the two conditions of writing emerges, confirming the hypothesis of a different modulation of performances as a function of the device used. Such difference may be attributed to the fact that the pen and the keyboard allow for idiosyncratic movements and correlated differences in the time writing is initiated and deployed. This is true in general, and in our experimental setting in particular, as in the touch screen condition participants had to move the hand from the central point to the indicated line while, in the keyboard condition, they had to move the hand(s), and the finger(s), from the bar space to the key corresponding to the first letter of the stimulus words.

Thus, the difference between the two different writing modalities may be ascribed to the different manual skills involved in each. Mangen and Velay (2010) point out that in handwriting we use only one hand and we are the maker of the shape of the letters but the keyboard involves both the hands without graphomotor movements implicated [10].

Another difference regards visual attention that is focused on one point in handwriting, whereas writing trough the keyboard split visual attention focused on the screen - and the haptic input focused on the keys.

It is plausible to hypothesize that these manual variances have different cognitive consequences. Several studies show that memory for figures, alphabetic characters, and pseudo-characters are improved thanks to previous handwriting experiences, given the directed experimentation of the shape of the proposed stimuli in comparison with both the previous keyboard or the sole visual experiences [11–13].

Furthermore, different cerebral structures mediating performance in the two writing modalities, and their links with other linguistic process - for example reading have been identified (e.g. [14, 15]).

These theoretical proposals can be seen in connection with the embodied cognition approach, as they assume that the perceptual mechanisms involved in language are bounded with motor actions, here the action of writing.

The second finding from the study regards the issue of spatial iconicity in writing. Our attempt to replicate in writing the results reported by, e.g. [1], was not successful. Our prediction that the congruency between the location of stimuli and the typical position of the corresponding referent would generate faster response times compared to incongruent condition is not supported by the data. This was not true even for the touch screen condition, the condition we thought was more likely to yield the effect because of the fewer constraints on movements and the greater overlap between meaning and movement. However, the results of the presented study show no difference between congruent and incongruent condition in either the devices.

This pattern suggests either that in linguistic production there is not a recovery of spatial iconicity of the produced words, irrespective of the writing tool or that structural constraints of the experiments, such as timing, may have prevented the effect to emerge. Further data are needed to disentangle this result.

An interesting, and unexpected, finding is the dissociation between the learning curve of the two writing conditions, with a continuous improvement in the touch screen condition but a flat performance after block 2 in the keyboard condition. This pattern may reflect a floor constraint for the latter condition, such that the planning and execution of the writing performance have reached maximum speed by the end of the first block. An alternative, not necessarily inconsistent, account would allow for the fine-grade adjustment for the trajectory movement in reaching the indicated line in the touch screen condition, adjustment that requires time to reach the optimal standard.

To conclude, we did not find a spatial congruity effect in writing in neither of the two devices used, the touch screen and the keyboard, in spite of significant time differences in performing the task in the two conditions, which may have allowed detection of the effect either at an early or a late onset. Interestingly, the speed to perform the task across the experiment varied markedly as a function of the devised used, with increased speed up to the last block for the touch-screen condition but an increase between the first and second block only for the keyboard condition.

References

1. Šetić, M., Domijan, D.: The influence of vertical spatial orientation on property verification. Language and Cognitive Processes 22, 297–312 (2007)
2. Pezzulo, G., Barsalou, L., Cangelosi, A., Fischer, M., McRae, K., Spivey, M.: Computational Grounded Cognition: A New Alliance between Grounded Cognition and Computational Modeling. Frontiers in Psychology 3, 612 (2013)
3. Estes, Z., Verges, M., Barsalou, L.W.: Head up, foot down: Object words orient attention to the objects typical location. Psychological Science 19, 93–99 (2008)
4. Pecher, D., Van Dantzig, S., Boot, I., Zanolie, K., Huber, D.E.: Congruency between word position and meaning is caused by task induced spatial attention. Frontiers in Cognition 1, 1–8 (2010)
5. Zwaan, R., Yaxley, R.: Spatial Iconicity Affects Semantic Relatedness Judgments. Psychonomic Bulletin and Review 10(4), 954–958 (2003)
6. Glenberg, A.M., Kaschak, M.P.: Grounding language in action. Psychonomic Bulletin and Review 9, 558–565 (2002)
7. Job, R., Treccani, B., Mulatti, C.: Perceptual and motor spatial representations in word recognition. In: Cadinu, M., Galdi, S., Maass, A. (eds.) Social Perception, Cognition, and Language, pp. 151–165. Cleup, Padova (2011)
8. Lakens, D.: High Skies and Oceans Deep: Polarity Benefits or Mental Simulation? Frontiers in Psychology 2, 21 (2011)
9. Jasmin, K., Casasanto, D.: The QWERTY effect: How typing shapes the meaning of words. Psychonomic Bulletin and Revie 19, 499–504 (2012)
10. Mangen, A., Velay, J.L.: Digitizing literacy: Reflections on the haptics of writing. In: Zadeh, M.H. (ed.) Advances in Haptics. InTech (2010) ISBN: 978-953-307-093-3, http://www.intechopen.com/articles/show/title/digitizing-literacyreflections-on-the-haptics-of-writing
11. Longcamp, M., Zerbato-Poudou, M.T., Velay, J.-L.: The influence of writing practice on letter recognition in preschool children: A comparison between handwriting and typing. Acta Psychologica 119, 67–79 (2005)
12. Longcamp, M., Boucard, C., Gilhodes, J.C., Velay, J.L.: Remembering the orientation of newly learned characters depends on the associated writing knowledge: A comparison between handwriting and typing. Human Movement Science 25, 646–656 (2006)
13. James, K.H., Atwood, T.P.: The role of sensorimotor learning in the perception of letter-like forms: tracking the causes of neural specialization for letters. Cognitive Neuropsychology 26, 91–110 (2009)
14. James, K.H., Engelhardt, L.: The effect of handwriting experience on functional brain development in pre-literate children. Trends in Neuroscience and Education 1, 32–42 (2012)
15. Longcamp, M., Hlushchuk, Y., Hari, R.: What differs in visual recognition of handwritten vs. printed letters? An fMRI study. Human Brain Mapping 32, 1250–1259 (2011)

Basic Study on Treating Tinnitus
with Brain Cognition Sound

Takeya Toyama[1], Daishi Takahashi[1], Yousuke Taguchi[1], and Ichiro fukumoto[2]

[1] Kitasato Junior College of Health and Hygienic Sciences, Niigata,
Japan
{t-toyama,d-taka,y-tagu}@kitasato-u.ac.jp
[2] Institute of Biomedical Engineering Nagaoka University of Technology, Niigata, Japan
ichiro@vos.nagaokaut.ac.jp

Abstract. This study aimed to develop a novel treatment method for tinnitus using phase-shift sound stimulation. We performed physical audio signal processing to create simulated sound stimuli resembling subjective tinnitus. The preliminary study utilized two tinnitus models representing different origins of tinnitus, and in each model the simulated tinnitus sound was presented simultaneously with a phase-shifted sound. We then measured audio brainstem response wave latencies; wave latency prolongation served as an evaluation index. The main study involved subjects with tinnitus but no underlying disease. To modulate the perception of tinnitus, an oscillator was used to identify tinnitus frequency and produce sound output that was then phase shifted. Preliminary study results indicated that excitation of the nerve impulse by an additional sound can modulate coding of preceding tinnitus information in the auditory brainstem. The experimental study demonstrated the reproducibility of time delays. Both results suggest the clinical usefulness of this treatment method.

Keywords: Phase-shifted sound, simulated tinnitus sound, auditory brainstem response.

1 Introduction

In Japan, approximately 10% to 20% of the population currently has tinnitus, and over 5% of them, including many with refractory tinnitus, have reported that tinnitus adversely affects their daily lives. Between 80% to 90% of patients treated at a hospital for tinnitus are reported to have some form of underlying hearing loss. For people whose tinnitus is caused by an underlying disease, the disease should be treated first. Many tinnitus cases are refractory because the underlying disease is refractory, of unknown origin, or accompanied by a psychological disorder such as stress. However, in some patients, tinnitus can persist even after the underlying disease has been successfully treated. Today, few effective treatments are available despite numerous ongoing studies investigating new or modified treatment methods. Moreover, many remedies for tinnitus are supported by little or no medical evidence.

T. Yoshida et al. (Eds.): AMT 2013, LNCS 8210, pp. 258–265, 2013.
© Springer International Publishing Switzerland 2013

2   The Nature of Tinnitus and Its Treatment

2.1   Clinical Definition

Tinnitus is defined as the perception of sound in the absence of external sound sources. Because acute tinnitus is the only symptom of sudden sensorineural hearing loss, if acute tinnitus develops, an early consultation with an otorhinolaryngologist is recommended. In Japan, tinnitus is commonly known as *miminari*, although *jimei* is the official medical term.

2.2   Acute and Chronic Tinnitus

Acute tinnitus is the sudden onset of sound in the ears and is often accompanied by sensorineural hearing loss. It is mostly transient and caused by acoustic trauma due to long exposure to loud noise or it is caused by sudden hearing loss due to various medical conditions. As generally defined, chronic tinnitus is tinnitus lasting more than three months after failed treatment of the acute symptom. Underlying diseases include inner ear injuries, acoustic trauma, Ménière's disease, cerebrovascular disease, cervical degenerative disc, temporomandibular joint disorder, diabetes, hypertension, and abnormal lipid metabolism.

2.3   Treatment Methods

Although the primary disease must be treated first, during the acute phase tinnitus can be successfully treated by therapy to improve blood circulation using, for example, steroid administration, intravenous infusion, or hyperbaric oxygen therapy. If the underlying disease is refractory or if tinnitus persists after successful treatment of the underlying disease, a determination is made whether drug therapy, cerebrovascular circulation treatment, physical and psychological therapy, acupuncture and/or moxibustion treatment is indicated.

The new method described in this study is similar to tinnitus masker therapy and tinnitus retraining therapy (TRT). Tinnitus masker therapy uses audiometry to determine the frequency band of the tinnitus sound and then a sound resembling the patient's tinnitus is used as a stimulus to mask the tinnitus. The actual masking effect lasts only as long as the therapy session, but because the cochlear nerve is fatigued, tinnitus can improve after the therapy session.

In TRT, individuals with tinnitus are trained to perceive tinnitus as an environmental sound that reaches the brain as auditory information, yet stays below the conscious level. TRT utilizes an inoffensive slight noise as a stimulus for a predetermined time period to help them expand their consciousness [1]. TRT does not mask the tinnitus as in tinnitus masker therapy.

Against this background, in this study, we performed physical audio signal processing to investigate whether simple and non-invasive forms of sound energy can be used to objectively assess tinnitus and improve a patient's perception of it [2]. We designed this method to mask tinnitus physiologically rather than conventionally as in tinnitus masker therapy. Our method might also reduce or mask tinnitus more effectively than TRT. Individuals with a primary complaint of tinnitus but no known underlying disease served as subjects in this study.

3 Experimental Methods

3.1 Development of Two Tinnitus Models

Reference and Phase-Shifted Sound Combination in the Cochlea of the Inner Ear. A bone conduction receiver was placed behind the ear, and the cochlea directly stimulated by a reference sound via the ear bone. This intracochlear tinnitus model was based on the hypothesis that tinnitus is induced by abnormal excitation of the sensory hair cells in the cochlea [3]. The reference sound was then phase shifted, and a receiver conducted the sound through air to stimulate the ipsilateral cochlea. This *Ipsi model* was so named because the sound combination was produced inside the same (unilateral) cochlea. We then investigated the effect of the sound combination on amplitude reduction.

Sound Combination in the Superior Olivary Complex in the Central Nervous System. A bone conduction receiver was placed behind the ear to directly stimulate the cochlea by a reference sound via the ear bone based on the same hypothesis described above. However, the contralateral cochlea was stimulated by a phase-shifted sound via air conduction using a headphone. This *Contra model* was so named because sounds from the other side were combined in the superior olivary complex, which is central to the cochlea. We then investigated the extent of amplitude reduction due to the sound combination in the superior olivary complex.

3.2 Physical Audio Technique to Alleviate Tinnitus

Effective phase-shift treatment depends upon detailed information on frequency of the tinnitus sound, however, characterizing subjective tinnitus accurately is difficult. To modulate the perception of tinnitus, we used an oscillator to identify the frequency of the tinnitus and produce sound output that was then phase shifted by time delay processing (Fig. 1). Specifically, an oscillator determined tinnitus frequency in the main frequency band, sound output was digitally converted, and a time delay of 1 to 30 ms added.

3.3 Audio Conversion System and Interface

LabVIEW was used to characterize and phase shift the frequencies, and the degree of conversion was visually checked on a color panel (Fig. 2).

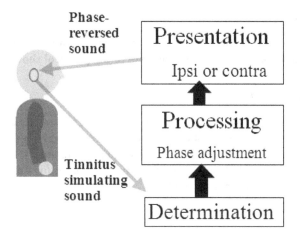

Fig. 1. Conceptual diagram of this novel tinnitus treatment method

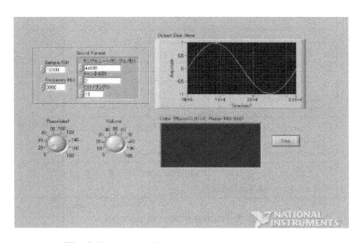

Fig. 2. Physical audio conversion system interface

3.4   Auditory Brainstem Response (ABR) Measurement Settings

The ABR measurement settings were as follows: number of additions, 1000; analysis time, 20 ms; low cut filter, 10 Hz; high cut filter, 3 kHz; electroencephalography electrode positions, Cz-A1 and Cz-A2 of the International 10/20 system; sound source, click 3 KHz or 10 Hz; tone burst, 500 Hz, 2 kHz, COND, RARE. Rise time, plateau time, and fall time were 10 ms, 50 ms, and 10 ms, respectively.

3.5   Tinnitus Evaluation Protocol

After obtaining informed consent, we conducted the main study with individuals with **tinnitus but no underlying disease according to the protocol shown in Figure 3.**

3.6   Evaluation Criteria

Tinnitus Handicap Inventory (THI). The THI is a 25-item clinical questionnaire that assesses the adverse effects of tinnitus in daily life. Based on responses using a 5-point scale, a total THI score from Grade I to V indicates tinnitus.

Visual Analogue Scale (VAS). A VAS for pain was used for subjects to evaluate, albeit subjectively, the severity of tinnitus.

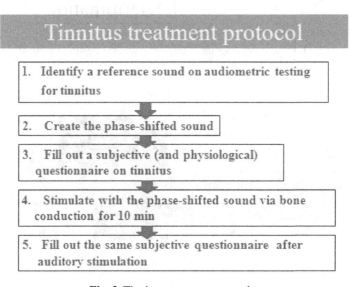

Fig. 3. Tinnitus treatment protocol

4   Results

4.1   Preliminary Study

Ipsi Model. Phase-shift sound stimulation prolonged ABR wave V latency only in the Ipsi model and not in the Contra model. Also, ABR wave V latency in both models was markedly prolonged in the "low tone" condition and shortened in the "high tone" condition.

Low Tone, 71.1%
(+) 0.28–0.64 ms, (−) 0.08–0.66 ms
High Tone, 0%
(+) 0.16–0.56 ms

Contra model
Low Tone, 43.3%
(+) 0.05–0.36 ms, (−) 0.07–0.86 ms
High Tone, 43.2%
(+) 0.04–0.68 ms, (−) 0.06–0.58 ms

4.2   Experimental Study with Subjects

Case 1. In a 47-year-old man with tinnitus in the right ear, standard pure-tone audiometry results were normal (Fig. 4)

Tinnitus assessment results

THI:	48, Grade 3
VAS:	36
Post-VAS:	28
Pitch match:	200 Hz
Loudness balance:	20 bB
Tinnitus tone:	250 Hz
Time delay:	6.0 ms

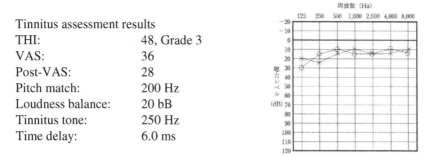

Fig. 4. Results of pure-tone audiometric testing in Case 1

Case 2. In a 34-year-old man with tinnitus in the left ear, standard pure-tone audiometry results were normal (Fig. 5).

Tinnitus assessment results

THI:	24, Grade 3
VAS:	26
Post-VAS:	24
Pitch match:	8000 Hz
Loudness balance:	20 dB
Tinnitus tone:	10200 Hz
Time delay:	29.0 ms

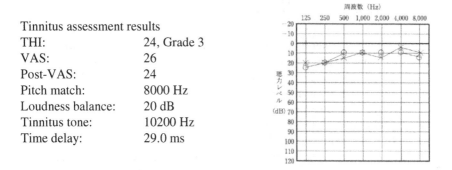

Fig. 5. Results of pure-tone audiometric testing in Case 2

Figure 5 shows the change in VAS score after the subjects with tinnitus were stimulated with phase-shifted sound (Fig. 5).

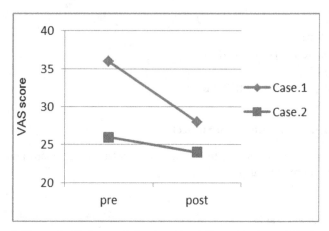

Fig. 6. Change in VAS score following stimulation with phase-shifted sound

5 Discussion

ABR wave V latency was markedly prolonged in both the Contra and Ipsi models using a low tone, which suggests that the combination of a reference sound mimicking tinnitus and a phase-shifted sound successfully canceled sound pressure in the cochlea. One possible explanation is that nerve impulses generated by basilar membrane vibration due to the sound stimulus were influenced by the phase-shifted sound in the endolymphatic sac. Also, the mechanically suppressed amplitude may have occurred while the phase-shifted sound was moving through the inner ear as sound energy. We developed these models in consideration of the above possibilities [4]. However, the extent of impulse suppression cannot be known based on these models, which are too similar in design. Also, mechanical sound conversion might have altered the sound pressure before it reached the inner ear. Because abnormal excitation of the cochlea nerve and outer hair cells is sometimes observed in patients with inner ear tinnitus, such patients, albeit limited in number, might benefit from this method.

ABR wave V latency was shortened, not prolonged, in the Contra and Ipsi models using a high tone, which result might be related to our difficulties in controlling the high tone range (cf. low tone range) when shifting the frequency of the reference sound. Thus, synchronization by phase shift might not have been completed.

Further study is needed to reveal the time difference between sound reaching the cochlea by bone conduction and sound reaching the outer ear and eardrum by air conduction. Moreover, by bone conduction, sound can be reduced up to a few decibels due to sound loss on the contralateral side, keeping in mind shadow hearing on this side. After revising the experimental protocol, we plan to perform additional experiments to carefully consider synchronization. The results in this study indicate that perception of preceding sound information in the brain is modified by nerve impulses generated by additional sound information. In the experimental study, we observed reproducibility in the amount of mechanical sound conversion that successfully alleviated subjective tinnitus.

6   Conclusion

ABR results in the preliminary study suggest the relevance of this method, and the experimental study showed that tinnitus was improved by time delay in sound stimulation and was eliminated by phase shift. These findings suggest the effectiveness of this novel treatment method.

7   Future Directions

Biofeedback (BF) has been used to effectively modify the phase of tinnitus sound. Although we used conventional tinnitus testing in this study, we plan to develop a method that uses BF to determine tinnitus frequency in real time. Amplitude modulation due to fluctuation at the time of frequency determination needs to be investigated and a new protocol should account for synchronization of test sounds and shadow hearing due to bone conduction. Because anticipation and anxiety can affect the outcome of tinnitus indicators (THI and VAS), other physiological indicators need to be included. Currently, we are considering the use of salivary chromogranin-A and low/high frequency heart rate variability on electrocardiograms as psychological stress indicators.

Acknowledgements. The authors thank students at Nagaoka University of Technology and Kitasato University for their participation in the preliminary study.

References

1. Toyama, T., Uchiyama, H., Fukumoto, I.: Novel physical audio processing-based treatment for tinnitus. Audiology Japan 55(5), 509–510 (2012)
2. Choy, D.S., Lipman, R.A., Tassi, G.P.: Worldwide experience with sequential phase-shift sound cancellation treatment of predominant tone tinnitus. J. Laryngol. Otol. 124(4), 366–369 (2010)
3. Meeus, O., Heyndrickx, K., Lambrechts, P., De Ridder, D., Van de Heyning, P.: Phase-shift treatment for tinnitus of cochlear origin. Eur. Arch. Otorhinolaryngol. 267(6), 881–888 (2010)
4. Schaette, R., König, O., Hornig, D., Gross, M., Kempter, R.: Acoustic stimulation treatments against tinnitus could be most effective when tinnitus pitch is within the stimulated frequency range. Hear. Res. 269(1-2), 95–101 (2010)

Designing Enhanced Daily Digital Artifacts Based on the Analysis of Product Promotions Using Fictional Animation Stories

Mizuki Sakamoto[1], Tatsuo Nakajima[1], and Sayaka Akioka[2]

[1] Department of Computer Science and Engineering, Waseda University, Japan
{mizuki,tatsuo}@dcl.cs.waseda.ac.jp
[2] Department of Network Design, Meiji University, Japan
akioka@meiji.ac.jp

Abstract. The *virtual forms* present dynamically generated visual images containing information that influences a users behavior and thinking. In a typical way, adding a display to show visual expressions or projecting some information on an artifact offers computational visual forms on the existing daily artifacts. Using *virtual forms* is a very promising way to enhance artifacts surrounding us, and to make our daily life and business richer and more enjoyable. We believe that incorporating fictional stories into *virtual forms* offers a new possibility for enriching user experiences. In particular, integrating fictional stories into our daily activities through transmedia storytelling is a promising approach. Transmedia storytelling enables *virtual forms* to be employed everywhere to immersively integrate fictional stories into our daily activities. If we can design attractive *virtual forms* in a structured way, it will become easy to enrich user experiences. Currently, the design framework for *virtual forms* is not well defined. The framework needs to take into account the semiotic aspect of a *virtual form*. One key factor, in particular, is how strongly we believe in the reality of a fictional story within the *virtual form*. In this paper, we show the extracted insights discussed in the workshops and present some design implications for designing *virtual forms* that integrate fictional stories into our daily activities.

Keywords: Fictionality, Product promotion, Reality, Augmented reality, Transmedia storytelling.

1   Introduction

Our daily life is becoming increasingly virtual as our surrounding daily artifacts become more intelligent [6,10]. We define a virtual object as something that does not really exist but that has a real effect on our daily life as if it exists. *Jean Baudrillard* explains our consumption behavior as consuming the symbols associated with things and not the things themselves [1]. Because the symbolization of things will accelerate by embedding computers in our lives, our virtual consumption will progress rapidly.

T. Yoshida et al. (Eds.): AMT 2013, LNCS 8210, pp. 266–277, 2013.
© Springer International Publishing Switzerland 2013

Virtual forms offer the potential to integrate virtuality into products and services to offer immersive experience to the users [13]. The *virtual forms* present dynamically generated visual images containing information that influences a users behavior and thinking. In a typical way, adding a display to show visual expressions or projecting some information on an existing daily artifact offers computational visual forms on the artifacts. In past studies, we have developed several *virtual forms*, and assessed their effectiveness [5,12,15,19]. The story already plays an important role in product advertisements because it increases the appeal of the target product [9]. Using fictionality to incorporate stories makes our experiences richer because the stories can more easily translate the meaning of the product. Fictional stories are particularly useful tools for enhancing our daily experiences to increase our buying impulse. Fictional stories can represent stories that do not exist in the real world or stories from the future. The stories can flexibly offer us a broad range of information using nonexistent artifacts such as magic. It is easy to embed ideological messages in these stories to make it possible to teach about various social issues. Additionally, the stories are useful to encourage a user to change their behavior and attitude because their positivity can be used to increase our self-efficacy.

Using fictional stories through transmedia storytelling [3] is a promising approach for enhancing *virtual forms*. Transmedia storytelling allows a fictional story to be harmonized into our real world by fragmenting the story into multiple media presented at various locations in our everyday life. In this case, *virtual forms* are installed everywhere to embed the fictional stories into the real word. The *virtual form* becomes a layer to enhance the real world through transmedia storytelling. The story embedded in the *virtual forms* virtualizes our real world and offers us additional enrichment. This approach offers the potential to enhance our experiences in our daily life, and the stories embedded in our world immersively encourage us to solve various social problems anytime, anywhere. For the successful integration of fictional stories, users will need to feel that the fictional stories are realistic. From our past experience, achieving reality is an important design criterion for successful integration. If a user does not feel the reality of the fictional stories, he/she will feel that the additional values offered by the *virtual forms* do not exist in the real world.

However, the semiotic aspect of designing *virtual forms* is an essential issue behind offering the reality to enrich user experiences. We need to investigate how the reality has been designed in the past media and extract design guidelines that can offer reality for the fictional stories presented in *virtual forms*.

In this paper, we analyze several case studies of products and services that are promoted using fictional stories to understand how reality should be achieved in fictional stories. We have conducted two workshops to extract useful insights for designing *virtual forms*. In respective workshops, about 20 participants attended and discussed for a couple of hours. The participants are of age 21-24. All of them are Japanese, and very familiar to recent Japanese animations/games and product promotions. After watching movies or playing games described in the following sections, the participants discussed how they feel the reality of the movies and games and what are major elements that they feel the reality. The movies and games used in each workshop were selected by two experts who are working on the media analysis research according to the main topic of the workshop. Finally, the experts concluded the discussions presented in the paper.

In the first workshop, we discussed the design patterns that allow a user to feel the reality of fictional stories when the stories are used for product and service promotions. It is essential that a user feels that the products exist in the real world and they are very attractive. In the second workshop, we discussed how the fictional stories can be seamlessly integrated into our daily activities. We show case studies that use fictional stories to enhance our daily activities. Finally, the paper presents how the insights extracted from the two workshops can be used to design future *virtual forms* and identifies the technological issues behind realizing them.

We believe that the discussions in this paper are essential to better design the integration of fictional stories into our real world; this integration is essential to make our lifestyle richer and more desirable. This makes us to develop attractive new services for marketing, community management and social activism.

2   Workshop 1: Product Promotion with Original Animated Stories

Animated movies are very useful tools because they make it easy to offer fictional worlds and characters. With animation, we can offer empathetic fictional creatures and marvelous futuristic worlds that are attractive to the user. It is easy to embed ideological messages that represent human dreams and expectations into animated stories. In addition, the typical Japanese animated stories are full of positive thinking, so the stories can be enjoyable while increasing our self-efficacy and helping us to overcome difficult problems.

Currently, typical product promotions use empathetic characters that appear in popular animated stories as tools to increase the attractiveness. For example, *Pokmon*[1] characters are widely used to promote foods for kids, such as a retort-packed curry food and bread. This animated movie is very popular with most Japanese children, so a promotion broadcast within the animated television movie is effective way to make these products appealing to these children. Although the promotion may increase the buying impulse of children, it does not convince them that the products are attractive. Thus, the products will be forgotten when they become adults.

2.1   Design Patterns to Promote Products with Fictional Stories

In this section, we present four design patterns for promoting commercial products by using fictional stories. The design patterns were extracted from discussions in the workshop, while watching several animated movies for product promotions.

---

[1] http://www.pokemon.com/us/

I. Offering vivid visual impact or surprise attracts us to the products:

Pepsi NEX uses *Cyborg 009* for their product promotion[2], and *Tiger & Bunny*[3] uses several Japanese company logos, such as *Softbank*[4] and *Bandai*[5], to promote their company brands. In the *Pepsi NEX* promotion, the cyborg heroes move very quickly, which becomes a metaphor for the *Pepsi NEX*s sharp taste. Additionally, a pretty heroine creates an empathetic feeling for the products. In the animated story of *Tiger & Bunny*, justice and heroics are commercialized commodities. Some people choose to become costumed superheroes, and each is sponsored by an actual major company, which is featured as an advertisement on the heroes uniforms. These superheroes are perceived as cool and become metaphors implying that the companies are also cool. These vivid expressions of an unusual daily life surprise people, which generates strong and memorable impressions of the target products.

II. Offering a non-fiction story that makes us believe the promotion:

TAISEI Corporation[6] promotes its brand image using an animated movie[7]. In the movie, a woman is working on the construction of the *Bosporus tunnel*. The non-fiction story explains that her work contributes to an amazing construction that will appear in the world map. The movie demonstrates that the company has achieved this amazing work, so the audience for the movie can own the positive feelings about the company through the amazingly realistic scenes of the tunnel. The promotion is very useful because most of us do not know that the *TAISEI Corporation* has built these amazing constructions all over the world.

III. Offering a reality that makes us believe the fictional story in the promotion:

*DOCOMO*s[8] promotion, *Xi AVANT*[9], shows a vision of the future mobile phone; this promotional movie uses several realistic landscape scenes of present-day *Barcelona*. The reality of the background landscape scene offers a feeling that the vision told in the fictional story will be realized in the near future real world. The *Tokyo Disney Resort*s promotion[10] reminds each person of a real memory from when she previously visited the *Tokyo Disney Resort*. So the story shows that she will enjoy visiting *Tokyo Disney Resort* even when she becomes old. Therefore, we believe that the happy memories from the *Tokyo Disney Resort* will be inherited by our children.

IV. Offering empathy that attracts us to the products in the promotion:

---

[2] http://009.ph9.jp/pepsi-nex/
[3] http://tigerbunny.wikia.com/wiki/Tiger_%26_Bunny_Wiki
[4] http://mb.softbank.jp/en/
[5] http://www.bandai.co.jp/e/
[6] http://www.taisei.co.jp/english/
[7] https://www.youtube.com/watch?v=OKoCl-3EOVw
[8] DOCOMO is one of the biggest mobile telecom companies in Japan.
[9] https://www.youtube.com/watch?v=IP5nAkG5lME
[10] https://www.youtube.com/watch?v=clFq7xwxV-Q

It is typical to use animated characters to create empathy in promotion videos. As described previously, *Pokmon* characters are used to promote various commercial products for kids. Additionally, the *Japan Racing Association* uses characters and giant humanoids in *Evangelion*, which is a very popular animated movie that is liked by many young Japanese adults[11]. The purpose of the promotion is to promote horse racing to young adults.

2.2   Analyzing Promotions that Use Original Animated Stories

Animated stories are preferred across generations in Japan, but each animation covers only a specific target generation. Thus, an animation that is of interest to a specific generation can be used to appeal to a different generation who may not have a strong interest in the target products. This section analyzes three promotional movies that use the original animated stories discussed in the workshop. The movies are both successful and unsuccessful at promoting products and company brands. We consider how the movies fit the design patterns explained in the previous subsection.

The first movie is *Toyotas Peace Eco Smile*[12], which promotes its brand using an original animated story. In the movie, a young male person who comes from outer space tries to learn about products and rules in our world. In particular, the story uses a love story to explain that the technologies developed by *Toyota* are very eco-friendly. However, the characters in the story are not empathetic enough because the characters are very new to most people and the story is not long enough to allow viewers to develop empathy for the characters. The story does not give us enough information to understand that *Toyotas* technologies are superior. In addition, the background scenes are not realistic enough, although there are a few impressive visual representations that provide us with metaphors showing *Toyotas* excellence. Thus, the promotion does not fit the proposed design patterns and has not been successfully accepted by most audiences.

The second is a promotional movie named *Mercedes-Benz Next A-Class*[13], which promotes *Mercedes-Benzs new A-Class* cars. In the story, the promoted car is depicted in a near-future world. The speed of the car is nicely shown in the movie, fitting with design pattern 1. Additionally, the movie shows that the car offers very high performance, which fits design pattern 2. In the animated movie, the presentation of the car is very realistic. Additionally, the story is about finding a legendary ramen noodle shop[14]. Finding a nice ramen noodle shop is a very popular activity for young Japanese adults. Thus, the story is particularly realistic for the young adults who are the target users for the *A-Class* cars, which fits design pattern 3. Finally, the movie adopts a character from *Evangelion*, and most young males know and like *Evangelion*. *Evangelion* is one of the most popular animations in Japan, and its characters are well known.

---

[11] https://www.youtube.com/watch?v=toEcz4inet8
[12] http://www.toytoyota.com/pes/
[13] http://next-a-class.com/
[14] Ramen is a Japanese noodle dish, which is of Chinese origin.

The people who like *Evangelion*s characters also like the characters in *Next A-Class*, which fits design pattern 4.

The third promotional movie is for *Subaru*, which is a Japanese automobile company. This promotional movie is named *Wish Upon the Pleiades*[15]. This promotion is very interesting because there are very few connections to Subaru in the movie. The name of the main heroine is *Subaru*, but the movie does not show any cars in the story. However, the characters in the story and the story itself are vividly attractive to many Japanese traditional animation fans. The story is based on a magic girls story, and many scenes in the movie are very typical of a magic girls animated movie. Thus, the movie creates empathy with many animation fans, and the characters in the movie have become very famous in the avid animation communities.

The movie makes the name *Subaru* famous, although the company name may not be popular among young adults. The company has also opened several public festival events using the characters. Many young adults who are interested in the characters visit the events and learn more about the company. The original story follows only design pattern 1 and 4, but design pattern 2 and 3, which are not satisfied in the story, are compensated by the reality offered in real world attractions in festival events. This strategy shows a possibility for using transmedia storytelling for effective promotions in the future.

Original stories need not take into account their consistency with promoted products. Thus, there is a very broad freedom for the promotion; however, it is expensive to create a movie from an original story[16].

3   Workshop 2: Fictional Stories in the Real World

Alternative Reality Games (ARG) are a promising approach for conveying messages to users using multiple channels. Fictional stories are embedded into a game that is played in the real world and that uses multiple channels [18]. The channels offered in the game are used to exploit the games fictional story. For example, in *Perplex City*[17], trading cards are used to introduce the characters and story. Web sites, emails, phone calls, and SMS messages are then cooperatively used to solve riddles in the mystery story. Because the media is tightly integrated with our daily activities, we feel that the fictional story is realistic. For example, social media such as *Twitter* and *Facebook* have become very popular. Thus, fictional stories that are embedded in social media increase the feeling that the stories are occurring in the real world.

Theme parks are entertainment attractions, rides, and other events in a location for the enjoyment of large numbers of people. A theme park is more elaborate than a simple city park or playground, usually providing attractions that are meant to cater specifically to particular age groups, although some are

---

[15] http://sbr-gx.jp/

[16] For example, designing a new attractive character needs great efforts, in particular, for making a short movie.

[17] http://perplexcitywiki.com/wiki/Main_Page

aimed towards all ages. A theme park is a typical way to integrate fictional stories into our activities in the real world. *Disneyland* is a typical and the most famous theme park. Many *Disney* characters appear in *Disneyland*. Each attraction at the park is constructed based on a *Disney* story. Because the stories are very popular with most people, we feel that the character exists in the real world when we are at *Disneyland*, and we feel that we can meet the *Disney* characters and enjoy being with them during our visit [20].

In this workshop, we discussed two case studies that use fictional stories in the real world. The first case study is a promotion for *Meiji fruit gummi candy*. We can communicate with one of characters in the story, and the story changes according to the real-time communication. Thus, we feel that the story is non fictional. The second case study is a recent trend for young people called *Seichi Junrei*. A user of a fictional story visits the place that the story is based on and expands the story by himself/herself. This ability increases his/her desire to join into the story.

The promotion movie for *Meiji fruits gummi candy* is *Megumi and Taiyo Tweet Love Story*[18]. In the story, a heroine *Megumi* eats a grape gummi candy when she needs to think deeply. When using a fictional story, it is not easy to know how the audiences feel the reality of the story through only its video movie. *Tweet Love Story* uses a social media service, *Twitter*, to make us feel the reality of the story. The audience can talk with the storys hero *Taiyo* via *Twitter*. *Taiyo* answers us when we give him advice on how to get closer to *Megumi*. *Megumi* and *Taiyo* also talk with each other on *Twitter* so that everyone can see their conversation. The audiences advice has a strong impact on the conversation between *Megumi* and *Taiyo*; therefore, the story changes its ending according to our advice to *Taiyo*, which makes the story feel non-fictional in our real world.

Taiyo works at a vineyard, and there are scenes showing healthy and delicious grapes. These scenes give information showing the products excellence. Finally, the story uses a character designer whose characters are currently very popular in many media. Thus, the audience easily feels empathy for the characters even though the story and its characters are original.

The case study shows that communication with virtual characters in a fictional story increases the users belief in the reality of the story. In particular, a recent trend in social media makes it easy to realize such communication because the communication can be automatically generated by computers. It is possible to realize this approach on *Facebook*. A *Facebook* page for a virtual character is a promising tool to allow interaction between us and a virtual character. If we become friends with the virtual character on Facebook, we can receive his/her feeds, and we can feel that the communication is realistic. Thus, we believe that the character exists in the real world.

Seichi Junrei is a typical geek culture activity in Japan, particularly related to Japanese animation, manga(comic) and games, in which people visit famous locations from animation, manga and games. "Seichi" means "Holy Land", "Junrei" means "Pilgrimage". Anime fans arrive at a location and take pictures with

---

[18] http://www.meiji.co.jp/sweets/candy_gum/fruits_gummi/part1/

the same screen/angle of the animation and upload them to their blogs. The most important aspect of Seichi Junrei is that something is brought from the fictional story to the real world. The fans create new stories using these pictures and the virtual characters from the fictional stories and co-construct the stories to share them within their community. This phenomenon is a very interesting example of harmonizing the real world and the fictional world. We believe that interactive pervasive games or social information services based on fictional stories are promising tools for increasing the reality of the fictional world, and the tools enhance the *Seichi Junrei* phenomena by realizing a tighter integration between the fictional and the real world. The experiences described in this paper will offer useful insights to help design tools that will realize new types of transmedia storytelling.

Recent animated movies use many scenes from the real world to increase the reality of fictional movies. A user of *Seichi Junrei* tries to find the scene that appeared in the movie in the real world and takes a photo of the scene. The user enjoys finding these scenes and visiting the location of the scenes. Then, he/she takes a photo, similar to a scavenger hunt. The photos are usually uploaded to the Web, and many other people enjoy looking at the photos.

In Japan, in particular, animated movies have recently been used to promote the local districts that are used in stories. Some districts plan to attract people who like the story by offering extra original new stories to those who visit. This strategy suggests that there is possibility to promote the districts with fictional stories. If real products that are famous in a district appear in the stories, then it will increase the feeling that the stories are realistic especially if the stories audiences are familiar to the products. If the stories are attractive, they will also promote the products to people who do not live in the district.

4   Design Implications and Current Status

In this section, we present how the discussions from the previous sections are related to the design of future *virtual forms*. We discussed two issues. The first issue is how to increase the reality of fictional stories. The second issue is how to seamlessly embed fictional stories in the real world. Designing future *virtual forms* to integrate fictional stories into the real world through transmedia storytelling is similar to the design of games embedded in the real world. The approach has been investigated within pervasive games [8,11], but *virtual forms* enable fictional stories to be more seamlessly integrated into the real world by using information technologies.

A pervasive game is designed with virtual and tangible objects. For example, *Pac-Man and Ghosts* consist of human players in the real world experiencing a computer graphics generated fantasy-reality by using wearable computers [8]. The basic concepts from pervasive games can be useful in realizing our goal. As shown in [2,4], it is essential to use tangible objects to increase human performance in the real world rather than using virtual objects. Our approach to enhancing a traditional pervasive game is to use transmedia storytelling. A fictional story is fragmented into multiple media, and they are integrated with our daily activities. We consider using the following three technologies for the integration.

The first technology is a *smart artifact* [5,6]. The key technical issue for developing *smart artifacts* is the ability to immersively sense our daily environment by using multiple *smart artifacts* [7]. The *smart artifacts* are deployed everywhere in our environment and connected by a network. The information sensed by them can be used to develop context-aware services that offer personalized and location-aware information to users. It is also important to maintain the functionality of existing artifacts so that a user feels their reality. In [14], we suggested that using virtual artifacts significantly reduces the feeling of reality associated with the existing artifacts such as paper-based trading cards. If a user is involved in a fictional story that uses tangible physical artifacts, his/her belief in the reality of the story will increase. In particular, if a typical artifact that is used in the story appears in the real world, and the user is reminded of the story by using the artifacts, the artifact is a promising source of media for enjoying the story. For example, in *live action role playing*(LARP) [11], a player plays a specific role from a fictional story. He/she wears clothes to represent the role and brings physical artifacts that the person in the role typically uses. *Smart artifacts enable us to influence the story when we wear the clothes and use the artifacts. This influence increases the feeling that we are involved in the fictional story.*

The second technology is a *markerless augmented reality* that superimposes some information on the real world. In [19], we use the technology to enhance existing artifacts to help the users decision making. To realize natural augmented reality, one important technology would provide the ability to sense the environment because adding markers to existing artifacts loses the natural integration between the real and the virtual world. Additionally, the natural integration of projected information into the real world is an essential technical issue. Our experience shows that the projected information should be a part of the existing artifact so that the user does not lose their belief in the reality of the existing artifacts. In addition, showing virtual characters is also useful because it reminds the user of the fictional story by projecting the characters into the users daily life. The reality is significantly increased if the characters movement is synchronized with the movement of a real person in the real world. However, as described in [14], the movement should be consistent with the typical features of the characters so that the reality of the character is not lost.

The third technology is a *persuasive ambient mirror* [12] that reflects the users current behavior into digital media. To realize *persuasive ambient mirrors*, the important technologies will provide the ability to sense human behavior. Another critical technical issue is the natural blending of the information that reflects the users current situation with the digital media. If the users behavior is reflected into a fictional story that is represented in the real world, the feeling of reality is significantly increased. In Section 3, we showed that communication with a virtual character via social media increases the belief in the reality of a fictional story. The *persuasive ambient mirror* is an opposite technology that allows the user to be involved in the story. However, as described in [12], the user needs to be aware that his/her behavior has an influence on the virtual world in the fictional story.

The above discussions show that these technologies are useful for designing *virtual forms* that integrate respective media in transmedia storytelling in the real world. We are currently designing a system called *micro-crowdfunding* [16]. In the system, a person designated as the mission organizer proposes a new mission to sustain our real world, and other persons, designated as mission investigators, offer funding to achieve the mission. If the amount of total funds exceeds the threshold determined by the mission organizer, the mission can be achieved by the person designated the mission performer. To encourage participants, we are extending the original system to incorporate a fictional story. Some of the missions will achieve sustainability in the real world, as described in [16]. A virtual character from a fictional story will appear in the real world through the use of augmented reality technology. Other missions for achieving sustainability in the virtual world will appear in the fictional story. A participants activity in the real world can be reflected in the virtual world through the use of a persuasive ambient mirror. In addition, the participants will be able to manipulate the virtual world by using smart artifacts within missions through the use of tangible interaction technologies. These missions represent typical scenes in the fictional story and the participant can use the typical artifacts that appear in the scenes.

This approach enhances three aspects of the original system. The first aspect is that a virtual character offers guidance on how to achieve a mission. Empathy for the character encourages the participants to join the mission. The second aspect is that a fictional story conveys an ideological message to the participants. Thus, their desire to achieve the mission is increased. The final aspect is that a mission to solve a real social problem increases the feeling of challenge. Visualizing the current situation by using *persuasive ambient mirrors* will be useful for encouraging participation.

The design patterns presented in Section 2 provide tools to increase the belief in the reality of a fictional world. For example, using a landscape from a persons current location in the real world can increase the belief in the reality of the fictional story. It makes the boundary of the fact and fiction blurry, in particular, if a story in which a fictional role is played is based on the same place when the person actually lives in. Similarly, if the fictional story uses a lot of recently popular accessories and fashions in the real world, his/her belief in its reality is also increased. Popular habitual behavior in specific generation is also an important factor to increase the belief in the reality. As described in Section 2, attractive vivid expressions can be effective for increase the belief in the reality of the fictional items and characters.

As described in Section 3, if the story is leaked into the real world, the participant feels that the fictional story is more realistic, and the positivity of the story increases the users self-efficacy and his/her ability to solve the problem in the real world. The advance of social media like *Facebook* and *Twitter* enables us to play a fictional role on the social media easily. Also, a festival event held in the real location used in a fictional story makes the fictional roles more realistic. If people play their fictional roles in the real event, the belief in the reality is significantly increased.

The most important insight that we found from our research is that we can feel the reality of a fictional story from even a small cue in the real world. This paper shows several examples of cues that were found in the case studies discussed in the workshops. If the cue is shown in a fictional story, then we feel that the story is more realistic. We still need to investigate the issue as to how we feel reality from even a small cue that represents a part of the real world. Our results are a first step towards building a more rigorous framework to design reality when a fictional story is integrated in the real world. This topic is very important for developing a more attractive enhancement of the real world.

5   Conclusion

Fictional stories can enhance the attractiveness of *virtual forms*, but good design guidelines do not exist to design richer user experiences. In this paper, we reported some insights extracted from two workshops conducted to discuss the integration of fictional stories into the real world to allow for the design of richer user experiences. The insights from these workshops can be used to design future *virtual forms* that embed fictional stories through transmedia storytelling. In the next step, we will design new case studies of *virtual forms* from experiences with the insights described in this paper. The insights are also useful for designing future media for ubiquitous computing environments.

Mixing multiple existing stories described in [17] is useful to expand our approach. For example, *MacDonald* sells *Happy Meals*, which include character goods from various stories. Currently, there is no interaction among the characters in different stories. However, it is impossible to develop a story that everyone likes. The approach offers a promising way to enhance the consumers experience by consistently mixing stories with multiple characters.

References

1. Baudrillard, J.: The Consumer Society: Myths and Structures. Sage Pub. Ltd. (1998)
2. Cuendet, S., Bumbacher, E., Dillenbourg, P.: Tangible vs. Virtual Representations: when Tangibles Benefit the Training of Spatial Skills. In: Proceedings of NordiCHI 2012 (2012)
3. Dena, C.: Transmedia Practice: Theorising the Practice of Expressing a Fictional World across Distinct Media and Environments. Dissertation Thesis, University Sydney (2009)
4. Esteves, A., Hoven, E., Oakley, I.: Physical Games or Digital Games? Comparing Support for Mental Projection in Tangible and Virtual Representations of a Problem Solving Task. In: Proceedings of the 7th International Conference on Tangible, Embedded and Embodied Interaction (2013)

5. Fujinami, K., Kawsar, F., Nakajima, T.: AwareMirror: A Personalized Display using a Mirror. In: Gellersen, H.-W., Want, R., Schmidt, A. (eds.) PERVASIVE 2005. LNCS, vol. 3468, pp. 315–332. Springer, Heidelberg (2005)
6. Hazenberg, W., Huisman, M.: Meta Products: Buidling the Internet of Things. BIS Publisher (2011)
7. Kawsar, F., Fujinami, K., Nakajima, T.: Deploy Spontaneously: Supporting End-Users in Building and Enhancing a Smart Home. In: The Tenth International Conference on Ubiquitous Computing (Ubicomp 2008) (2008)
8. Magerkurth, C., Cheok, A.D., Mandryk, R.L., Nilsen, T.: Pervasive Games: Bringing Computer Entertainment Back to the Real World. ACM Computer in Entertainment 3(3) (2005)
9. Mattila, A.S.: The Role of Narratives in the Advertising of Experiential Services. Journal of Service Research, 35–45 (2000)
10. Marzano, S., Aarts, E.: The New Everyday View on Ambient Intelligence. 010 Publisher, Rotterdam (2003)
11. Montola, M., Stemros, J., Waern, A.: Pervasive Games - Theory and Design. Morgan Kaufmann (2009)
12. Nakajima, T., Lehdonvirta, V.: Designing Motivation in Persuasive Ambient Mirrors. Personal and Ubiquitous Computing 17(1), 107–126 (2013)
13. Sakamoto, M., Nakajima, T., Alexandrova, T.: Digital-Physical Hybrid Design: Harmonizing the Real World and the Virtual World. In: Proceedings of the 7th International Conference and Workshop on the Design & Semantics of Form & Movement (DesForm 2012) (2012)
14. Sakamoto, M., Alexandrova, T., Nakajima, T.: Analyzing the Effects of Virtualizing and Augmenting Trading Card Game based on the Players Personality. In: Proceedings of The Sixth International Conference on Advances in Computer-Human Interactions (2013)
15. Sakamoto, M., Alexandrova, T., Nakajima, T.: Augmenting Remote Trading Card Play with Virtual Characters used in Animation and Game Stories Towards Persuasive and Ambient Transmedia Storytelling. In: Proceedings of the 6th International Conference on Advances in Computer-Human Interactions (2013)
16. Sakamoto, M., Nakajima, T., Liu, Y., Alexandrova, T.: Achieving Sustainable Society through Micro-level Crowdfunding. In: Proceedings of CHI 2013 Extended Abstracts in Human Factors in Computer Systems (2013)
17. Sakamoto, M., Nakajima, T.: An Analysis of Composing Multiple Fictional Stories and Its Future Possibility. In: Proceedings of 15th International Conference on Human Computer Interaction (2013)
18. Szulborski, D.: This Is Not a Game: A Guide to Alternate Reality Gaming. Lulu. Com. (2005)
19. Yamabe, T., Nakajima, T.: Playful Training with Augmented Reality Games: Case Studies towards Reality-Oriented System Design. Multimedia Tools and Applications 62(1), 259–286 (2013)
20. The Imagineers: Walt Disney Imagineering. Disney Book Group (2010)

Task Context Modeling for User Assist
in Organizational Work

Masashi Katsumata

Department of Computer and Information Engineering, Nippon Institute of Technology, Japan
katumata@nit.ac.jp

Abstract. E-mail-based communication and collaboration are important to organizational work. In order to help the multi-tasking knowledge worker, a task management-based software environment requires a support mechanism of an automated user operation in addition to a support function that manages task resources. In this paper, we propose a task context model that manages task-related e-mail messages and their resources for the purpose of reusing them. In addition, we describe a task context model-based user assist functions that allows users to send or reply to e-mail quickly and efficiently and that automatically extracts data from e-mail messages. To validate the task context model, we implement the prototype system and describe its experimental results.

Keywords: User Assist, Task Model, Organizational Work, Ontology.

1 Introduction

E-mail based communication and collaboration are important for organizational work. Therefore, e-mail communication among group members for a given task is required to achieve effective task management for reusing an e-mail message and its related resources (schedule, attached file, contact list, etc.). Office workers are able to efficiently search and use e-mail messages and their related resources by organizing them according to individual tasks [1], [14]. For this reason, multi-tasking office workers often set up automatic filtering into folders and manually move e-mail messages into folders. Recently, enhanced existing task management systems and task-centric mail clients have been used for this purpose, and several research studies have performed that support the discovery of e-mail messages and related resources by adding metadata to e-mail messages and their related resources [2-5].

In this paper, we propose a task context model that manages task-related e-mail messages and their resources for the purpose of reusing them. In addition, we describe a task context model based prototype system with functions that allows users to send or reply to e-mail quickly and efficiently and that automatically extracts data from e-mail messages. In order to realize the concept of the task context model, we build an ontology-based semantic representation model of associations between task management and task-related e-mail processes. Using the ontology-based task context model, the prototype system can realize the user assist function for task users by sharing data

T. Yoshida et al. (Eds.): AMT 2013, LNCS 8210, pp. 278–289, 2013.
© Springer International Publishing Switzerland 2013

between task management and task-related e-mail processes. We also describe the effectiveness of task context model from experimental results.

This paper is organized as follows. Section 2 discusses the task context model and user assist. In Section 3, we address the system implementation based on the task context model. In Section 4, we describe the validation of the system based on the task context model. In Section 5, we review related work. In Section 6, we provide conclusions and future research directions.

2   Task Context Model and User Assist Functions

2.1   Task Context Model for User Assist

In this study, we create an ontology-based semantic representation model that represents the associations between the task management and e-mail processes. We call this task context model. In this model, e-mail messages for a task in an organization are handled as a task unit, and resources related to the e-mail are managed under the same task unit. The e-mail messages and their related resources for the task unit are called task context data. The created files and schedule data for the task are also handled as task context data. In addition, data extracted from e-mail messages by automatic retrieval function are also handled as task context data. The automatic retrieval function is considered later in this paper. These task context data are utilized in the user assist function to accomplish a task.

Semantic representation of task context model is represented by Resource Description Framework (RDF) [6]. RDF is a collection of triples, each of which consists of a resource, a property, and a literal. A set of such triples is called an RDF graph (see Fig. 1). The left part of Fig. 1 shows the conceptual model that represents associations between a task and its related attribution information for task context management; task context data are managed under this model.

The concept of task context management in the task context model represents task related data that include frequently used information contained in an e-mail communication for organizational work. Task context is represented as a property of a "Task" resource. We define the conceptual associations of the task context's attribution as follows: Task is "Subject," task context is "Predicate," and the value of the task context denotes "Literal."

In the task context model, the type of task context is defined as a file, schedule, participant, memo, etc. A file can be any file (e.g., Microsoft Word document) on a user's personal computer or a URL on the Internet. The Subject value of the File property is the path of the existing file. In an organization, data in a reply e-mail are collected frequently, and thus, we also define the concept of frequently used exchanged data in e-mail communication for the purpose of the task work. We call this concept an "Action." There are three types of Action: Event Notification, Questionnaire Request, and File Collection. An Action resource is defined as the hierarchies of resources. The conceptual model of the mail process represents the associations of an e-mail message, its related data, and mail form for Action. E-mail messages are represented as resources.

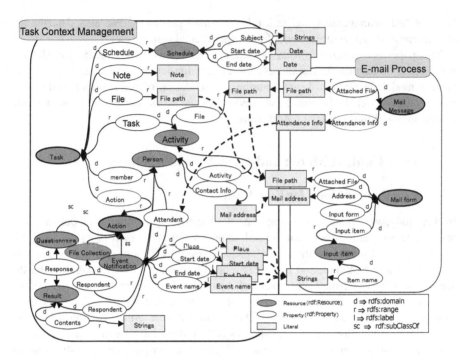

Fig. 1. Example of the task context model

Attached file and form data contained in the message body are represented as properties. The value of property is mapped to the value of property in the task context for linking two conceptual models (task context management and e-mail process). On the other hand, mail forms are represented as resources, and attached files and mail addresses are represented as properties. The task context model shown in Fig. 1 is created for the prototype system described in Section 3.

Task Context Management. The task context model manages the task context's property and its value under each task. The property of the task context represents a task-related file, member, contact information, and schedule data, as shown in Fig. 1. The value of the task context is automatically retrieved from an e-mail message on the basis of the task context model when the e-mail arrives on the mail server implemented in this study (see Fig. 2). Figure 3 shows how task context is assigned to property in the task context model. When the task owner registers the task for group work in the client, RDF/XML format data based on the task context model are automatically generated and managed by the task context server. The upper part of Fig. 3 shows the RDF model of the task context model, and the lower part of Fig. 3 shows task context data that are generated according to the task context model. Figure 3 shows how this model can concurrently handle multiple tasks such as "meeting" and "lab party."

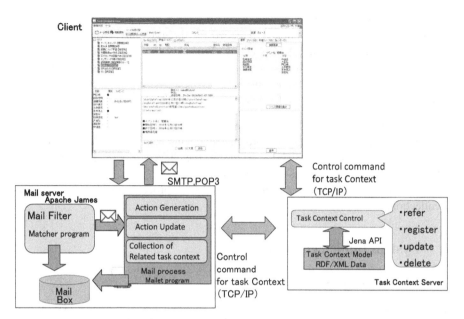

Fig. 2. Configuration of the prototype system

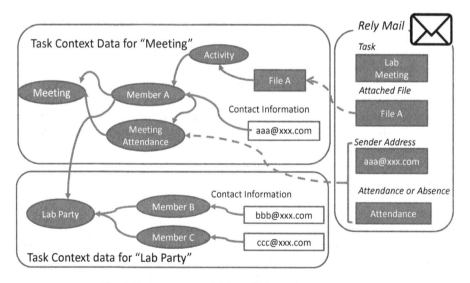

Fig. 3. Task context model-based information management

E-mail Process. The aims of the conceptual model of e-mail process are to support 1) creation of e-mail forms and 2) collection of the data contained in reply e-mails. The support for creating e-mail forms is performed when a task owner creates an e-mail form for a task request and a task member creates an e-mail form for replying to a task request. On the other hand, the support for collection of data contained in reply e-mail is performed when a task owner receives a reply mail from a task member. In this study, the user assist function is intended for three actions: Event Notification, Questionnaire Request, and File Collection. We considered that these actions frequently occur in e-mail collaboration in an organization. The user assist function in e-mail processes provides service that supports both creation of e-mail forms and organization and retrieval of data from reply e-mail for these mail actions.

1. Event Notification. This action represents the concept of a mail process for the target of attendance confirmation toward an event being held in an organization.
2. Questionnaire Request. This action represents the concept of a mail process for the target of questionnaire request and collection.
3. File Collection. This action represents the mail process for the attached file collection.

3   System Implementation

3.1   System Overview

In order to validate the task context model, we implemented a prototype system that performs the user assist functions. Prototype system is composed of the following three systems: task context server, mail server, and mail client (see Fig. 2). The task context server manages task context and its value, which is generated according to the RDF/XML data format.

In the task context server, task context (file path, schedule, contact information, etc.) is managed under each task. We implement a task context server that can connect to the mail server and the mail client. The task context server can accept a request command (create, refer, update, and delete) for the task context model from the mail server and client via TCP/IP. By accepting a request command, the task context server can perform the model according to the RDF/XML format by the Jena application programming interface [7].

In this study, the mail server in our prototype system is built on Apache James [8]. In addition to general mail transport and storage functions, Apache James is a mail application platform that enables users to program the code of custom applications for e-mail processing. Apache James provides e-mail filtering in a function called Matcher and provides e-mail processing in a function called Mailet.

In this paper, we introduce extended e-mail headers for realizing the user assist functions (see Table 1). After Matcher refers to the extended e-mail headers, Mailet can perform according to the purpose of extended e-mail headers.

The client can connect with both the mail server and the task context server. In addition to general e-mail operations, the client provides the user interface that manages task context data. E-mail message submitted by client is automatically added to the extended e-mail header. In our prototype system, the client displays the structured mail form by referring to the extended e-mail header. This client can also

Table 1. Types of extednded mail headers

Header name	Role
X-Task-Name	Indicates the name of task
X-Task-Owner	Indicates the task owner
X-Action-Model-Type	Indicates the request of generating Action and the type of Action
X-Action-Update-Type	Indicates the request of updating the status of Action and the type of Action
X-Action-Retrieve-Type	Indiscates the request of retrieving task context data and the type of Action

receive other e-mail messages according to the RFC2822 [9]. This client's user interface is composed of six main areas (task member, file, schedule, action, message, and form).

3.2   E-mail Process and Task Context

The aim of the user assist function is to provide a platform that is useful for managing and operating the task-related information in e-mail processes. In this section, we describe the role of the task context in the e-mail process.

E-mail Form Creation Support. When a task owner selects the type of Action on the client, an e-mail form for the selected Action is displayed. In the e-mail form for the Action, task-related data are provided as a list of suggestions of possible input. As a result, the task owner spends less time typing and querying for information related to the task. When the task client receives the e-mail from the task owner, the replying form is displayed by referring to the extended e-mail header X-Action-Model-Type. Displayed input fields on the replying form are the elements of Action type corresponding to "replying form creation" as shown in Table 2. The task member can type the value according to the displayed input field on the replying form.

Table 2. Support functions corresponding to each action type

Function	Event Notification	Questionnarie Request	File Collection
Request mail form composition	Schedule form (Start date, End date, Place, Event name)	Questionnaire form (Support of form composition)	File name and stored folder name
Reply mail form composition	Attendace form(Yes, No)	Questionnaire form (reply form)	Selection of attached file
Automated extraction	Attendance data	Questionnarie response data	Attached file

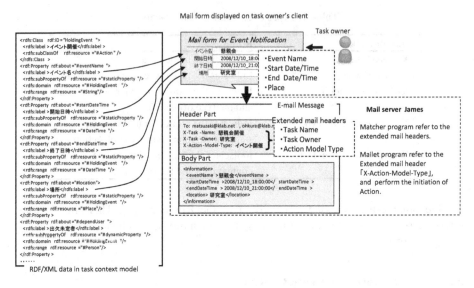

Fig. 4. Example of the generation process of Action (Event Notification)

Automated Process for Reply E-mail. When the mail server receives an e-mail according to the type of Action, the contents of the reply e-mail are automatically retrieved as task context. In the task context server, retrieved task context data are managed as RDF/XML format data that are based on the conceptual model for Action. When the task context server receives a reply e-mail, the value of task context in Action is updated, and the state of Action is displayed on the client's State panel. In this way, the task owner can confirm the state of the task intuitively without checking each reply e-mail. Automated processes such as generation or update of Action are performed via Mailet according to the value of the extended mail header. In Fig. 4, the generation process of Action instance is represented. When mail server received the e-mail from task owner, the Matcher refers the extended e-mail header, and request command for generation of Action to Mailet. Mailet generates task context model according to given Action on the task context server. When the mail server receives the reply e-mail message from the task member, Matcher program refers the extended e-mail header X-Action-Update-Type, and the Mailet updates the state of Action (see Fig. 5). Mailet updates the value of state by referring the text element of XML format contained in the body part of the e-mail message. The above procedure is also followed for reply e-mails from other task members and updates to the value of the attendance for Action. The above procedure is also followed for reply e-mails from other task members and updates to the value of the attendance for Action. Attendance data and questionnaire data can be written to a Comma Separated Value (CSV) file on the assumption that task context data might be used by a spreadsheet application (e.g., Microsoft Excel). This function provides the task owner with a useful environment that allows him to easily shift the work with collected data on mail

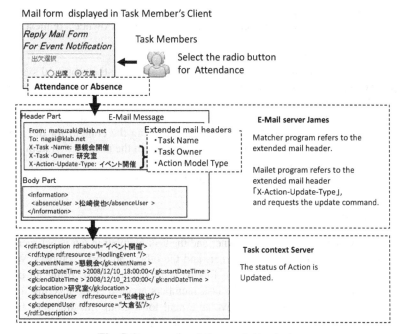

Fig. 5. Process of updating an Action instance

server from task-based mail operation environment. In the case of File Collection, files are automatically renamed according to a predefined file name from the attached file in the task member's reply e-mail.

4 Experiment

4.1 Experiment Objective and Conditions

We report qualitative data compiled from 26 university students (male, 21 – 22 years old) who used the prototype system for two months. The prototype system was set up in our laboratory. The work for experiment was held by two ways: using an existing e-mail client and using our pro-posed e-mail client. In this experiment, we made the following several assumptions:

1. The case of an existing e-mail system
   (a) Task owner's usage condition and work
       (i) *The e-mail addresses of the task owner and task member are registered in the e-mail client.*
       (ii) The date of the group meeting is provided to the task member.
       (iii) Indication of attendance confirmation, questionnaire request, and deadline of the research report are handled by e-mail, and the task owner checks the reply e-mail.

(b) Task member's usage condition and work
    (i) The replying e-mail is created according to the task owner's request e-mail.
2. The case of the proposed system
    (a) Task owner's usage condition and work
        (i) The group member's e-mail address is registered in the proposed system.
        (ii) The schedule date of the group meeting is registered in the scheduler of the client system.
        (iii) Indication of attendance confirmation, questionnaire request, and deadline of the report are handled by e-mail. Collecting the content for the reply e-mail is performed by the user assist function in the proposed system.
    (b) Task member's usage condition and work
        (i) The replying e-mail is created according to the task owner's request e-mail.

4.2   Experimentation Overview

Here, we describe the role of the task owner and the task member in our experiment. One of the 26 people is elected as a task owner, and the others are task members. We performed the experiment according to the following procedure. First, using an existing e-mail client, the task owner notifies the task members about a group meeting for a research report. When task members receive an e-mail for a group meeting, they reply to the e-mail for attendance of meeting to task owner. After the group meeting, the task owner requests from the task members the completed questionnaire and the meeting report for a research presentation. The task owner checks and confirms the reply e-mails from the task members and collects the content data of the reply e-mails. Next, in this work process, a similar experiment was performed in an environment using our proposed system. In the similar way, the experimentation was conducted changing the person assigned to the role of task owner. In our experiment, this work procedure (using the existing e-mail system and using our proposed system) was performed four times two months.

4.3   Experimental Result and Discussions

We investigated the results of the questionnaire about the prototype system's superiority compared with the existing e-mail client. Table 3 shows the results of the questionnaire for the following user assist functions: Event Notification, Questionnaire Request, File Collection and Task context management.

Table 3. The results of the questionnaire on usage of the prototype system (Excellent = 5, Good = 4, Fair = 3, Poor = 2, Very Poor = 1)

User Assit Functions	Task Owner		Task Member
	Request mail form composition	Collection of Reply mail	Reply mail composition
Event Notification	4.0	5.0	4.7
Questionnarie Request	3.7	5.0	4.0
File Collection	3.7	3.5	4.0
Task context management	4.5		3.7

Consideration of user assist function toward the Actions can be described as follows:

1. **Event Notification.** The task owner rates 4.0 and 5.0 on average for request mail form composition and confirmation of attendance for the reply e-mail, respectively. We believe that this assessment of the user assist function for the task owner was given a high valuation because of the automated function for the task owner. In confirmation of attendance, the following opinion was obtained: "It is good because of possibility to check the state of the task on the client". On the other hand, task member rates 4.7 on average for the reply e-mail composition. We think that the assessment of the work necessary for the reply e-mail attained a high valuation.

2. **Questionnaire Request.** The work of the task owner and task members in this Action is as follows: The task owner creates a questionnaire form. After submitting the questionnaire e-mail to the task members, the task owner collects the reply e-mails for the questionnaire. The task owner rates 3.7 and 5.0 on average for questionnaire mail form composition and collection of the reply e-mail for the questionnaire, respectively. A comment on how to improve the system from the task owner's perspective is as follows: "Cannot change the assessment value for the questionnaire item." On the other hand, the task member rates 4.0 on average for reply mail form composition. We found that a customized form provided by performing assist function was highly evaluated.

3. **File Collection.** The work of the task owner in the experiment is to request the submission of a report for the presentation and to collect the file attachments from the task members. The task owner rates 3.7 and 3.5 on average for request mail composition for file collection and collecting the file attachments of reply e-mails, respectively. We think that this low rating is not expected to result due to not enough participants in experimentation. On the other hand, the task member rates 4.0 for reply mail composition. Although superiority for the proposed system was highly evaluated, the following opinion was put forth: "It is difficult to judge superiority for the pro-posed system toward adding the file to the task without drag and drop on the client." We think that attaching the file using the proposed system is comparable to performing that operation with the existing e-mail client. In fact, we created the function with consideration for the work necessary to continuously accomplish the task, because file management is too burdensome for searching for and saving file attachments.

4. **Task Context Management.** Handling of task context data for task owner and task member was verified. Task owner's rating was higher than task member's rating. We found that user interface's view for task owner was effective for organizing task related information. Moreover, we also found that burden operation for organizing task related information decreased by performing user assist function. Therefore, task member rating was not highly evaluated compared with existing e-mail client.

5   Related Works

In using e-mail for organizational work, a vast amount of task-related information has to be handled. Therefore, research on supporting resource management of each task has been studied.

TaskMaster [2] enhances an e-mail client to function as a task management system and manages the resources as e-mail messages and file attachments for each task. In addition, a useful user interface with both browsing and operating resources is also provided. This system makes it easy to search through the resources of a task. TV-ACTA [3] provides prestructured containers that are created inside the e-mail folder hierarchy to support personal information management. Specialized subfolders called "Components" within each ACTA Activity automatically organize and present information appropriately for aspects of the activity at hand. ACTA is designed to create a more efficient personal information management environment with the ultimate goal of providing context metadata for machine learning and automation techniques.

KASIMIR [10] and OntoPIM [11] are ontology-based personal task-management systems. These systems provide semi-automated functions for retrieving and registering the task-related information within e-mail messages according to an ontology-based model. Activity Explorer [4] support knowledge workers with context switching and resource rediscovery by organizing and integrating resources, tools, and people around the computational concept of a work activity. Support functions in their research did not apply to reusing the managed data for the task. In this study, we focused on the associations between the related resources and the e-mail form and aimed toward realizing the support of reusing the managed data for accomplishing the task. In [12], [13], e-mail-related attribution information (group, project, member, etc.) is modeled by ontology, and a system that provides a user interface with visualized and grouped formal concept analysis is reported. By this user interface, searching approach is reported.

6   Conclusion

In this paper, we verified the proposed task context model by implementing a proto-type system based on it. We examined whether the assist function can help a task user by having study participants use the system and then fill out a questionnaire about their experience. The results confirm a high valuation for assist functions such as processing e-mail and managing task resources in e-mail communications for a given task. In addition, we obtained some beneficial comments in order to improve the proposed system for practical use from the point of view of the user. Finally, we need to investigate the influence of organizational work as well as the system's functional aspects.

References

1. Ducheneaut, N., Bellotti, V.: Email as a Habitat: An Exploration of Embedded Personal Information Management. ACM Interactions 8, 30–38 (2001)
2. Bellotti, V., Ducheneaut, N., Howard, M., Smith, I.: Taking Email to Task: The Design and Evaluation of a Task Management Centered Email Tool. In: Proceedings of the SIGCHI Conference on Human Factors in Computing Systems, pp. 345–352. ACM, New York (2003)

3. Bellotti, V., Thornton, J.D., Chin, A., Schiano, D., Good, N.: TV-ACTA: Embedding an Activity-Centered Interface for Task Management in Email. In: Proceedings of the Conference on Email and Anti-Spam (2007)

4. Geyer, W., Muller, M.J., Moore, M.T., Wilcox, E., Cheng, L.T., Brownholtz, B., Hill, C., Millen, D.R.: Activity Explorer: Activity-Centric Collaboration from Research to Product. IBM Syst. J. 45, 713–738 (2006)

5. Cozzi, A., Farrell, S., Lau, T., Smith, B.A., Drews, C., Lin, J., Stachel, B., Moran, T.P.: Activity Management as a Web Service. IBM Syst. J. 45, 695–712 (2006)

6. Candan, K.S., Liu, H., Suvarna, R.: Resource description framework: metadata and its applications. ACM SIGKDD Explorations Newsletter 3(1) (2001)

7. Jena API, http://jena.sourceforge.net

8. James, A.: http://james.apache.org

9. Resnick, P.: Internet Message Format. RFC Editor (2001)

10. Grebner, O., Ong, E., Riss, U.V.: KASIMIR - Work Process Embedded Task Management Leveraging the Semantic Desktop. In: Proceedings of Multikonferenz Wirtshaftsinformatik, Workshop Semantic Web Technology in Business Information Systems, pp. 715–726 (2008)

11. Lepouras, G., Dix, A., Katifori, T., Catarci, T., Habegger, B., Poggi, A., Ioannidis, Y.: OntoPIM: From Personal Information Management to Task Information Management. In: Proceedings of SIGIR Workshop on Personal Information Management, pp. 78–81 (2006)

12. Eklund, P., Cole, R.: Structured Ontology and Information Retrieval for Email Search and Discovery. In: Hacid, M.-S., Raś, Z.W., Zighed, D.A., Kodratoff, Y. (eds.) ISMIS 2002. LNCS (LNAI), vol. 2366, pp. 75–84. Springer, Heidelberg (2002)

13. Brendel, R., Krawczyk, H.: E-mail User Role Identification Using OWL-Based Ontology Approach. In: Proceedings of the 1st International Conference on Information Technology, pp. 18–21 (2008)

14. Krämer, J.-P.: PIM-Mail:Consolidating Task and Email Management. In: Proceedings of the 28th of the International Conference Extended Abstracts on Human Factors in Computing Systems, pp. 4411–4416 (2010)

Selection of Core Technologies from Scientific Document

Myunggwon Hwang[1], Jangwon Gim[1], Do-Heon Jeong[1,*], Jinhyung Kim[1],
Sa-kwang Song[1], Sajjad Mazhar[1,2], Hanmin Jung[1], and Jung-Hoon Park[1]

[1] Korea Institute of Science and Technology Information (KISTI)
245 Daehak-ro, Yuseong-gu, Daejeon (305-806), South Korea
[2] Korean University of Science and Technology (UST)
217 Gajung-ro, Yuseong-gu, Daejeon (305-350), South Korea
{mgh,jangwon,heon,jinhyung,esmallj,
ms,jhm,jhpark}@kisti.re.kr

Abstract. Extraction and management of technical terminologies become an
important process in the business intelligence. To do this, historic methods have
a focus on calculating weight values and selecting top n terminologies according to the values for the cores that represent given scientific documents. These
terminologies selected through those methods can be used as important clues
for business intelligence services such as technology trend analysis, potential
market discover, and so on however the terminologies extracted from the documents do not mean the technologies of the organizations publishing the documents. Therefore, our research is based on a fundamental that there are only a
few technologies an organization participates in directly even though a scientific document of the organization contains various technical terminologies. In this
paper, to enhance the quality of business intelligence services, we propose a
method to select core technologies of an organization and utilize semantic networks of technical terminologies of a given scientific document and we suggest
its possibility through simple experimental evaluation.

Keywords: core technology, technology ontology, elementary technology,
scientific document.

1 Introduction

Nowadays, business intelligence (BI) is getting important [1] and text mining from
scientific documents becomes a key process to extract valuable information and to
suggest insightful words. The scientific documents contain papers, patents, and web
articles mainly and each one deals with many technical terminologies in the document. Each terminology is meaningful itself and its value is determined by interrela-
tions with other terminologies in the same context. To calculate the value, the existing
works are mainly depending on TF-IDF which is a representative method based on
statistics or shallow semantics [2] based on simple ontologies. These research works
have a theoretical background that all terminologies are basically important while

---

* Corresponding author.

T. Yoshida et al. (Eds.): AMT 2013, LNCS 8210, pp. 290–295, 2013.
© Springer International Publishing Switzerland 2013

have different values in a given document. This background is useful for the general search aspect which returns not exact documents but related documents by calculating values of queries inputted by users as well as the calculated value is applied to BI aspect to figure out technology trend [3].

Organizations create various types of scientific document for some objectives such as property protections, technology/product promotions, user feedbacks and so on. This means that the technical terminologies described in the documents are connected to the organizations, however, when we consider BI aspect deeply, the all terminologies are not directly participated in (or similar activities such as manufactured, patented, sold, and so on) by the organizations even though the documents contain many technical terminologies. For example, let us assume that one organization is specialized for a technology "LED Display" and they have an intellectual property right such as a patent type. And the patent document is containing "LED Display applicable to smart phone, navigation system, and any other portable devices." In this case, the historic method may extract all terminologies (four technologies in this example, LED display, smart phone, navigation system, and portable device) and give different weight values. However, we cannot say the organization makes smart phone, navigation system, and portable device except led device. In the example, if we use the terminologies simply extracted from scientific documents to compare relative levels of organizations for competence analysis and technology transfer analysis, this may return a list including many organizations unrelated to the technology "LED Display."

Therefore this research work suggests a method to select which technologies are actually related to the organizations which own the right of publications. To do this, we employ KISTI (Korea Institute of Science and Technology Information) ontology, which was automatically constructed by text mining and Semantic Web techniques, to analyze various relation patterns discovered among technical terminologies. And the method contained in this paper is a tentative version to check the possibility of the main idea and we describe examples and results of using only elementary relation of technologies.

This paper is organized as follows: Section 2 describes the technology ontology utilized for analyzing semantic network of technologies. In section 3, we propose our method for core technologies and evaluate the method. Finally, we summarize our research with a few suggestions in the fourth section.

2   Technology Ontology of KISTI

We will select core technologies by using semantic network that can be constructed among technologies extracted from scientific documents. In constructing the semantic network, we use technology ontology [4, 5] and Figure 1 shows its relation schemes of objects.

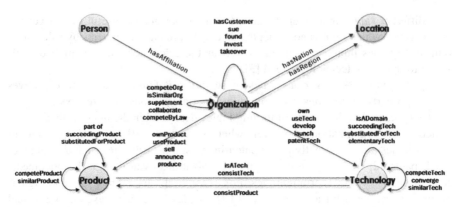

Fig. 1. Relation architecture of technology ontology (excerpted and redrawn from [4] and [5])

To fill the ontology as shown in the figure 1 with instances, they analyzed various scientific documents with high reliability such as papers, patents, Web articles, Wikipedia, and so on collected from 2001 to 2012 and have constructed the ontology with about 498millions triples. Table 1 shows defined objects and its statistics.

Table 1. Statistics of Technology Ontology

Object categories	Statistics
Technologies	43,201,941
Products	62,327,156
Persons	25,110,360
Organizations	40,993,708
Locations	50,350,884

Among the objects and the relation types in the table 1 and figure 1 respectively, this work uses the technology category and the relation "elementTech" which means elementary technology.

3   Selection of Core Technologies and Evaluation

3.1   Selection of Core Technologies

As shown in figure 1, there are total 7 relation types among technologies and we use only *elementaryTech* relation ($^{\rightarrow E}$). If a sentence "Smart phone includes touch device" is found in a document, the triple analyzer makes "smart phone $^{\rightarrow E}$ touch device" triple from the sentence. Our work starts from an assumption that core technologies can be selected through analyzing a few patterns which consist of various relations and we estimate its possibility in this paper. To select core technologies, this work defines next two characteristics.

1. Concentration: If there are core technologies among terminology list collected from a scientific document, the cores have a role as a bridge to connect other terminologies.
2. Relation direction: Even though some terminologies have a role of the bridge, its concentration can be delivered to another one according to relation direction.

Figure 2 shows an example which depicts these characteristics and a part of semantic network using terminologies extracted from a patent document[1].

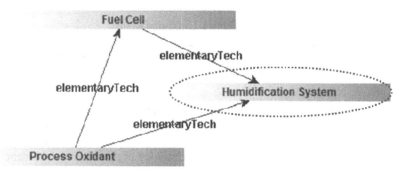

Fig. 2. An example of visualizing[2] technical terminologies

In figure 2, we can figure out that "humidification system" is related to "fuel cell" and "process oxidant" with $\xrightarrow{E}$ as well as "fuel cell" is connected to "process oxidant." In here, the concentration of "fuel cell" can be transitive to "humidification system" according to the characteristics defined above. Our tentative study is simple like this example, we select core technologies which have high $\xrightarrow{E}$ concentration from other technologies with considering its directions.

3.2   Experimental Evaluation

We selected 10,000 technical terminology sets (test set) and each set was consisted of one or more technologies extracted from one patent document of NDSL[3]. And we checked each test set whether the set contains multiple technical terminologies because one technology cannot make relation and made totally 5,419 sets that have technologies more than one. And we made 1,411 (about 26.04%) evaluation sets through constructing semantic network with $\xrightarrow{E}$ from each set. For the evaluation, we selected just 100 evaluation sets randomly and evaluated the performance of

---

[1] The example shows a part of technologies extracted from a patent "Humidification system for a fuel cell power plant":
http://patent.ndsl.kr/patDetail.do?cn=WPA2001010106588
[2] Java Graph Visualization and Layout: http://www.jgraph.com/
[3] National Discovery for Science Leaders – Patent:
http://patent.ndsl.kr/index.do

selection of core technologies automatically selected by our system. In here we analyzed original patent documents manually of the evaluation sets. If the system selected one or more technologies as the core candidates and the candidates contained right answer, we decided that the result was right. For example, a case of one patent has {thermal storage, motor vehicle, air conditioning, ventilation system, ...} as technologies which expresses its contents and our evaluation system selects "air conditioning" and "thermal storage" as the candidates. In this document, real core technologies is "air conditioning" and we evaluate this case is right. The following table 2 summarizes our evaluation result.

Table 2. Evaluation Result

Total count	Right cases	Wrong cases
100	57	43

When evaluating this work by using the number in the table 2, we may simply decide that the work is inefficient. However, we think this research can be extended with positive possibility because we used only one relation. And the evaluation set contained 10 sets that do not have core technology of the original document. This was caused from text mining that is out of range of this work. The wrong cases include these sets. Based on the experimental result, we confirmed semantic network could be efficient for selection of core technologies. In the future, we will continuously keep this work with various pattern analyses of semantic network and extra clues.

4   Conclusion

To provide more adequate business intelligence service, this paper proposed a method for detecting the core technologies from scientific documents based on semantic network. We selected the core technologies based on the semantic network features such as concentration and relation direction, and figured out its positive possibility through the simple experiment. However this work should be extended with analyses under more various views. The technology ontology has 7 relation types between technologies. This means that the diversity causes many patterns and we need to define its meanings in the future work.

Acknowledgement. This work utilized scientific and technical contents constructed through 'Establishment of the Sharing System for Electronic Information with Core Science and Technology' project (K-13-L02-C01-S02) and we thanks to Hwanmin Kim, the head of Department of Overseas Information for providing the contents.

References

1. Watson, H.J.: What's new and Important in Business Intelligences. In: Proceedings of International Conference on Information Technology Interfaces, pp. 23–24 (2009)

2. Kong, H., Hwang, M., Hwang, G., Shim, J., Kim, P.: Topic Selection of Web Documents using Specific Domain Ontology. In: Gelbukh, A., Reyes-Garcia, C.A. (eds.) MICAI 2006. LNCS (LNAI), vol. 4293, pp. 1047–1056. Springer, Heidelberg (2006)
3. Kim, J., Hwang, M., Jeong, D.-H., Jung, H.: Technology Trends Analysis and Forecasting Application Based on Decision Tree and Statistical Feature Analysis. Expert Systems with Applications 39(16), 12618–12625 (2012)
4. Hwang, M.-N., Seo, D., Lee, S., Cho, M., Song, S.-K., Lee, J., Hong, S.-C., Choi, S.-P., Jung, H.: Ontology Model of Technical Knowledge for Analytics. In: Proceedings of International Conference on Smart Media and Applications, pp. 13–14 (2012)
5. Hwang, M.-N., Lee, S., Cho, M., Kim, S.Y., Choi, S.-P., Jung, H.: Ontology Construction of Technological Knowledge for R&D Trend Analysis. The Journal of the Korea Contents Association 12(12), 35–45 (2012)

Integration System for Linguistic Software and Data Set: uLAMP (Unified Linguistic Asset Management Platform)

Jung-Ho Um, Sung-Ho Shin, Sung-Pil Choi, Seungwoo Lee, and Hanmin Jung

Dept. of Computer Intelligence Research
Korea Institute of Science and Technology Information
245 Daehakno, Yuseong-gu, Daejeon, 305-806, Korea
{jhum,maximus74,spchoi,swlee,jhm}@kisti.re.kr

Abstract. Numerous linguistic resources are readily available in area of expertise due to the development of wireless devices such as smart-phones and the internet. To select useful information from the massive amount of the data, many systems using semantic web technologies have been developed. In order to build those systems, data collection and natural language processing are essential. However, most of those systems do not consider software integration and the data required by the processes used. In this paper, we propose a system, entitled uLAMP which integrates software and data related to natural language processing. In terms of economics, the cost is reduced by preventing duplicated implementation and data collection. On the other hand, data and software usability are increasing in terms of management requirements. In addition, for the evaluation of the uLAMP usability and effectiveness of uLAMP, a user survey was conducted. Through this evaluation, the advantages of the currentness of data and the ease of use were found.

Keywords: Linguistic Asset Management, Natural Language Processing, Contents Management System.

1 Introduction

Currently, various types of software are developed using web technology to confirm the usefulness of information published on the internet [1-2]. These types of software offer the function of natural language processing. To build the required system, it is necessary to recognize the entities, their relationships, and to ensure mapping to an ontology scheme. However, there are few systems which combine all components of natural language processing.

Currently, Yahoo pipe[1] and U-compare[2] offer a combination of software and linguistic data sets. Yahoo pipe offers an environment in which a user can create a new application by combining legacy web applications and web resources. U-compare also offers an environment of specified bio-data management. It creates work-flow

---

[1] http://pipes.yahoo.com/pipes/
[2] http://nactem.ac.uk/ucompare/

T. Yoshida et al. (Eds.): AMT 2013, LNCS 8210, pp. 296–300, 2013.
© Springer International Publishing Switzerland 2013

through the combined extraction of bio-document systems. However, neither can deal with all linguistic data sets and the related software.

Therefore, if a system combining the essential software and data resources for knowledge extraction (natural language processing) processes to be built, the system, to support QoS, will be built considering the management of a vast amount of information. This system can show increased efficiently by preventing the duplicated development of software or the duplicated collection of data. Moreover, it can utilize the reusability of software and data resources for improved management [3].

Thus, we proposed overall architecture of a system called uLAMP (unified Lingusitic Asset Management Platform) in [4]. It combines software and data resources related to language processing. To do this, in this paper, we design a meta-data structure for the management of user software and data sets, and we formulate the data structure of the collected data. We also design the registration, storage, usage and search functions. Finally, we discuss this system in terms of user evaluations.

2   Proposed uLAMP System Architecture

There are two characteristics of the proposed system. First, the system consists of a storage structure and functions enabling the use of other users' registered applications. In the storage structure, full-text entries, entities, and triples are stored, where the full-text entries are from web articles and paper and patent sources, for instance. Next, the system offers complete sets of software and data resources; a user can register and use software, allowing considerable amount of knowledge from users to be handled.

The system has two advantages. First, it can be used for data verification because it provides full texts, extracted entities and triple resources for language processing. Second, the system can share software and data through its registration environment for users' data and software. Finally, the cost of developing the software and the data collection costs are reduced through the scheme for managing software and data content related to knowledge acquisition.

The proposed system handles language resources such as extracted entities, triples, dictionaries, corpuses, and ontology-related items. The system also handles software related to the recognition of knowledge. Figure 1 shows the overall system architecture of the proposed unified Linguistic Asset Management Platform (uLAMP).

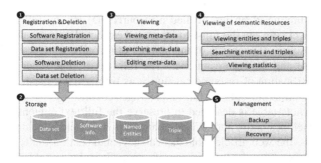

Fig. 1. Overall system architecture

2.1   Collection of Linguistic Asset

There are two means of collecting the data. First, web articles, papers and patents are collected to build semantic resources. Next, corpuses, dictionaries, and linguistic software are provided from a linguistic experts group. Tables 1 and 2 present the number of entities and triples of collected web articles, Wiki pages, papers, and patents. The classifications of the entities are as follows: technology, product, person, organization, and location.

Table 1. The number of entities

	Technology	Product	Person	Organization	Location
Web Article	293,742	544,167	265,845	1,255,730	4,333830
Wiki	1,183,919	910,228	2,031,990	11,936,827	6,368,304
Paper	2,018,990	3,943,222	1,031,365	5,725,623	1,919,186
Patent	1,922,472	2,536,421	134,195	2,619,301	1,089,637
Total	5,419,123	7,934,038	3,463,395	21,537,481	13,710,957

Table 2. The number of triples

Web Article	Wiki	Paper	Patent	Total
1,396,681	12,841,106	7,366,030	4,155,124	25,758,941

2.2   Design of Resource Storage Scheme

The uLAMP system stores semantic resources (entities and triples), linguistic software and data sets, as well as meta-data for the software and data sets. First, semantic resources can be classified with a full-text storage scheme and with entity and triple storage schemes extracting from the full-text items, as shown in Figure 2. In this case, uLAMP uses unique identifiers (called a resource ID) for the full-text, entities and triples to connect between the full-text items and extracted objects such as entities and triples. The resource ID consists of the number of resource categories, the code name of the database and the record number. The full-text storage scheme is designed to include a classification of resource categories as web articles, papers and patents, for instance. Figure 2(a) presents the full-text storage scheme.

The web article table is designed to preserve the full-text contents. Thus, the web article table includes body, author, and publication data fields. In addition, to classify the news cases, the web article table has category and section fields. The paper table is designed to recognize the information of papers. It includes fields for the publisher, journal name, ISSN or ISBN, author, keywords, subject, and other such information.

The patent table consists of meta-data and the body of the patent to present patent information such as the application number (AN), inventor(s) (PAE), patent name, claims and applied country fields, as well as other pertinent information. On the other hand, entities and triples extracted from full-text storage and their frequency tables are designed as shown below in Figure 2(b). To store the data and the type of semantic resources, the entity table consists of the entity name and its type fields, while the triple table has subject name, subject type, relationship name, object name and object type fields. Additionally, joins between entity/triple tables and frequency tables can be measured to determine the weight and the history of the data.

The meta-data of the software and linguistic data set includes the name, type, manual, number, location, and other pertinent information. The meta-data guides the user to know the details about the software and linguistic data sets. Also, users can use upload and download functions through the uLAMP system.

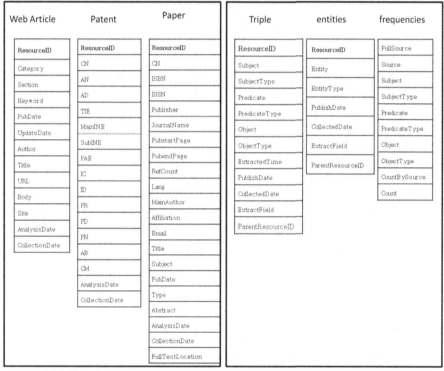

(a) full-text storage scheme   (b) semantic resource storage scheme

Fig. 2. uLAMP storage scheme

3   User Evaluation

To evaluate the proposed system, we surveyed 11 experts in the area of natural language processing. The evaluation measured the usability and effectiveness of the proposed system. The questions for the user evaluation are referenced from the survey of DCMS at KISTI [4]. Regarding the database, the questions are related to correctness, completeness, currentness and concurrency. There are also questions related to searching, easiness of use, customer support and the cost of the system. There are eighteen questions in total, and answers are given on a Likert-type five-point multiple-choice scale. Because there are eighteen questions, the scores for each question on this scale range from 1.5 to 5.5. To measure the satisfaction of the users, Table 3 was used.

Table 3. User Satisfaction Rate (USR)

Level	Point Interval	description
Level S	USR > = 90	Both effectiveness and usability are excellent
Level A	70<=USR<90	Both effectiveness and usability are good
Level B	50<=USR<70	Both effectiveness and usability are not good
Level C	30<=USR<50	Both effectiveness and usability are poor
Level D	USR<30	System can't be maintained

The average score was 84 points, indicating that both effectiveness and usability were rated highly. High scores were also noted for the concurrency and currentness of the database and for easiness of use. Low scores were given for completeness of the database and visibility of the system.

4   Conclusion

In this paper, uLAMP managing software and data resources related to all components of language processing are proposed. It was found though a user evaluation that uLAMP has advantages in terms of easiness of use and the currentness of the database. As future work, we will study how to represent full-text information related to extracted entities or triples. In addition, we will continue to collect user software and language data resources so that we can contribute to the research on natural language processing and related areas.

References

1. Sung, W., Jung, H., Kim, P., Kang, I., Lee, S., Lee, M., Park, D., Hahn, S.: A Semantic Portal for Researchers Using OntoFrame. In: 6th International Semantic Web Conference (2007)
2. Song, S.-K., Oh, H.-S., Myaeng, S.H., Choi, S.-P., Chun, H.-W., Choi, Y.-S., Jeong, C.-H.: Procedural knowledge extraction on MEDLINE abstracts. In: Zhong, N., Callaghan, V., Ghorbani, A.A., Hu, B. (eds.) AMT 2011. LNCS, vol. 6890, pp. 345–354. Springer, Heidelberg (2011)
3. Jakubička, M.: Software asset management. In: IEEE International Conference on Software Maintenance (ICSM) (2010)
4. Shin, S., Um, J., Song, S.-K., Choi, S.-P., Jung, H.: uLAMP: unified Linguistic Asset Management System. In: JIST 2012 (2012)

Scalable Visualization of DBpedia Ontology Using Hadoop

Sung-min Kim, Seong-hun Park, and Young-guk Ha[*]

Department of Computer Science and Engineering
Konkuk University, Seoul, Republic of Korea
{allmax75,wolfire,ygha}@konkuk.ac.kr

Abstract. Existing visualizing methods for big ontology data have many problems. To solve the problems and visualize big ontology data efficiently, we used Hadoop framework, which is for distributed processing across clusters for handling large dataset. The system that we devised is made up of three parts-a data server, a visualization server, and user devices. First of all, The data server preprocesses big data, and the visualization server processes the outputs for visualizing them and transform the outputs to match web standard. The data server and the visualization server use Hadoop framework. User devices have web browsers. Through web browsers, users can be provided with the visualization results by the visualization server We processed DBpedia ontology and visualized the data. In this paper, we will introduce a method for processing and visualizing DBPedia ontology. And we will show the performance of the method by measuring execution time and the experimental results of the visualizing process.

Keywords: Ontology visualization, RDF visualization, Big data visualization, Hadoop, DBpedia Ontology.

1   Introduction

Big ontology data processing is one of burning issues recently. Engineers in various fields do a lot of researches and suggested solutions to process big ontology data. The big ontology data processing is important. And it's also important to show big ontology data. Handling big ontology data visualization using existing visualization method has many problems like slow processing speed, too many processing steps. To solve the problems, we processed big ontology data using Hadoop framework.

Hadoop framework is the one that allows us to perform distributed processing easily by using a simple programming model that it provides [1]. Hadoop is widely used for big data processing because more data can be handled by distributed processing at the same time. These advantages help us to visualize big ontology data efficiently.

Our visualization system consists of a server part and a client part. The server part is divided into a data server and a visualization server. The data server and the

---

[*] Corresponding author.

T. Yoshida et al. (Eds.): AMT 2013, LNCS 8210, pp. 301–306, 2013.
© Springer International Publishing Switzerland 2013

visualization server use Hadoop. The data server processes big data and generates processing the result. The visualization server process output data of the data server to extract information which is used in the client part. And then, the visualization server provides information to web services. We used web services to provide cross platform to users. The client part receives information that user want to get from the visualization server through a web browser.

The outline of this paper is organized as follows: Section 2 explains DBpedia and DBpedia ontology datasets. Section 3 explains detail of our propose system and section 4 explains method of DBPedia ontology visualization. In Section 5, we are showing experiments. Finally, Section 6 will give conclusions and future works.

2   DBpedia Ontology

DBpedia is a crowd-sourced community effort to extract structured information from Wikipedia and make this information available on the Web. DBpedia allows you to send sophisticated queries against Wikipedia, and to link the different data sets on the Web to Wikipedia data.

The DBpedia Ontology is a shallow, cross-domain ontology, which has been manually created based on the most commonly used infoboxes within Wikipedia. The ontology currently covers 359 classes which form a subsumption hierarchy and are described by 1,775 different properties. And it currently contains about 2,350,000 instances [2], [3]. DBpedia provide DBpedia ontology and DBpedia datasets on web site. The dataset is composed of some files that are variety combination of class, instance, property and others.

3   Propose System

Figure 1 shows overview of the proposed system. The system consists of the data server, the visualization server and user devices. The data server and the visualization server are divided logically, which means that they could be in one physical machine.

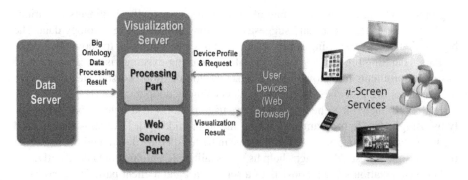

Fig. 1. System overview and steps of big data visualization

A single machine has a scalability problem about to visualize big data. The single machine is slowed depending on growing of data size. The problem is overcome by using the Hadoop.

Data Server
The data server preprocess big ontology data in order that the system visualizes big ontology data or processing result. The data is stored in HDFS (Hadoop Distributed File System) in order that the visualization server uses the data at any time.

Visualization Server
The visualization server consists of a processing part and a web service part. This structure is similar to some other systems [4]. The two parts work together. The processing part processes outputs of the data server to use Hadoop. The processing part generates visualization results for web format using the HTML5 and Javascript. The result is serviced by a web server in the web service part. To provide cross platform to user, we used web format in visualization result.

User Devices
User devices should have a web browser supporting HTML5. Users request the desired information through the Web and confirm the result. Users are provided with information from the visualization server regardless of device.

4   DBpedia Ontology Visualization Method

Our goal is to extract data that users want to get and to visualize the data. For achieving that goal, we had to know what data in files in DBPedia ontology, how they are defined, and what relationships are among them. Each file in DBpedia dataset is made up of triples that have the same meaning. For example, the file "instance_types.nt" is a set of triples like <Instance, Type> and defines types for each instance. We defined meanings that each file represents and schema [5].

Defined data and actual DBpedia dataset are processed by Hadoop framework. The system receives data format in which users want to receive information [6]. If there is the data format in the defined triple schema, the system runs a MapReduce program. If the data format doesn't exist in the schema, the system make a set of triple schema to generate the data format in which users want to receive information, and runs the MapReduce program. And all the information generated by the system is suited to JSON format.

5   Experiment

Our goal is to generate useful information from DBPedia ontology and to show the information on web browsers. Our system's structure is shown below(Table 1).

Table 1. System Environment

Hadoop Environment			
Hadoop Version	Hadoop 1.0.4		
Node	NameNode	2ndNameNode	DataNode
	1	1	30
Node Environment			
CPU	3.10 GHz quad-core		
Memory	16 GB		
Operating System	Ubuntu Server 12.04 LTS (64-bit)		
Web Server Environment			
Web Server	Tomcat 6.0		

Table 2. Experiments Result

First Experiment	
Use the files	dbpedia_3.8.owl (800KB) instance_types_en.nt (1.9GB)
Output	175 MB
Hadoop Map / Reduce	Map 29 : / Reduce : 30
Execute Time	42 sec
Second Experiment	
Use the files	First Experiment Result (175MB) infobox_properties_en.nt (8.8GB)
Output	
Hadoop Map / Reduce	Map : 121 / Reduce : 30
Execute Time	54 sec

We conducted two kinds of experiments. The first experiment is to show relationships between classes and instances as a list. In the list, classes which are in DBPedia's namespace match all the instances. Each class has a list that shows its own instances, and the lists are visualized through being suited to web format. The second experiment is to show major properties for each class. For figuring out major properties, the numbers of properties of all the instances in each class are counted and ranked. The ranking result is shown as a list. Table 2 shows files used for each experiment, the results, and performance. We used DBPedia 3.8 as a dataset for the experiments.

Big Ontology Data processed by the experiments are 2GB and 9GB respectively. The two executing time do not show a big difference, which means that our system is scalable enough to process big data.

We visualized the output results through a web browser. Figure 2 is the first experiment's result, which shows a list of instances of the class "Airport".   Figure 3 is the second experiment's result, which shows a list of 20 properties that appear most frequently in the class "Infrastructure".

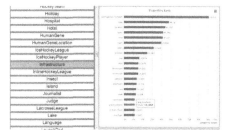

Fig. 2. First Visualization

Fig. 3. Second Visualization

6   Conclusion

In this paper, we proposed a big ontology data visualization system. Our visualization system consists of a data server, a visualization server and user devices. The data server and the visualization server used Hadoop. In the visualization server, the distributed processing of the Hadoop is used by the system for visualization because Hadoop is effective enough to process big data. And the visualization server provided cross platform service through web.

As a future work, we will continue to carry out the research to improve our system. And we will devise a tool to show many different kinds of information for users.

Acknowledgment. This research was supported by MSIP(Ministry of Science, ICT&Future Planning), Korea, under the ITRC(Information Technology Research Center) support program (NIPA-2013-(H0301-13-1012)) supervised by the NIPA (National IT Industry Promotion Agency).

References

1. The Apache™ Hadoop, http://hadoop.apache.org
2. DBpedia, http://dbpedia.org
3. Papantoniou, A., Loumos, V., Poulizos, M., Rigas, G.: A Framework for Visualizing the Web of Data: Combining DBpedia and Open APIs. In: 2011 15th Panhellenic Conference on Informatics (PCI), Kastonia, pp. 240–244 (2011)

4. Zhu, X., Wang, B.: Web service management based on Hadoop. In: 2011 8th International Conference on Service Systems and Service Management (ICSSSM), Tianjin, pp. 1–6 (2011)
5. Lama, M., Vidal, J.C., Otero-Garcia, E., Bugarin, A., Barro, S.: Semantic Linking of a Learning Object Repository to DBpedia. In: 2011 11th IEEE International Conference on Advanced Learning Technologies (ICALT), Athens, GA, pp. 460–464 (2011)
6. Ochs, C., Tian, T., Geller, J., Chun, S.A.: Google Knows Who is Famous Today – Building an Ontology from Search Engine Knowledge and DBpedia. In: 2011 Fifth IEEE International Conference on Semantic Computing (ICSC), Palo Alto, CA, pp. 320–327 (2011)

Content and Expert Recommendation System Using Improved Collaborative Filtering Method for Social Learning

Kyungsun Kim, Kyounguk Lee, and Jinwoo Park

Diquest, R&D Center
5F, Kolon Villant 2, 222-8, Guro 3 dong, Guro-gu, Seoul, Korea
{kksun,arp1710,jwpark}@diquest.com

Abstract. Social Learning as a new concept of learning model emphasizes an individual's activity and formation of relationships with other people. On the contrary, traditional recommendation system provides a target user with the appropriate recommendation information after analyzing a user's preference based on the user's profiles and rating histories. These kinds of systems need to modify recommendation algorithm; these traditional recommendation systems are limited to only two attributes - user profiles and rating histories – that includes the problem of recommendation reliability and accuracy. In this paper, we present a user-context based collaborative filtering (UCCF) using user-context and social relationships. The UCCF analyzes user-context and social relationships, and generates a similar user group which uses the user's recommendation score from similar user groups. The UCCF reflects strong ties of users who have similar tendency and improves reliability and accuracy of the content and expert recommendation system.

Keywords: Item-Based Collaborative Filtering, Social Network Analysis, Semantic Ontology, Compound Knowledge, Case-Based Reasoning.

1   Introduction

The concept of social learning has been spread by the social media as a new concept of sharing information and knowledge generated in a social network service [1]. In accordance with the formation of the concept, intelligent recommendation technology development that can provide every learner with the best learning environment is required [6, 7, 8]. The main purpose of the learning environment analyzes contents effectively created in a social network and diverse correlation among participants so it can recommend customized learning contents and experts to learners. The relationship of users who share information and exchange opinion in SNS can point out that the relationship merely has no similar tendencies and reflects a stronger weight of closeness and credibility [4, 5, 9]. This paper is about content and expert recommendation system for social learning. To share educational knowledge and information through various SNS, we have developed social platform. On the platform, we have analyzed

T. Yoshida et al. (Eds.): AMT 2013, LNCS 8210, pp. 307–313, 2013.
© Springer International Publishing Switzerland 2013

numerous information and knowledge such as experts, contents and learners. We also have developed content and expert recommendation system using an improved collaborative filtering that reflects user-context and social relationships. These elements are stored as types of compound knowledge in ontology [10, 11, 12].

2   Related Study

Previous recommendation system is using collaborative filtering and user preference for the system is classed as implicit and explicit.

-. Explicit rating for Collaborative Filtering [2, 3, 5]
Explicit ratings are when the user herself explicitly rates the item. One example of this is the thumbs up / thumbs down rating on sites such as Pandora and YouTube. Amazon has been adopting 5-Star rule as the explicit rating.

-. Implicit rating for Collaborative Filtering [4, 6]
For implicit ratings, we don't ask users to give any ratings—we just observe their behavior. An example of this is keeping track of what a user clicks on in the online New York Times. Amazon is also using the implicit rating. Consider what information Amazon can gain from recording what products a user clicks on in Amazon. This information is used to display the items in the section "Customers who viewed this also viewed." Another implicit rating is what the customer actually buys. Amazon also keeps track of this information and uses it for their recommendations "Frequently Bought Together" and "Customers Who Viewed This Item Also Bought".

3   Construction of Compound Knowledge for Recommendation System

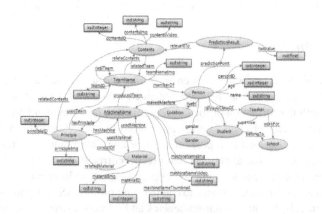

Fig. 1. Construction of Compound Knowledge

3.1   Data Collection of User-context and Social Relationship

This system builds knowledge based on ontology. Ontology which consists of scientific principles and materials, and focuses on Mathematics and Science are defined by Goldberg contest. In addition, LMS (Learning Management System) based on SNS aims at people who have interest in this contest and contextualizes content information with user-context, social relationship weight and ratings of content. In conclusion, compound knowledge consists of form of RDF triple. (Subject, Predicate, Object).

-. User Context
  Gender, Age, Area, Education, etc ...
-. Ratings of content
  User's preference data is user's ratings on 1 or more item and the range is 1~5.
-. Content information
  Writer of content, level (K1~K12), domain (Science: Physics, Chemistry Mathematics: arithmetic operation, statistics)
-. Social Relationship Weight
  Grasp of structure 'Like', 'Comment' on content and assign social relation weight among users.

3.2   Case Based Reasoning and User Similarity Computation

Based on the analysis of the data collected from 3.1, the analyzed data is stored as a form of Semantic Ontology Triple to find set of learners who have a similar history or feature. For the sake of creating learner group that has similar preference and strong ties, user similarity is computed with the degree of consistency of user-context and social relationship weight that are stored in triple repository by case-based reasoning.

3.2.1   Consistency Degree of User-context

Gender	equal: 1	No equal: 0
Level of Education (LOE)	1- |LOE of himself – LOE of target| / N (N: 12, K1~K12)	
Area	equal: 1	No equal: 0
Social Relationship Weight (SRW)	Inverse of Depth N(1~6) SRW: 1/Depth (1, 1/2, ..., 1/N)	
Materials (M)	(M of himself ∩ M of target) / (M of himself M of target)	
Scientific Principle (SP)	(SP of himself ∩SP of target) / (SP of himself SP of target)	

3.2.2   Social Relationship Weight

Social Relationship Weight is SNS depth, meaning, SNS depth presents connection depth which links a user's post with the other users' comments. If user B leaves a comment on user A's post, SNS depth of users A and B will have 1. If user D gives a comment on user C's post, who has N depth with user A, SNS depth of users D and A will be N+1. In conclusion, social relationship weight traces a maximum of 6 depths and uses inverse number of SNS depth. In other words, if SNS depth is 1, social relationship weight is 1 and if the SNS depth is 2, social relationship weight is 1/2.

3.2.3   User Similarity

Similarity (U1, U2) = (Whether Gender equal or not + Age gap + Education gap + whether area equal or not + Consistency degree of service context + social relationship weight) / 6

Similarities	User1	User2	User3	.....
User1	1	Sim(U1, U2)	Sim(U1, U3)	
User2	Sim(U2, U1)	1	Sim(U2, U3)	
User3	Sim(U3, U1)	Sim(U3, U2)	1	
......				1

As can be seen above, user context and social relationship weight is included in computing the similarities of users that would make it more accurate.

4   Social Relationship Weight and UserContext-Based Collaborative Filtering for Content Recommendation

For content recommendation, the first step is to organize user-content matrix. The second step is to apply item-based collaborative filtering algorithm of Weighed Slope-One to user-content matrix and predict the score of an unknown rating. The last step is to recommend learning content based on the standards that have been illustrated above. As shown, since UCCF algorithm includes ratings of users who have higher similarity, the target learner is able to get a recommendation about the learning content which improves accuracy and reliability.

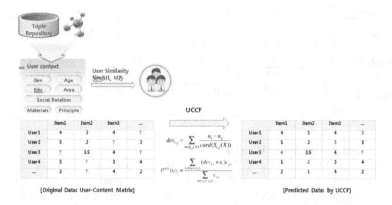

Fig. 2. UCCF (User-Context based Collaborative Filtering) relationship model

```
PREFIX rdfs: <http://www.w3.org/2000/01/rdf-schema#>
PREFIX rdf: <http://www.w3.org/1999/02/22-rdf-syntax-ns#>
PREFIX sole: <http://www.diquest.com/sole#>

select ?con ?team ?conid ?video where {
sole: potential energy sole:relatedContents ?con .
?con sole:relatedTeam ?team .
?con sole:contentsID ?conid .
?con sole:contentsVideo ?video .
}
```
Ex) Recommendation Sqarql for learning video content related to potential energy

5   Expert Recommendation

For expert recommendation, content information is contextualized with user-context, social relationship weight and ratings of content and stored as a form of RDF triple. Users are able to record 'Like' or 'Comment' in LMS based on SNS and these attributes are considered as important factors for inferring user's degree of recommendation and interest. The writer who has a higher degree of recommendation could be referred to as an expert in the field.

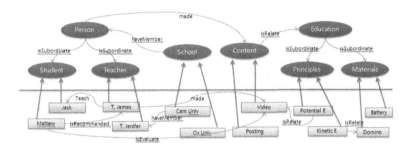

Fig. 3. UCCF relationship model for expert recommendation

5.1   Expert extraction Using Ratings on Content

Expert extraction goes through 4 steps as below.

[Step1. User-Content matrix]

Ratings	Content1	Content2	Content3	.....
User1	5	0	3	0
User2	2	5	0	0
User3	0	2	5	0
......	0	1	0	5

[Step2. Content-Writer matrix]

Posting Owner	Content1	Content2	Content3	.....
Writer	User1	User10	User35	

[Step3. Content-Writer Ranking matrix]

Ranking	Content1	Content2	Content3	.....
User1	5	3	3	2
User2	2	5	1	2
User3	4	2	5	3
......	3	1	3	5

[Step4. Sum of ranking and recommend expert]

Expert	Content10	Content27	Content35	.....
Expert	User10	User23	User54	

6   Social Learning Platform and Recommendation System

We have been performing experiments for UCCF algorithm, as introduced in this paper, which are aimed at participants of Goldberg contest with the use of social learning platform and recommendation system.

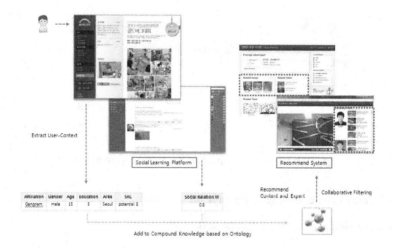

7   Conclusion

In this paper, we present social network analysis that applied semantic technology such as OWL ontology, RDFS inference, and Sparql. In addition, we have developed content and expert recommendation system using collaborative filtering with user-context and social relationship. The degree of satisfaction evaluation are based on the result of the recommendations given by FGI(Focus Group Interview) targeting Goldberg contest. Moreover, we are going to distribute data which have been collected from Goldberg contest for social learning recommendation system. Though recommendation system suggested in this paper is for education, our improved collaborative filtering algorithm can be applied to diverse domain like movies, shopping and so on.

Acknowledgments. This paper is supported by Korean Industrial Technology Development Program –"Develop Social Learning Platform with intelligent recommendation and social matching technology for self-paced learning: Project No. 10042794"

References

[1] Banduar, A.: Social-Learning Theory of Identificatory Processes, Stanford University
[2] Linden, G., Smith, B., York, J.: Amazon.com Recommendations, Item-to-Item Collaborative Filtering
[3] Sarwar, B., Karypis, G., Konstan, J., Riedl, J.: Item-Based Collaborative Filtering Recommendation Algorithms
[4] A Programmer's Guide to Data Mining, http://guidetodatamining.com/
[5] Lemire, D., Maclachlan, A.: Slope One Predictors for Online Rating-Based Collaborative Filtering (February 7, 2005)
[6] An Expert Recommendation System using Ontology-based Social Network Analysis
[7] Wasserman, S., Faust, K.: Social Network Analysis: Methods and Applications. Cambridge University Press, Cambridge (1994)
[8] Mika, P.: Flink: Semantic Web Technology for the Extraction and Analysis of Social Networks. Journal of Web Semantics (2005)
[9] Chee, S.H.S., Han, J., Wang, K.: Rectree: An efficient collaborative filtering method. In: Kambayashi, Y., Winiwarter, W., Arikawa, M. (eds.) DaWaK 2001. LNCS, vol. 2114, pp. 141–151. Springer, Heidelberg (2001)
[10] http://www.w3.org/TR/owl-features/
[11] http://www.w3.org/TR/rdf-schema/
[12] http://www.w3.org/TR/rdf-sparql-query/

Automatic Animation Composition System
for Event Representation
— Towards an Automatic Story Animation Generation —

Yusuke Manabe[1], Takanori Ohsugi[2], and Kenji Sugawara[1]

[1] Chiba Institute of Technology
{ymanabe,suga}@net.it-chiba.ac.jp
[2] Kyorindo Co. Ltd.

Abstract. This paper proposes an automatic animation composition system based on six databases, which are defined through analyzing a Japanese folktale animation movie. The system can automatically translate a text-based event representation (simple case frame) into an animation script (TVML). We show that our proposed system can compose several TVML scripts, which can represent animations of events as parts of a story.

1 Introduction

The concept of narrative includes several aspects such as story, discourse and representation. This study focuses on story and its visual representation [1–6], especially animation movie. Our final goal is to develop a system that can automatically generate an animation from a text-based story representation such as case frame [7]. In order to convert a text-based story into a visual image such as an animation, it is necessary to build many and various databases or knowledge bases for bridging between a text-based story representation and a visual representation including image resources, cinematic techniques, actions of characters, screen layouts and so on. However no such knowledge base or database exists. Actually some contributions, which try to generate animation movies automatically, face the same drawback that inhibits a realization of automatic animation generation [1–4, 6].

Generally we can illustrate the content of text-based story as visual image through complementing visual scene, character appearance and time background even if the sufficient visual information about a story is not given. This implies that we make up for the lack of visual information through knowledge of many stories that we have already read. From this viewpoint, conventional animation movies include informative animation resources, such as screen layouts, cinematic techniques, actions etc. If such resources can be used as knowledge for generating a new animation, we can automatically generate various animations by a computer program and knowledge bases (databases).

Therefore, in this study we define six databases through analyzing a Japanese folktale animation and developed an automatic animation composition system

T. Yoshida et al. (Eds.): AMT 2013, LNCS 8210, pp. 314–323, 2013.
© Springer International Publishing Switzerland 2013

for event representations. Our developed system can convert a text-based event representation into an animation script, TVML[1].

2   Analysis of a Japanese Folktale Animation and Database Definition

In order to make up for the lack of information for bridging a text-based and a visual-based story representation, we analyze a Japanese folktale animation and define some databases for composing animation.

We extract many image resources from the original Japanese folktale animation and make a TVML script, which realizes as same animation with the original animation as possible. The TVML script can be regarded as a template of animation composition. These elements are classified and stored in each database to be used as knowledge bridging a text-based and a visual-based story representation.

2.1   Animation Material and Analysis Steps

Animation material is a DVD animation, "Fleas are medicine (Nomi Wa Kusuri)", in a series of *Japanese Folktales For Reading to Children* [9]. The reason why we choose this title is that it is composed by minimal animation techniques; the number of characters is two, all of actions are expressed by simple 2D operations such as switching, rotation, transition and scaling of images.

Analysis steps are shown as follows;

1. To make a list of composition elements for all of scenes of the folktale animation.
2. To capture image resources such as characters and backgrounds from the animation.
3. To make a TVML script using captured image resources.

Step 1: Making a List of Composition Elements. We made an analysis list with respect to 14 elements, which are shown as Figure 1. We extracted 49 cuts from whole of a movie (time is 5m45s).

Step 2: Capturing Image Resources. We captured some scene images from all of cuts and clipped many character images and background images. The number of character images is 96, the number of background images is 37. The former are used as templates of character and action representation, the latter are used as templates of location and layout representation. In actual operational phase, all of image resources are replaced by other resources which are free from copyrights, although they are used for making a TVML script in the next step.

---

[1] TVML stands for TV program Making Language and it is a text-based language developed by NHK Science and Technical Research Laboratories [8]. TVML can be interpreted as a 2D/3D CG animation (OpenGL) in real time by TVML Player.

Table 1. An Example of the Analysis List (Translated in English)

Cut #	1	2	5
Time (start-end)	0:00 - 0:15	0:15 - 0:20	0:29 - 0:34
Narrations, Scripts		とんと昔・ある村に	ある年の夏、仕事で街へ出かけ、
Camera Movements	Move(Top --> Bottom)	Fix	Move(Bottom --> Top)
Characters	Man	Man	Man
# of Characters	1	1	1
Postures of Characters	[Man] left side view, whole body, Kimono, lean tree, get lying down, cross arms claspe hands behind head, cross legs (right leg over left leg)	[Man] left side view, whole body, Kimono, lean tree, get lying down, cross arms claspe hands behind head, cross legs (right leg over left leg)	[Man] back view, whole body, kimono, put a woven hat on, carry something to right shoulder
Activities/Transitions of Characters		[Man] Activity(Nod, Shake right leg)	[Man] Transition(Bottom --> Top)
Objects	[Man] chew a grass	[Man] chew a grass, loop a rope around left foot	
Object Movements		[a grass] rotate reef	
Semantics of Scenes	Show the title	peaceful noon?	[Man] man goes to village
Backgrounds	Sky, Cloud, Sun, Tree, Mountain, Hill?	Sky, Cloud, Tree, Road(Hill)	Tree, Cloud, Mountain,, Sky, Road, Homes
Screen Layouts	[Sun]TopCenter [Cloud]TopLeft, MiddleLeft [Hill]BottomCenter,BottomRight.Midd leRight [Tree]TopRight, MiddleRight [Mountain]BottomCenter [Man]MiddleRight	[Cloud]TopLeft [Tree]TopRight,MiddleRight [Road(Hill)]Bottom [Sky]TopCenter,MiddleLeft,MiddleCen ter	[Road]MiddleCenter, BottomCenter [Homes]:MiddleCenter, BottomCenter [Mountain]MiddleLeft, MiddleCenter, MiddleRight [Cloud]TopRight, TopLeft [Tree]BottomCenter [Man]BottomCenter --> MiddleCenter
BGM/Effects	BGM: Music	Nothing	Effect: Cicadas are shrilling

Step 3: Making a TVML Script. Using the clipped image resources, we made a TVML script, which realizes as same animation with the original animation as possible. That is, this is a reconstruction of the original animation by TVML. We employ prop commands in TVML for image composition as the following list;

prop: position() it can be used for positioning,
prop: visible() it can be used for image switching,
prop: transform() it can be used for rotation, transition and scaling.

The total number of line of the TVML script we made is 1841 excepting comments and blank lines. This reconstruction does not include narrations, speech conversations and sounds but include actions (character's performances), screen layouts and cinematic techniques.

Here is an example of a TVML script, whose content is a cut that two characters talk. Figure 1 shows six image resources are assigned by prop: assign() because each character is configured by the three image resources. Figure 2 shows each character is properly configured by prop: position(). Figure 3 shows a 'talk' action can be expressed by prop: transform() and prop: visible().

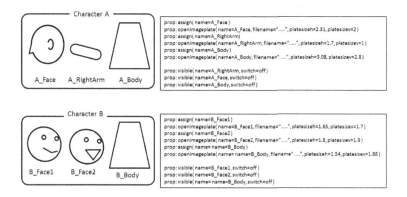

Fig. 1. Assigning Character Image Resources

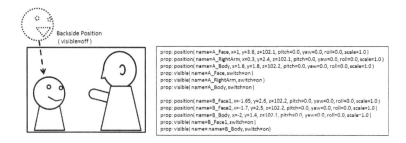

Fig. 2. Positioning Two Characters

As the result, some animations (i.e., fade-in/fade-out, concurrent actions, synchronization between a camera and a character) could not be realized due to the limitations of TVML and TVML Player although we could reconstruct most of the original animation.

2.2 Database Definition

After the above analysis steps, we classified the TVML script, the extracted image resources and the analysis list. As the result, they are stored in six databases; character image, location image, layout, action, transition and activity. In this subsection, we describe about main columns in each database.

Character Image DB. This database consists of 16 columns. `RPosX`, `RPosY`, `RPosZ` are relative coordinate values for composing a character by some image resources. `RotX`, `RotY`, `RotZ` are rotation degrees with respect to each axis. `Posture`, `Direction1&2`, `Embodiment` are properties of character image. `ActivityID` includes several IDs of activity database and it denotes that a relevant character image is used for expressing some activities tied with ID.

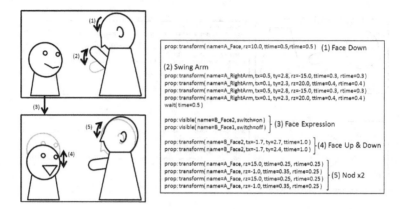

Fig. 3. Action Script

Location Image DB. This database consists of 11 columns. `initCamTVML` is a TVML script to capture properly a background image by a camera. `setLocTVML` is a TVML script to position a location image. `Posture`, `Direction1&2`, `Embodiment` are constraint properties for positioning character image. `LayoutID` includes several IDs of layout database and it denotes desirable positioning for character image. `TransitionID` includes several IDs of transition database and it denotes that a relevant location image allows to express some transitions tied with ID.

Layout DB. This database consists of 4 columns and it stores positioning information of character image.

Action DB. This database consists of 5 columns. The most important column is `TransitionID`, which is time sequence of activities and actions. Figure 4 shows an example of composition column in an action, 'come'. This figure shows 'come' action is composed of 2 transitions (MOVE1 and MOVE2) and 3 activities (ACT2CACT3CACT4). The meanings of another descriptor are shown as Table 2.

Transition DB. This database consists of 3 columns. The most important column is `setTransTVML`, which is TVML templates for transition of character image.

Activity DB. This database consists of 7 columns. `CharaID` is a character image ID for composing the activity. Thus an activity should be composed of four character images at a maximum. `setActTVML` is TVML templates for activity of character images.

X@MOVE1 #1,X@ ACT2 ,X@ MOVE2 #2,X@ ACT3
X@MOVE1 #3,X@ ACT2 ,X@ MOVE2 #4,X@ ACT3
X@MOVE1 #5,X@ ACT2 ,X@ MOVE2 #6,X@ ACT4

Fig. 4. An Example of Composition Column in 'come'

Table 2. Descriptors in Composition Column

Symbols	Interpretation
X[Y]	Character Identifier (X and Y in case # of Actor is 2.)
ACT(Number)	ID in Activity Database
MOVE(Number)	ID in Transition Database
(Number) after #	Transition Times
@C#	Segmentation Symbols

3   Automatic Animation Composition System

3.1   Processing Steps

Based on the previous results, we developed an automatic animation composition system, which can generate a TVML script based on text-based event representation. In this study we define an event as a part of a story. That is, a story is composed by a time sequence of many event representations.

Our proposed system is composed of 3 processes; input process, decision process and generation process.

Input process is to set a text-based event representation through GUI operation. The event representation is composed of four elements; Agent(AGT), Counter-Agent(C-AGT), Location(LOC) and Action(ACT). Namely our proposed system postulates that a text-based event representation is not natural language but a formal data like simple case frame.

Decision process is to assess whether animation composition can be realized by the databases or not. The system examines the following six conditions.

1. Whether the location DB has location images corresponding to a text-based input.
2. Whether the character DB has character images corresponding to a text-based input.
3. Whether the action DB has action representations corresponding to a text-based input.
4. Whether the action representation refers to the transition DB, the activity DB or the both.
5. Whether the location image DB has location images to allow some transition representations because an action representation depends on location images when it includes the transition representations.
6. Whether the character image DB has character images to allow some activity representations because an action representation depends on character images when it includes the activity representations.

When it is decided that an animation composition can be realized, this system can automatically generate a TVML script.

Generation process is to determine all of image resources that the system needs and to calculate each positions of resources, on-off control of image display, image operations such as transition, rotation and scaling. Figure 5 shows the workflow in generation process.

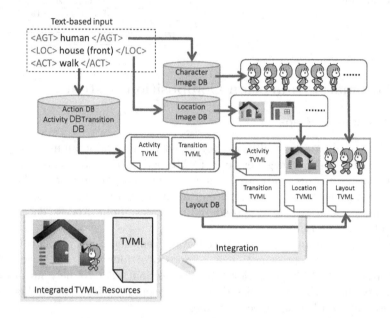

Fig. 5. Workflow in Generation Process

Mainly the system executes the following processing.

1. To get candidates of character images and location images based on a text-based input.
2. To get candidates of transition TVML and activity TVML for composing action representations.
3. To determine transition TVML and activity TVML from their candidates.
4. To determine location images from its candidates and get a location TVML.
5. To determine character images from its candidates.
6. To get a layout TVML for locating character images.
7. To integrate all of TVML and image resources.

Finally a composed TVML script is interpreted as a CG animation by the TVML Player.

3.2   Implementation

We developed the system by C# language under Microsoft Visual Studio 2010. Figure 6 shows a screen shot of the developed system. The menu items in the left side are used for input of an event. Setting an event and pressing a generation button, a TVML script is automatically written in the text area. The output button is used for output of a TVML script, which can automatically interpreted as an animation by TVML player.

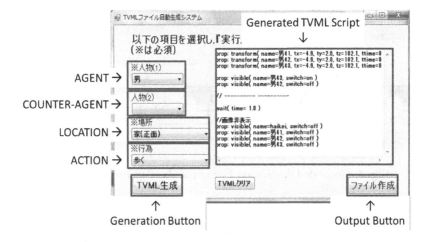

Fig. 6. Screen Shot of Developed System

To design six databases, we employ a database management system, MySQL. The number of data samples is 70 rows in the Character DB, 30 rows in the Location DB, 59 rows in the Layout DB, 9 rows in the Action DB, 4 rows in the Transition DB and 16 rows in the Activity DB.

4   Feasibility of Automatic Animation Movie Generation by Proposed System and Databases

In order to discuss a feasibility of automatic animation generation, we automatically composed some TVML scripts by our proposed system and databases. Then we evaluated animations based on the automatic composed TVML scripts.

In this paper, we show results of 3 animations based on 3 events as follows;

1. <AGT>man</AGT> <C-AGT>old woman</C-AGT> <LOC>room</LOC> <ACT>talk</ACT>
2. <AGT>man</AGT> <LOC>house(front)</LOC> <ACT>pass</ACT>
3. <AGT>man</AGT> <LOC>house(front)</LOC> <ACT>come</ACT>

The first event is equivalent to an event in the original animation movie but the image resources are changed as simple cartoon pictures. The second event and the third event do not exist in the original animation movie. The 'pass' action (from right side to left side by a walk) and the 'come' action (from left side by a walk) are newly defined based on 'come' action (from right side by a walk) in the original movie. The former corresponds to a distance change and the latter corresponds to a direction change. Image resources are changed as simple cartoon pictures or clip-art pictures[10, 11].

Figure 7, 8 and 9 show some screen shots of the animations based on the generated TVML scripts. Although it is hard to judge each animation by static screen shots, we could confirm that each animation has a sufficient quality of an event representation.

Fig. 7. Generated Animation by Simple Cartoon Pictures ('talk')

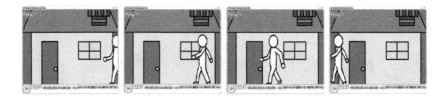

Fig. 8. Generated Animation by Simple Cartoon Pictures ('pass')

Fig. 9. Generated Animation by Clip-art Pictures ('come')

5   Conclusion

In this paper we analyzed a Japanese folktale animation and defined six databases for composing animation in order to make up for the lack of information for bridging a text-based and a visual-based story representation. Based on the six databases, we developed an automatic animation composition system, which can automatically translate text-based event representation (simple case frame) into animation script (TVML). Finally we found that our proposed system can compose several TVML scripts, which can represent an event animation as parts of a story.

As future works, we would like to deal with the following items;

1. Making a uniform approach to prepare image resources.
2. Making a camera database for controlling cinematic techniques.
3. Refining each database, especially a design of a hierarchical action database.

References

1. Pérez, R., Sosa, R., Lemaitre, C.: A computer model for visual-daydreaming. In: AAAI Fall Symposium on Intelligent Narrative Technologies, pp. 102–109. AAAI Press, Washington D.C. (2007)
2. Miyazaki, K., Nagai, Y., Nakatsu, R.: Architecture of an Authoring System to Support Interactive Contents Creation for Games/E-Learning. In: Pan, Z., Aylett, R.S., Diener, H., Jin, X., Göbel, S., Li, L. (eds.) Edutainment 2006. LNCS, vol. 3942, pp. 70–79. Springer, Heidelberg (2006)
3. Miyazaki, K., Nagai, Y., Wama, T., Nakatsu, R.: Concept and construction of an interactive folktale system. In: Ma, L., Rauterberg, M., Nakatsu, R. (eds.) ICEC 2007. LNCS, vol. 4740, pp. 162–170. Springer, Heidelberg (2007)
4. Hoshina, K., Noguchi, T., Sugimoto, T., Enokidzu, H.: A trial of story comprehension simulation: Through automatic generation of animation from story text. In: Proc. of the 29th Annual Meeting of the Japanese Cognitive Science Society, vol. P1–7, pp. 204–210 (2012) (in Japanese)
5. whiteonwhite: algorithmicnoir, http://www.rufuscorporation.com/wowpr.htm
6. Lu, R., Zhang, S.: Automatic Generation of Computer Animation. LNCS (LNAI), vol. 2160. Springer, Heidelberg (2002)
7. Manabe, Y.: How to Automatically Generate Animation Movie of Story — Towards Action Representation Ontology—. In: Proc. of Literature and Cognition/Computer II, vol. 15G-6, 10 pages (2008) (in Japanese)
8. TV program Making Language. NHK Science and Technical Research Laboratories (2013), http://www.nhk.or.jp/strl/tvml/index.html
9. Fleas are medicine (Nomi Wa Kusuri). In: DVD Japanese Folktales For Reading to Children (Yomikikase Nihon Mukashi Banashi), vol. 1. Columbia Music Entertainment, Inc. (2006)
10. doroid-chan.net (Doroid Chan Matome Site), http://droid-chan.net/ (in Japanese)
11. materialand, http://www.material-land.com/ (in Japanese)

Narrative on the Road

Hitoshi Morita

University of Nagasaki
morita@sun.ac.jp

Abstract. This research discusses the generation method of the narrative text linked with geographic space data. First, analysis methods for the existing folk tale text are described. It is clarified that there is geographic bias in the narrative text. Next, the technique for using geographic space data for the sightseeing tour is described. The sightseeing tour is an action for the tourist to touch the narrative text in the local area while moving. We maintained the place that related to the text as geographic space data. The text including place information code can be plotted on the digital map. We propose a new expression technique of the story text by integrating these methods.

Keywords: Japanese Folktale, Tale Type, Motif Index, ITS for Sightseeing, Smart Phone Navigation, LRT, GPS, NFC.

1 Classical Part

In this part, we describe analysis methods for the existing Japanese famous folk tale text.

1.1 Introduction

The purpose of this part is to examine whether a new story can be created by analyzing the legend. The tool of folklore like the tale type and the motif indexes ([1], [2], [3], [4], [5], [6]) are used for the analysis. The result of analysis is structurized by the technique of the natural language processing like the story grammar and the discourse analysis. And, the computation model of the story [7] is assumed to be a hypothesis and the possibility of the text generation is examined. We clarified the following two points by this research. 1) Various social factors transformed the story. 2) If an existing motif is calculated, it becomes possible to generate a new story.

1.2 Method and Matrial

The feature of this part is to consider not only the text of the legend but also the relation between the text and the society. The title of material for this research is "Momotaro" (Peach Boy).

"Momotaro" (Peach Boy) is the most famous folk literature in Japan. However, all "Momotaro" is not necessarily the same. More than 680 kinds of varieties circulate

T. Yoshida et al. (Eds.): AMT 2013, LNCS 8210, pp. 324–332, 2013.
© Springer International Publishing Switzerland 2013

this story in Japan. Among these varieties, 63 samples are compared according to our original scale.

This story was a rejuvenation type until Edo period. We examine the factor converted from the rejuvenation type to the abnormal birth type from the outside of the text (social factor). The factor is the following four points. 1) Prosperity of popular literature. 2) Enactment of elementary school textbook by nation and leveling of story. 3) Diversion of folklore to nationalism. [8] 4) Fight uplift animation in World War II made from Momotaro

We reflect our original scale and show the structure of the story from the following points. { Feature }, { hero's name }, { attendant }, { motivation }, { act }, and { return } (see Table 1). And, we brought the structure of this story together from the viewpoint "Subtraction of the story" and "Addition of the story". "Subtraction of the story" is to be subtracted some motifs from the story. "Addition of the story" is to be added another motif to the story. It is two kinds of motifs that subtracted from "Momotaro" in Edo period. "Motif at rejuvenation" of birth from aged woman. And, "Motif of the bride removing" of having brought the bride from demon's island. "Motif at rejuvenation" and "Motif of the bride removing" are in the co-occurrence relation. On the other hand, the motivation "Why do you go demon extermination?" being talked in "Momotaro" is rare. Most cases are based on tacit consent "Because it is a bad demon, the demon is punished". The Japanese animation in World War II avoided clarifying the motivation and the reason for the invasion. Here, "Momotaro punishes the demon without the motivation" becomes the same motif.

Table 1. Structure analysis on "Momotaro"

	abnormal birth type	*rejuvenation type*
Circumstances of birth	born from the peach	old woman gave birth.
Who	Momotaro	Momotaro
With whom	dog, monkey, pheasant	crab, mortar, cows pie, rope, others
Purpose	N/A	N/A
What was done	Demon extermination	Demon extermination
Episode of the closing	took the treasure home	brought the bride

Table 2. Structural comparisons of Japanese 5 great folktales

Title	*Momotaro*	*Fights between Monkey and Crab*	*Tongue-Cut Sparrow*	*Magic Ashes make a Cherry Blossom*	*Kachi-Kachi Mountain*
Tale Type	AT[1] 513A(S)	AT 9C(S), AT 210(Ts)	AT 480(S), AT 1143A(Ts)	AT 1655A(S)	AT 1087(Ts)
	Ikeda [2]302	Ikeda 210	Ikeda 480D	Ikeda 503F	Ikeda 176
	Tsuukan [3] 127	Tsuukan 522A	Tsuukan 85	Tsuukan 364A	Tsuukan 531

Table 2. (*continued*)

	Shuusei [4]187	Shuusei 59	Shuusei 271A	Shuusei 268	Shuusei 78
	Taisei [5]143	Taisei 24	Taisei 191	Taisei 190	Taisei 32
Cha-racters	old man, old woman, Momotaro, dog, monkey, pheasant, demon	monkey, persimmon, crab, chestnut, bee, cow pie, mortar	old woman, old man, sparrow, cattleman, horse rear-ing	old woman, dog, old man, the next old man, mortar, ash, lord	old man, asian raccoon, old woman, rabbit
Com-mon epi-sode	Old woman obtained the object at the riverside.	N/A	Old woman obtained the object at the riverside.	Old woman obtained the object at the riverside.	N/A
Motif [6]	B391	B296	B222	B153	B437.4
	B421	B481.3	B222	B182.1	B511
	B441.1	B762	B350	B211.1.7	B512
	B469.10	D1273	B451.7	B335	B857
	D981.2	D2157.2	C321.2	B421	G61
	F601.7	F1025	F127	D1174	G85
	F611.3.2	F601.7	F127	D1263	J1706
	G500	F811.19	F379.1	D1561.1	J1706
	H1221	H1228.2	F379.1	D1561.1	J1706
	N538.2	K1161	F379.2.2	D1571.1	J2171.1.3
	T543.3	K170	F379.2.2	D422.2.1	K1014
	T550.2	K171.9	H1049	D866	K1941
	T596	W111	H1049	D931.1.2	K2310
	T615		J2411	D950	K2310
	Z293		J2415	E631	K2345
			L210	F811.19	K581.1
			N820	J2401	K735.2
			N820	J2411	K800
			Q272	J2411	K800
			Q285.1.1	J2411	K800
			Q285.1.1	J2415	K800
			Q3	J2415	K910
			Q3	J2415	K926
			S167	N511	R210
			Z293	N550	
				Q272	
				Q272	
				Q272	
				T615	
				T615	

1.3   Some Versions of Folktale

As a result, it becomes possible for the speaker and the receiver to supplement the motivation and the reason even if it misses in the main body of the story. Generally, attendants with whom Momotaro connects the relation of master and servant on the way arrive at demon's island are set as the dog, the monkey, and the pheasant. However, there is a case where the attendant doesn't appear in the talk at all. Moreover, there is a case where the bee, the mortar, and the cow pie, etc. become attendants. It is inosculated as other stories.

1.4   Narrative Data Processing

We clarified how the story "Momotaro" was influenced from the society. Next, we examine whether it becomes possible to generate the text by processing "Momotaro". We compared Japanese 5 great antiquity including "Momotaro" stories by the tale type and the motif (see Table 2). We are able to derive the operation possibilities of the following legends from this comparative study result as two hypotheses.

─ Calculation possibility of Narrative text

("Momotaro" + "Fights Between Monkey and Crab")   "Kachi-Kachi Mountain" =

"Tongue-Cut Sparrow" --> "Magic Ashes make a Cherry-Tree Blossom"

─ Narrative text linked with geographical data

There was a regional variation in the distribution of a different text. Especially, there was a regional trait in the attendant's composition. This feature can be related to the geographic data.

2   GIS Part

In this part, we describe the technique for using geographic space data for the sightseeing tour. The sightseeing tour is an action for the tourist to touch the narrative text in the local area while moving.

2.1   Tourist Support System in Sightseeing Town Nagasaki

Nagasaki LRT Navigation Promotion Council provides the Service named DOKONE. It is a system that distributes the position information of the low-floor vehicle operated by Nagasaki Electric Tramway to the mobile terminal. This system detects tram and user's positions by using GPS, and transmits data via mobile network (3G or LTE). Users can determine the position of the tram by the map displayed on the screen of the mobile phone. In addition, they can reserve getting on from a specific stop. This service started in October, 2011, and reached 40,000 by the number of accesses in one year. Our Promotion Council is composed by the university, private companies, and local governments (City and Prefecture). We developed a new navigation system for sightseeing that used Ucode and the NFC tag based on DOKONE in 2012. This article reports on the process of the evolution of the navigator.

2.2   Nagasaki City and Tram

Nagasaki is a beautiful tourist town on the edge of the west of Japan. This city has prospered as a window of Western culture in Japan until the 18th century. And, World War II ended by the atomic bombing to Nagasaki. It has been used as familiar transportation by the citizens and the tourist today though the tram received big damage in the war. Nagasaki Electric Tramway operates the tram of 11.5km in the total extension in the Nagasaki City. The number of stops is 39 places, and the number of vehicles is 79. Among these there are 5 low-floor vehicles. Because the driving spacing is very short, Nagasaki Electric Tramway doesn't maintain the timetable. The user doesn't understand when the low-floor car arrives at the stop from such circumstances. As a consignment business of the Ministry of Land, Infrastructure and Transport we started the location system of the low-floor car for the wheelchair and the stroller user at October 7, 2011 named DOKONE.(see Figure 1) This word is a meaning "Where are you?" in the Nagasaki dialect. We have formed the council the three months ago. Nagasaki LRT Navigation Promotion Council is composed by University of Nagasaki, Nagasaki Electric Tramway, Ougiseiko Co. Ltd. (System Integrator), and local governments (Nagasaki City and Nagasaki Prefecture).

2.3   System Overview

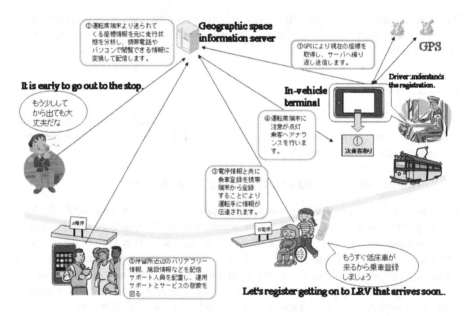

Fig. 1. Service image of DOKONE

1. The tablet terminal mounted to the low-floor vehicle collects the location data by GPS and it transmits to the server.
2. The server is converted from the location data into information that can be inspected and notice assignment works to user's mobile phone and personal computer.
3. User confirms the operation situation of the low-floor car with a terminal such as mobile phones. In addition, the getting on intention can be registered on the same system.
4. The getting on intention of the user who needs supporting is transmitted to the in-vehicle tablet terminal on the driver side. The driver requests consideration to the necessity of support to the passenger.
5. Barrier information and sightseeing information, etc. in the vicinity of the stop are offered to the portable terminal and the support service that can act smoothly when getting on and off tending the most much to be confused is provided.
    1) Pedestrian's position: It is specific according to reading ucodeQR that buries GPS or place information code with built-in portable terminal.
    2) The method of using place information code: Acquisition ucodeQR label is set up, and place information code in which the stop is shown is used to fix the stop where the pedestrian is located.

2.4   Navigation by Position Information System

We developed a new navigation system for sightseeing (DOKONE II) that used Ucode and the NFC tag based on DOKONE in 2012. The blue line of Figure 2 is a railway track of the tram, and a red line is a stroll road for sightseeing. The index board of Figure 3 laid 45 places on a red line underground is place information encoded and the sensor is installed. And, we constructed a new navigation system that used three kinds of sensors described to Table1 and enabled the transit of the stroll road and the tram. Figure 4 shows the operation image of the developed application program. It is scheduled to connect with the integrated tourism information platform in Nagasaki Prefecture. [9]

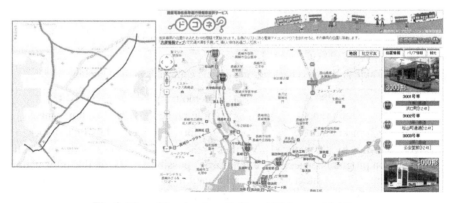

Fig. 2. Map of Service Area (left) and Web page (right)

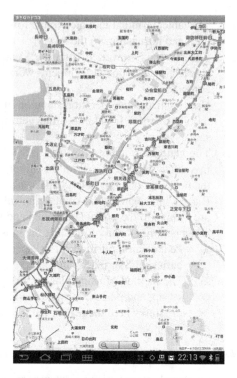

Fig. 3. Nagasaki-index(above) & those road →

Fig. 4. Bluetooth Sensor(left) and NFC tag(right)

Table 4. Various characteristics of field communication technology

Sensor	GPS （long range: 20m-km more）	Bluetooth （middle range: 1-20m）	NFC （short range: less than 0.1m）
Usage	Verifies by the application program connected with the integrated tourism information platform. [9]	Verifies the notice assignment possibility of information in the surrounding, and the Navigation to facilities.	Verifies the possibility of the information service corresponding to the user.

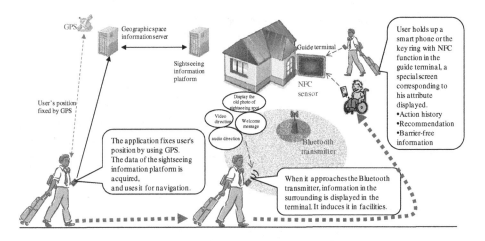

Fig. 5. Service image of DOKONE II

3   Geographical Text

We propose two kinds of evolutionary types in place of the conclusion at the end.

Figure 6 is a record of the collection ground of the folk tale in the Ureshino region on the map. The footprint where the folk tale moved on the line of the root 34(the national road) is recorded. Signs of distribution and the movement of the folk tale can be recorded in the map.

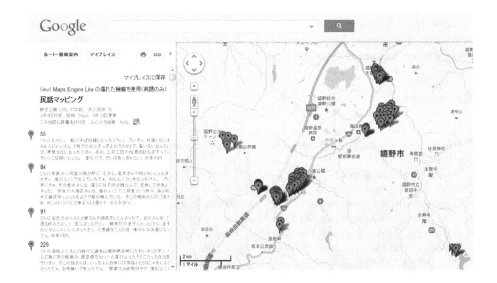

Fig. 6. Ureshino Folktale Map

【Cooperation of data and contents that make geographic space information common term】

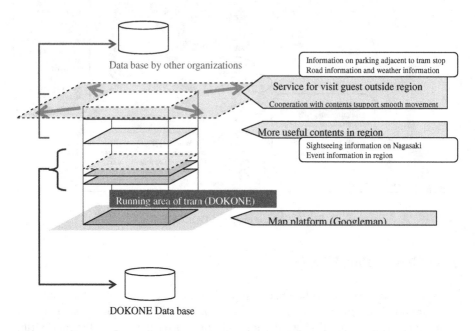

DOKONE Data base

Fig. 7. Expandability of geographic information database

Figure 7 is drawing of the expandability of the geographic information database. The possibility of linking with various texts that exist in the society is shown.

These two kinds of extension methods will be integrated. The mode of expression of contents that assume the node of the real world and the text to be geographic space information is scheduled to be developed.

References

1. Aarne, A.: The types of the folktale. Suomalainen Tiedeakatemia, Helsinki (1964)
2. Ikeda, H.: A type and motif index of Japanese folk-literature. Suomalainen Tiedeakatemia, Helsinki (1971)
3. Inada, K.: Nippon Mukashibanashi Tsuukan, General survey and analysis of Japanese folktales, vol. 28. Dohosha, Tokyo (1988)
4. Seki, K.: Nippon Mukashibanashi Shuusei. Kadokawa Shoten, Tokyo (1950-1958)
5. Seki, K.: Nippon Mukashibanashi Taisei. Kadokawa Shoten, Tokyo (1978-1980)
6. Thompson, S.: Motif-Index of Folk-Literature. Indiana Univ. Press, Bloomington (1955)
7. Morita, H.: Literary Hypertext. Yushodo, Tokyo (2006)
8. Yanagida, K.: Birth of Momotaro. Sanseidou, Tokyo (1933)
9. Watabe, K., et al.: Start of Regional Tourism ITS Contents & Service System toward Realization of "Driving Tours of the Future" in Nagasaki EV & ITS. In: Proceedings of the 45th Infrastructure Planning. Japan Society of Civil Engineers, Kyoto (2012)

Methods for Generalizing the Propp-Based Story Generation Mechanism

Shohei Imabuchi and Takashi Ogata

Iwate Prefectural University, Takizawa, Iwate, Japan
g231k005@s.iwate-pu.ac.jp, t-ogata@iwate-pu.ac.jp

Abstract. This paper discusses some methods for generalizing our Propp-based story generation mechanism. The mechanism has the following aspects: as the development of a system based on the literary theory by Propp and as the use in a more comprehensive architecture of narrative generation called the integrated narrative generation system. Considering the latter aspect especially, the generalization beyond the restriction of Propp's theory will become an important issue for the future development. The first half of this paper will introduce overviews of the Propp-based mechanism and the integrated narrative generation system. Then in the latter half, we will present four methods for the generalization.

Keywords: Integrated narrative generation system, Propp's theory, Propp-based story generation mechanism.

1   Introduction

Our integrated narrative generation system aims to organically combine a variety of mechanisms or modules dependent on our previous partial and fragmentary research results into one organic mechanism. Ogata and Kanai [1] described its ultimate philosophy and overall design, and Akimoto and Ogata [2] introduced the current status of its actual implementation. This system consists of the following phases of story generation, discourse, and surface expression. The first phase generates a story's conceptual structure including a temporal sequence of events from a set of fragmentary input information. The discourse phase can transform a story structure generated by the above phase into a variety of discourse structures to be expressed by the next phase. The last phase transforms a discourse structure into various types of surface expressions including language, image, and music. This integrated system is a kind of knowledge-based architecture in which several types of conceptual dictionaries and various knowledge bases are used. We have developed conceptual dictionaries [3] for noun, verb, adjective, and adjective verb concepts. The latter includes a content knowledge base for the story generation, a language notation dictionary, etc.

On the other hand, we have applied such literary theories as Genette, Jauss, and inter-textuality to the integrated system as one of the important characteristics of our research. Propp's theory [4] in this paper is also one of the basic literary theories relevant to some significant functions. We have had two types of approaches for the literary theory by Propp: the first is developing a story generation mechanism using

T. Yoshida et al. (Eds.): AMT 2013, LNCS 8210, pp. 333–344, 2013.
© Springer International Publishing Switzerland 2013

Propp ("Propp-based story generation mechanism") as comprehensively as possible and the second is introducing it as one of the modules into the integrated system [5].

The topic to be discussed in this paper is the "generalization" of Propp's theory. Propp was a Russian folklorist and his theory was, in spite of the generality, based on the analysis of the restricted number of Russian fairy tales. Although its faithful imitation is also an interesting topic, this paper aims to generalize toward the two directions. In particular, we use the mechanism as one of the modules or a group of functions in the integrated system and introduce more general and expanded knowledge into the mechanism. Table 1 presents generalization methods from these view points and the corresponding sections. This paper will describe a current comprehensive framework toward the generalization of the Propp-based story generation mechanism.

Table 1. Methods for generalizeing the Propp-based story generation mechaism

The type of "generalization"	The method of "generalization"	In this pepar
Using the Propp-based story generation mechanism as one of the functions or modules in the integrated narrative generation system	The Propp-based mechanism as a group of story techniques	Section 3.2
	The integration through a state management mechanism	Section 3.3
Introducing more general and expanded knowledge into the Propp-based story generation mechanism	The automatic extraction of generative rules using the state management mechanism	Section 4
	The automatic acquisition of scripts	Section 5
	Changing the order of functions	Section 6

2   The Current Propp-Based Story Generation Mechanism

We summarize the previous version of the Propp-based mechanism based on [5]. The focal system is a kind of story grammar as a systematic organization of Propp's theory. Although we have called it "Propp-based story grammar", we change the term "story grammar" into "story content grammar" since "story grammar" may cause misunderstanding for the following reason. Traditional story grammar provides a formal definition of a story's structure and the description contains generative rules for a part of a story. The point is that the rules are described for defining consistently formal developments in a story. The content knowledge for a "problem", for example, must be described outside the story grammar. By contrast, each element in the Propp-based story content grammar is equal to a definition for specifying the content information and the rules includes such elements for indicating a concrete event in a story as "villainy", which is one unit in 31 "functions" to be described in the next part.

Propp proposed the concept of "function", which means an action seen from the result and the principle of the sequential arrangement of 31 "functions" [1] as the most important theoretical idea. The "function" of "villainy", for example, means that a

---

[1] 1.Absentation, 2.Interdiction, 3.Violation, 4.Reconnaissance, 5.Delivery, 6.Trickery, 7.Complicity, 8.Villainy or lack, 9.Mediation, 10.Beginning counter-action, 11.Departure, 12.First function of donor, 13.Hero's reaction, 14.Provision or receipt of a magical agent, 15.Guidance, 16.Struggle, 17.Branding, 18.Victory, 19.Liquidation, 20.Return, 21.Pursuit, 22.Rescue, 23.Unrecognized arrival, 24.Unfounded claims, 25.Difficult task, 26.Solution, 27.Recognition, 28.Exposure, 29.Transfiguration, 30.Punishment, 31.Wedding.

character causes the damage of a different character. There is a variety of specific actions to actualize it, since the "function" means an abstract definition. We call more specified actions for actualizing each of the "functions" "sub-functions". Propp described a variety of examples corresponding to the sub-functions. The Propp-based story content grammar is based on the framework of Propp's theory with "functions" and sub-functions as one of the chief concepts. The detailed description will be presented in the next section. As the difference of "functions" and sub-functions apparently shows, it is a kind of blending of theoretical description and case description.

On the other hand, the Propp-based mechanism is designed to be combined with the integrated narrative generation system. Its primary medium was conceptual dictionaries including several hierarchical structures for the concepts of verb, noun, etc. These dictionaries have been developing to be used commonly in the integrated narrative generation system. Fig. 1 shows the overall formation of the Propp-based mechanism. A story as the output information is equivalent to a sequence of events bound hierarchically using "story relations" including "causal relation", "continuation relation", etc. The form of an event as a basic component in a story is a frame representation, which is formed by a verb concept and the corresponding case elements for such noun concepts as agent, object, and location. The verb concepts and noun concepts are associated with each corresponding element in the conceptual dictionaries. In Fig. 1, the structure generation mechanism and the event concept generation mechanism are respectively corresponding to generative mechanisms and the Propp-based story content grammar and the conceptual dictionaries are the parts of knowledge to be referred by the generative mechanism. As these generative parts and the parts to be referred are mutually independent, the redundancy of processing lessens.

2.1   The Propp-Based Story Content Grammar

The Propp-based story content grammar is divided into five hierarchical levels. The level 1 is the highest level to determine the macro or overall structure of a story. The level 2 is the layer for grouping 31 kinds of "functions" into the several higher parts and these "functions" are generated in the next level 3. The level 4 generates 198 kinds of sub-functions from the "functions". The story content grammar also implements "pairs of sub-functions" which mean co-occurring sub-functions. For this mechanism, we have defined 85 kinds of pairs among sub-functions. In the last level 5, each generated sub-function is derived to a tangible action or a sequence of actions (238 kinds), in which each action is described in a case structure that has a verb concept and the necessary noun concepts. Sub-functions and case structures at the level 4 and 5 have been described in hand according to the description of examples in [4]. In addition, an agent in a case structure under each sub-function is corresponded to one of the seven types of roles including hero, villain, victim, helper, dispatcher, donor, and false hero in narrative characters defined by Propp. In these levels, the level 3 is on the basis of the theoretical definition of "functions" and the level 4 is based on the case description of sub-functions. The level 1 and 2 are convenient categorizations

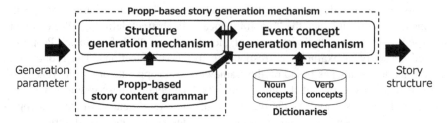

Fig. 1. An overview of the Propp-based story generation mechanism

according to the description by Propp. And the level 5 deals with a more general method for the operation of events to be basic units in a generated story.

Fig. 2 shows the part. We call each of the components with a right term and a left term a "generative rule" because a left term is transformed to a right term or a right term is generated from a left term. For example, the generative rule from the line seven in Fig. 2 means that the "function" of "interdiction" can be expanded to the two sub-functions of "interdiction" and "order or a suggestion". For the "order or a suggestion", though we described based on the Propp's description constantly, the actual program independently processes the "order" or "suggestion". Each of the right terms at the level 5 is equal to a case structure which is the framework of an event or event concept to be generated by an "event concept generation mechanism".

```
(setq *Propp-level1-list* '((Russian_folk_tale (P-Problem P-Trial P-Solution))))
(setq *Propp-level2-list* '((P-Problem (Reserve_portion Beginning))
                            (P-Trial (OR (Reserve_trial Battle_and_victory) (Reserve_trial Task_and_solution)))
(setq *Propp-level3-list* '((Reserve_portion (OR (01_Absentation 02_Interdiction 03_Violation)
                            (04_Reconnaissance 05_Delivery 06_Trickery)
(setq *Propp-level4-list* '((01_Absentation (OR bet-1_Going_out1 bet-2_Death bet-3_Going_out2))
                            (02_Interdiction (OR gam-1_Interdiction gam-2_Order_or_suggestion))
(setq *Propp-level5-list* '((bet-1_Going_out1 ((Go_out (1) (agent Parents) (object Business))))
                            (bet-2_Death ((Dead (1) (agent Parents)))) ···
(setq *Propp-function-pair-list* '((gam-1_Interdiction del-1_Violation)
                            (gam-2_Order_or_suggestion del-2_Order_execution) ···
```

Fig. 2. A part of the Propp-based story content grammar

2.2   The Event Concept Generation Mechanism

The integrated narrative generation system uses conceptual dictionaries [3] including four types of hierarchies. We shall detaile the dictionaries for especially noun and verb concepts in section 3.1. Each event concept in a story structure is represented by a case structure containing a verb concept and several necessary noun concepts. When events are generated at the level 5 in the story content grammar, the Propp-based story generation mechanism and the integrated narrative generation system are organically combined through its process. To be precise, the Propp-based mechanism, at that time, operates as a part of the integrated system. First, it searches for a verb concept described in the level 5 to get the case structure, for example, "order 1" for "interdiction". Next, it inserts the case information described in the level 5 into this case structure. At this time, if there is a description like "!hero" in the seven types of characters' roles, it assigns the real name like "Ivan" to the variable. On the other hand, if there is a description like "human" enclosed by double quotes, it refers to the noun conceptual dictionary to get the lower level's concept like "boy". Since the description of an event like "event 1" contained in a case structure means a nested event concept, it inserts a new event concept into the case structure.

2.3   Structure Generation Mechanisms in the Propp-Based Mechanism

The Propp-based story content grammar is opened to different types of operations including top-down, bottom-up, and hybrid. As mentioned above, in the all operations, the event concept generation mechanism commonly generates concretized events using the conceptual dictionaries. The top-down process is the standard method to simply expand the story content grammar from the highest level to the lowest level according to the user's input information including each name of the characters and the noun concept corresponding to the seven types of roles. This noun concept means such attribute of a character as a "brave" for a "hero" and a "snake" for a "villain". When the processing reaches to a case structure at the level 5, the event concept generation mechanism generates the event concept. In the bottom-up processing, the user inputs a case structure which is corresponding to an element of the lowest level in the story content grammar and other several types of information. The mechanism makes a story structure based on the inputs by using the hierarchy and pairs of subfunctions. The inputs have the following three types of possibilities: (1) The types, names, and noun concepts as attributes of one or more characters in the seven types of roles, (2) event concepts from 1 to 5, or (3) the combination of the above (1) and (2). This processing goes up the hierarchy from all case structures in which a type of the input information is contained. This mechanism puts together generated partial story structures into a whole. The last hybrid method is a mixed processing of the above two types of mechanisms and carries out a bottom-up processing to the hierarchy which the user designates as one of the inputs. Then, in contrast, it executes a top-down processing to generate a sequence of events at the lowest hierarchy.

3   Generalization (1): Combing with the Integrated System

Of the new version of the Propp-based mechanism, the combination with the integrated system is majorly conducted dependent on its application as a group of story techniques and its connection with a "state management mechanism". Before these explanations, we sketch an overview of the integrated narrative generation system.

3.1   An Overview of the Integrated Narrative Generation System

As mentioned in section 1, the integrated narrative generation system consists of three types of generation phases: story, discourse, and expression. On the other hand, from the type of processing, the system architecture has the following parts. The first "narrative techniques" are divided into "story techniques" and "discourse techniques" to respectively generate conceptual structures of stories and discourses. The next part has several mechanisms relating to knowledge to be referred by the part for generating narratives. Four types of conceptual dictionaries are mainly used in the event generation and the story techniques also use the content knowledge including scripts, causal relations, etc. stored in a story knowledge base. The other mechanisms include techniques to generate several types of surface representation and mechanisms to control or adjust the use and the timing of narrative techniques. The parts except for the conceptual dictionaries and the content knowledge bases are currently defined with about 550 original functions by Common Lisp.

For the conceptual dictionaries [3], we have been developing systems for concepts of verbs, nouns, adjectives, and adjective verbs with the respective hierarchical structures according to "is-a" relation. We differentiate a word's meaning or concept from the word itself as the fundamental policy of the development. In particular, one or more meanings relating to a noun are described in the noun conceptual dictionaries as one or more noun concepts, and the word to be used normally and notations for these concepts are described in another linguistic dictionary. The noun conceptual dictionary contains 115765 terminal concepts and 5808 intermediate concepts. The verb conceptual dictionary has 12174 terminal concepts and 36 intermediate concepts. A terminal verb concept in the hierarchy has the following information: a basic sentence pattern for natural language generation, one or more case structures, and constraints for each of the cases. A case structure defines several cases for nouns required in the verb. The structures of the adjective and adjective verb conceptual dictionaries are same as the verb conceptual dictionary, and 739 adjective concepts and 1372 adjective verb concepts are contained under 36 intermediate concepts in common with the verb conceptual hierarchy. In addition, knowledge for managing relations among events and mutual relationships between events & states are composed on the basis of the conceptual dictionaries for verb and noun concepts.

Although the Propp-based story generation mechanism can be used as an independent story generation system, its integration into the phase of story generation in the integrated narrative generation system increases the diversity and flexibility in story generation. A story is described, in the story generation phase, as the form of conceptual representation including a temporal sequence of events at the bottom level in a hierarchical tree structure with several types of relations as the intermediate nodes. Each of the events is described as a specific case structure which has a verb concept and the necessary instantiated noun concepts linked to the conceptual dictionaries. Fig. 3 shows a story structure and a case structure of an event within the story structure. The story generation mechanism generates story structures having a sequence of events based on structural operation techniques using the dictionaries and the other narrative knowledge. We call a procedure for expanding the tree structure of a story

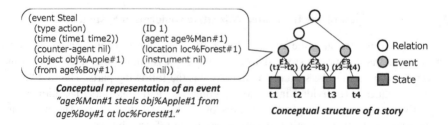

Fig. 3. The structures of a story and an event

using the story knowledge base, which stores concrete narrative fragments or structures for such various types of narrative relations as scripts and causal relations, a narrative technique for story generation or directly a story technique. The input to a story technique is an event, the whole of a story structure, or a sub-structure in a story structure and the output is chiefly the expanded story structure that is one of the possible various structures. They can treat from micro story structures to macro ones.

3.2   The Propp-Based Mechanism as a Group of Story Techniques

As stated above, a story is represented as a tree structure in which some events are bound under a relation, and is gradually expanded according to story techniques including various types of structural transformation functions. Each of the story techniques uses specific content information in the story knowledge base to be operated by the technique. The Propp-based mechanism can also be a kind of story technique. A group of story techniques are used to operate such micro structures as causal relations for forming partial structures in a story. In contrast, the Propp-based mechanism can construct a macro story structure in which names of the higher levels are corresponded to names of the relations for binding events in the lowest level. Considering the relation between story techniques and the story content knowledge, a set of the Propp-based grammar give an element in the story knowledge base, while the three types of developing methods are equivalent to a group of story techniques to return each tree structure. Since a set of the grammar means a sort of static data to be referred by the story techniques, it is classified into a type of story content knowledge.

3.3   The Integration through a State Management Mechanism

We stated that the Propp-based story generation mechanism can be combined with the integrated narrative generation system through the conceptual dictionaries and as a group of story techniques connecting with the story knowledge base in the previous sections. Additionally, the "state management mechanism" in the integrated system can also be a medium for the integration. Overall, a chief function of this state management mechanism is managing and complementing semantic consistency at the most micro level in each generated story structure. Semantic consistency in a story is obviously a significant issue in the Propp-based mechanism too. Since both of the Propp-based mechanism and the integrated system use a same form of story representation, the state management mechanism can commonly be applied to the part of processing based on the Propp-based mechanism.

A state in the integrated system is equivalent to the description of the attribute information of instances for characters, objects, and places at each time point between an event and the next event in a story structure. A state at a time point in which "the location of Ivan (the name of an instance of a young man) is a house" can, for example, be changed to the different state of "the location of Ivan is a wood" through the medium of the event of "Ivan goes to the wood from the house". The mechanism manages the consistency of a sequence of events, and at the same time the states in a story are equal to a set of knowledge to store the background information of events.

For the form, a state management mechanism uses a "state-event transformation knowledge base" to generate states from a sequence of events [6]. It contains many rules called "(state-event) transformation rules" and each of them consists of "changing content(s)" and "precedent condition(s)". Each of the changing contents defines the change of states caused by an event and each of the precedent conditions describes the requirements which an event is necessary to have for the realization. It repeats the following same three procedures from the first event to the last one: (1) Based on the verb concept and the case structure in the current event, the mechanism acquires a transformation rule from the state-event transformation knowledge base. (2) It checks whether the previous state of the targeted event satisfies the precedent condition(s) in the transformation rules acquired in the (1). If it does not satisfy all of the precedent condition(s), it generates a state in which the attribute slot violating the precedent condition is rewritten to the value described in the precedent condition to add (complement) the state to the next position of the previous state of the targeted event. (3) According to the changing content(s) described in the transformation rule acquired in the (1), it rewrites values of the attribute slots in the previous state of the targeted event or the state generated in the (2) to generate the state in the next position of the targeted event. In this way, a conceptual structure for a story or a partial story is generated according to the application of the Propp-based story techniques, and a set of states are made by the state management mechanism to expand the story structure.

4   Generalization (2): The Extraction of Generative Rules

The generation ability of the Propp-based story generation mechanism is limited in story generation in the style of a Russian fairy tale because it is intended to organize Propp's theory as constantly as possible. It means not only formal limitation, but also the limitation of content. The reason is that we used the knowledge of the content of Russian fairy tales based on the organization of sub-functions and more specified events according to actual examples by Propp himself. Our objective in this section is making the Propp-based story content grammar generate a variety of stories beyond the range of Russian tales by Propp's theory. This direction provides another method which is different from the combining with the integrated narrative generation system to generalize the Propp-based story generation mechanism. In particular, we present a method using the above state-event transformation knowledge base to extract new rules for some hierarchical levels under "functions" in the story content grammar.

In the current Propp-based grammar, one or more sub-functions are defined under each "function". The "interdiction", for example, has two sub-functions of "interdiction" and "order or a suggestion" based on examples by Propp consistently. However, if we consider the general property of a "function" which allows all the realization forms, we will be able to increase the species and number. In fact, Tosu [7] defined the structure of a Japanese folk tale, "The Grateful Crane"[2], dependent on the same "functions" and the partially different sub-functions. In this proposal, an episode or scene, *"when she prepared to make fine cloth in her room, she said to her parents.*

---

[2] http://www.geocities.co.jp/HeartLand-Gaien/7211/

"From now I'll make fine cloth. While I am working, promise me never to look into my room, Never. Never."", is equivalent to the sub-function of "interdiction" which is a particular method of the "function" of "interdiction". This example uses the same sub-function as Propp defined. However, although a scene, *"he found a crane in a trap. The more it moved, the harder the trap caught it. The old man felt very sorry for it. "Stop, don't move. I'll help you." He helped it from the trap. It flew away toward the mountain."*, is corresponded to the "functions" of "difficult task" and "solution", "the request of rescue" and "rescue" described in this scene do not appear in Propp's sub-functions. Our purpose is to automatically acquire new sub-functions which do not appear in the description by Propp to insert new rules corresponding to the sub-functions into the story content grammar. The automatic acquisition of case structures for the specified events also has a similar effect. Extracting rules automatically will be the first step toward the general constitution of the Propp-based grammar.

Sub-functions and case structures are extracted based on a comparatively simple method using the state-event transformation knowledge base. Fig. 4 shows the general framework. If the A and B are respectively a "function", the transformation process is handled by a sub-function. In such case, since a "function" means defining the change from the initial state A to the goal state B, if the definition can adequately be described, we will be able to call many specific sub-functions from the A to the B which meets the condition. The acquisition of a set of methods which causes the adequate change from the A to the B is equal to increase the diversity of rules at the level of sub-function. In the same fashion, this method also enables to extract case structures at the level under the sub-function. Generally speaking, if we can define the semantic conditions and effects of such corresponding objects as "functions" and sub-functions, we will be able to extract a variety of concrete methods to transform an initial state A to a goal state B. It is corresponded to expand the state-event transformation mechanism to the levels of "function" and sub-function, but may not necessarily be a sufficient automatic acquisition function since much pre-defined description is necessary. However, this attempt will contribute to make a framework of the advanced narrative knowledge automatic acquisition in the future. In addition, a semantic classification of verb concepts may be caused by the abstract definition of the transformation of states.

Currently, we are experimentally making an automatic extraction mechanism of case structures according to the description of initial states and goal states for several sub-functions. Based on the state-event transformation knowledge base, for example, the sub-function of "killing" in the "function" of "villainy" is defined as below: the first state is "hero-health: fine" and the goal state is "hero-health: death". On the basis of the definition, the mechanism searches for such verb concepts as "kill", "hold an execution", "blip off", and "bang" through which the attribute of "health" is changed to "death" from "fine" form the knowledge base. These extracted elements can be used as rules to generate case structures inserted into the right side at the level 5. Fig. 5 shows sentences from sub-functions extracted by this mechanism.

A Initial state　Extraction of rules　C　B Goal state

Fig. 4. The framework of the extraction of rules from an initial state and a goal state

In this experimental attempt, we respectively extracted 63, 415 and 15 case structures dependent on the state changes of "health", "possession", and "perception" in the three types of sub-functions of "killing", "acquisition", and "transformation". However, quite a lot of results were not usable because of the inadequate relationships among the verb concept and the noun concepts. The reason is that we do not use constraints in the verb concepts which we have defined in the current version. In the future, some sort of "functions" and sub-functions in the current definition used in the integrated system are necessary to define very abstract description for initial states and goal states. For the initial and goal states of the "function" of "villainy", for example, such abstract definition as the lowering of the victim's physical or mental state may be necessary. Finally, this attempt does not include the automatic extraction of 31 kinds of "functions" since we regard them as the substantial elements for deciding the general structure of stories. Treating the level of "functions" in the context of automatic knowledge acquisition is a topic to be discussed in the future. But alternatively we are considering a way for changing the arrangement sequence (section 6).

··· The snake <u>killed</u> the princess in the forest. Ivan came to the king's palace from ···

··· The snake <u>killed</u> the princess in the forest. Ivan came to the king's palace from ···
··· The snake <u>assassinated</u> the princess in the forest. Ivan came to the king's palace from ···
··· The snake <u>held on execution</u> the princess in the forest. Ivan came to the king's palace from ···
··· The snake <u>banged</u> the princess in the forest. Ivan came to the king's palace from ···
··· The snake <u>blip off</u> the princess in the forest. Ivan came to the king's palace from ··· (and more.)

Fig. 5. Sentences using extracted case structures of "killing"

5   Generalization (3): The Automatic Acquisition of Scripts

Since events generated at the level 5 in the Propp-based grammar are according to the sub-functions, the scope is restricted to the style of a Russian fairy tale. As mentioned above, one of the generalization methods was acquiring and using new events concepts. On the other hand, the use of "scripts" is also the way. In the current integrated system, we have prepared about seventy scripts by hand with the story technique of script to transform a more abstract event into a more concrete or detailed sequence of events. It has various roles. First, the script technique complements the semantic consistency between two events and gives the embodied image to the skeleton of a story. Second, the use of the script technique in the Propp-based mechanism contributes to emancipate the world of stories from Propp-like world. As the above automatic extraction, currently we are making a start on the automatic acquisition of scripts.

As the first attempt, we are implementing a mechanism using N-gram for natural language processing by [8] which proposed an analysis of sentence patterns based on N-gram. A sequence of verbs composing of N verbs is defined N-gram. For example, the sequence of verbs of "have-eat" is 2-gram (bi-gram). The automatic script acquisition process is composed of the next three phases. The targeted texts are novels in Aozora Bunko (the version of modern notation of kana letter)[3]: (1) Acquiring the

---

[3] http://www.aozora.gr.jp

frequency table of N-gram from the texts: The mechanism extracts all verbs according to the appearance order by a paragraph using the tool "MeCab"[4] to acquire all possible N-grams from the extracted verbs. For example, from a sequence of verbs, "have-eat-satisfy", we can acquire "have-eat" and "eat-satisfy" (2-gram) and "have-eat-satisfy" (3-gram). These are stored in frequency tables of N-gram. (2) Acquiring script candidates from the frequency tables: From the frequency table of N-gram, the mechanism can select the N-grams with the highest appearance frequency in the N-grams in which a verb is the starting point. For instance, if there are "eat-satisfy" (appearance frequency: 5) and "eat-drink-sleep" (appearance frequency: 7) in the N-grams of "eat", the latter is selected. (3) Giving a case structure to each of the selected verbs: It makes verbs in each of the selected N-grams into the case structures based on the description in the verb conceptual dictionary. The final form of a script is a sequence of case structures corresponding to the N-gram as shown in Fig. 6.

```
(script (persuade1 (elect3 like1 take1 pay14)) ((event presuade1 (agent (&v age1)) (counter-agent (&v age2)))
        ((1 (event elect3 (agent (&v age1)) (counter-agent (&v age2))))
        (2 (event like1 (agent (&v age1)) (counter-agent (&v age2))))
        (3 (event take1 (agent (&v age1)) (counter-agent (&v age2)) (to (&v loc1))))
        (4 (event pay14 (agent (&v age1)) (counter-agent (&v age2)))))))
```

Fig. 6. An example of scripts acquired

6   Generalization (4): Changing the Order of "Functions"

Several researchers have been trying to apply Propp's theory, especially the "functions", to the other genres of stories. Tosu [7] presented an example of the analysis of Japanese folk tales. This research has also showed that "The Grateful Crane" can be described according to the sequence of "functions" of "lack-difficult task-solution-liquidation-interdiction-violation-lack". Although this research uses the same "functions" as Propp theory, the order of the application is different from the original sequence by Propp. This sort of example indicates a general characteristic by Propp.

Using the changed order of "functions" is a method to generalize Propp's theory. We have experimentally substituted the next description in the level 1 and 2 for the description in Fig. 2: "(The_grateful_crane (00_preliminary_part Story))" and "(Story (08a_lack 25_difficult_task 26_solution 19_liquidation 02_interdiction 03_violation 08a_lack))". The result showed that stories with global story lines and different realization in events were generated. The method in section 4 extracts partial rules inserted into the particular levels in the grammar, while the method here means the substitution of patterns based on the "functions" and sub-functions to be treated as the expansion of the grammar.

The technique enabled the generation of stories which have the global structure of "The Grateful Crane" but differ in the details. The scene of "*a young man met an injured crane*" ("difficult task" in the analysis by Tosu [7]) was, for example, developed as the event of "*a middle-aged man posed an unsolvable riddle for a young man.*". The original scene of "*the crane interdicted opening the door to the young*

---

[4] https://code.google.com/p/mecab/

man." according to the "function" of "interdiction" was developed as the event of "*A grandfather gave an order to the young man sending breakfast into his brother.*". On the contrary, in the future, to generate stories with details similar to the original story, we need inserting narrative knowledge or rules dependent on "The Grateful Crane" into the lower levels in the Propp-based story content grammar.

7   Conclusions

We have discussed methods to generalize the Propp-based story generation mechanism from the next view points: as the development of a system based on Propp's theory and as the use in a more comprehensive architecture of the integrated narrative generation system. Regarding the latter especially, the generalization beyond the restriction of Propp's theory will be an important issue in the future. The first half of this paper has introduced the Propp-based mechanism and the integrated system. The latter half has presented the following methods: its integration with the integrated system based on the conceptual dictionaries and the state management mechanism, the automatic extraction of rules using the state management mechanism, the automatic acquisition of scripts from outside texts, and the changed orders of "functions". Finally, Propp's theory and narratology essentially have a general characteristic. Although the theory of "function" has extensibility and flexibility, the other various types of methods described in [4] also cover many parts of story generation. Narratological approaches to narrative generation still possess great potentials.

References

1. Ogata, T., Kanai, A.: An Introduction to Informatics of Narratology. Gakubunsha (2010) (in Japanese)
2. Akimoto, T., Ogata, T.: Macro Structure and Basic Methods in the Integrated Narrative Generation System by Introducing Narratological Knowledge. In: Proc. of the 11th IEEE International Conference on Cognitive Informatics & Cognitive Computing, pp. 253–262 (2012)
3. Oishi, K., Kurisawa, Y., Kamada, M., Fukuda, I., Akimoto, T., Ogata, T.: Building Conceptual Dictionary for Providing Common Knowledge in the Integrated Narrative Generation System. In: Proc. of the 34th Annual Meeting of the Cognitive Science Society, pp. 2126–2131 (2012)
4. Propp, V.: Morphology of the Folktale. University of Texas Press (1968)
5. Imabuchi, S., Ogata, T.: A Story Generation System based on Propp Theory: As a Mechanism in an Integrated Narrative Generation System. In: Isahara, H., Kanzaki, K. (eds.) JapTAL 2012. LNCS, vol. 7614, pp. 312–321. Springer, Heidelberg (2012)
6. Akimoto, T., Kurisawa, Y., Ogata, T.: A Mechanism for Managing the Progression of Events by States in Integrated Narrative Generation System. In: Proc. of the 2nd International Conference on Engineering and Applied Science, pp. 1605–1614 (2013)
7. Tosu, N.: Cultural linguistics. Keisou Shobou (1988) (in Japanese)
8. Nagao, M., Mori, S.: A New Method of N-gram Statistics for Large Number of n and Automatic Extraction of Words and Phrases form Large Text Data of Japanese. In: Proc. of the 15th Conference on Computational Linguistics, vol. 1, pp. 611–615 (1994)

GA-Based Method for Optimal Weight Design Problem of 23 Stories Frame Structure

Takao Yokota, Kiyoshi Tsukagoshi, Shozo Wada[1],
Takeaki Taguchi, and C. Tarn[2]

[1] Department of Systems and Information Engineering and
Department of Architecture Ashikaga Institute of Technology,
Ashikaga 326-8558, Japan
[2] Department of Medical Informatics, Takasaki University of Health and Welfare,
Takasaki, 370-0033, Japan

Abstract. In this paper, we formulate an optimal weight design (OWD) problem of a 23 stories frame structure for a constrained relative story displacement as a statically indeterminate structure problem and solve it directly by keeping the constraints based on an improved genetic algorithm (GA). We discuss the efficiency between the proposed method and the discredited optimum criteria methods.

Keywords: Frame structure, Genetic Algorithm.

1   Introduction

Generally, the optimal design of a framed structure is executed by combining structure analysis with an optimization method and then the iterative retrieval method is used as the search technique.

This method, though, has a large number of design variables which increases the amount of calculation and hence increases the analysis time. One of the methods proposed to solve the above problem is the sub optimization method for frame structure analysis of multimember.

Homma, *et al.* [1] introduced the scaling $I^{1/4}$(I:second moment of inertia) into the cross-sectional property of the H shaped steel beam, in an attempt to improve calculation efficiency.

Uchimura, *et al.*, [2] on the other hand, proposed a steel structure improved method focusing on lateral drift using the unit load method. Here the concept of displacement participation factor, which should be distinged from sensitivity analysis, is obtained from the unit load method.

The application of genetic algorithms (GA) (which uses the stochastic method) effectively solves the combinatorial problem making it a remarkable new tool for solving these kind of problems[3].

In this paper, we formulate an optimal weight design OWD problem of 23 stories frame structure for a constrained relative story displacement of frame member as a statically indeterminate structure problem, and are able to get a global solution and solve it directly by keeping the constraints by using improved GA [4]-[5]. As a result, the number of decision (design) variables does not increase and easily gets the best compromised solution.

T. Yoshida et al. (Eds.): AMT 2013, LNCS 8210, pp. 345–351, 2013.
© Springer International Publishing Switzerland 2013

2   An Optimal Weight Design Problem of 23 Stories Frame

Let us consider an OWD problem of 23 stories frame with a minimum weight which is a full stress design without the deflection as shown in Fig. 1[2].

We formulate the optimal design problem with a satisfied allowable relative story displacement with a minimized 23 stories frame weight $W(A)$. It is the following linear programming (LP) problem:

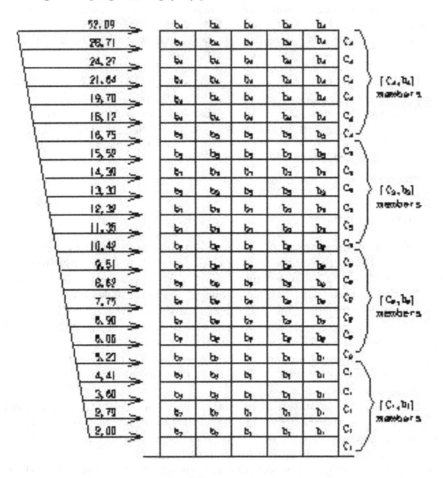

Fig. 1. An OWD problem of 23 stories frame

$$\min \quad W(A) = \rho \sum_{i=1}^{253} l_i A_i \tag{1}$$

$$\text{s. t.} \quad G_{1i} = \sigma_i \le b_{1i}, (i = 1, 2, .., 253) \tag{2}$$

$$G_{2i} = \theta_i \le b_{2i}, (i = 1, 2, .., 253) \tag{3}$$

$$\sigma_i^L \le \sigma_i \le \sigma_i^U, \quad (i = 1, 2, .., 253),$$

$$\theta_i^L \leq \theta_i \leq \theta_i^U, \quad (i = 1, 2, .., 253),$$

where

$$\sigma_i = M_{max(i)}/Z_i, (i = 1, 2, .., 253) \tag{4}$$

Where $W(A)$ is the weight function, ρ is density of frames' members, σ_i/θ_i $(i = 1, \ldots 253)$ are stress/angle of relative story displacement of frame's members, b_{1i}/b_{2i} are allowable stress/ angle of relative story displacement of frames' members, σ_i^L/σ_i^U and θ_i^L/θ_i^U are lower/upper limits of allowable stress and angle of relative story displacement. $M_{max(i)}/Z_i$ are maximum bending moument/section modulus of frame's members.

3   Representation and Evaluation Function

Yokota et al. [5] reported an IGA to cope with such a problem. That is, for the constraint condition, the following measure is introduced:

$$d_i = \begin{cases} 0; \\ G_i \leq b_i, \ i = 1, 2, \cdots, 253; \ \text{otherwise} \end{cases} \tag{5}$$

The measure d_i adopted here indicates how much the left-hand side of the expression corresponding to the constraint condition exceeds the right-hand side, and the evaluation function is defined as follows:

$$eval(V_k) = W(A_1, A_2, \cdots, A_{253}) \left(1 - \frac{1}{253} \sum_{i=1}^{253} d_i \right) \tag{6}$$

This IGA includes information about unfeasible chromosomes as near as possible to the feasible region in the evaluation function.

We can select the best chromosome with the following equation:

$$V^* = \arg \min_{V_k} \{eval(V_k)\} \tag{7}$$

$$mineval = \min\{eval(V_k)\} \tag{8}$$

Where the "arg" of argmin is an abbreviation for argument and argmin means that we adopt the chromosome whose evaluation returns the lowest value.

4   Algorithm

The procedure for solving the OWD problem of 23 stories frame by genetic algorithms is proposed in this section.

Step 1: Set population size *pop_size*, crossover
probability P_c, mutation probability P_m, maximum generation *maxgen*, initial generation *gen*=0, initial fitness value *maxeval*=0.
Step 2: Generate initial population $V_k(k = 1, \ldots, pop_size)$ randomly
Step 3: Caluculate each chromosome's fitness value $eval(V_k)$ and set *gene* = *gen* + 1.
Step 4: If *gen* < *maxgen*, goto Step 5. If *gen* = *maxgen*, output *maxeval* and terminate.

Step 5: Reproduce new chromosomes during arithmetic crossover and mutation process and perform the Elitest Selection.

step 6: Select any one chromosome in the chromosome population, Calculate the fitness of the chromosome.

Step 7: From equation (7) determine the best chromosome, calculate the fitness, register them, and return to **Step 3.**

Table 1. Coefficient value of frame members

No.	column c_{1-4}	beam b_{1-4}
1	□450 × 450 × 28	H 700 × 350 × 16 × 28
2	□450 × 450 × 25	H 700 × 350 × 16 × 25
3	□450 × 450 × 22	H 700 × 350 × 16 × 22
4	□450 × 450 × 19	H 700 × 350 × 16 × 19
S_1	$l_{1,24,47,70,93,116} = 450$	$l_{139-253} = 800$
S_2	$l_{2,25,48,71,94,117} = 400$	fx_{1-23}: lateral loads
S_3	$l_{3-23,26-46} = 360$	$\theta \leq 0.05,$
$\wr$	$l_{49-69,72-92} = 360$	$\theta \leq 0.06$
S_{23}	$l_{95-115,118-138} = 360$	

5   Numerical Example

We solve the OWD problem by using our proposed algorithm with coefficient values as shown in Fig. 1 and Table 1.

We set up the following initial parameters for the numerical example:

$$pop\_size = 20, \ P_c = 0.4, \ P_m = 0.1, \ maxgen = 100$$

The following results are obtained:

$$V^*_{1-253} = [4 \ 8 \ 4 \ 8 \ 4 \ 8 \ 4 \ 8 \ 4 \ 8 \ 4 \ 4 \ 8 \ 4 \ 8 \ 4 \ 8 \ 4 \ 8 \ 4$$

```
8 4 4 8 4 8 4 8 4 8 4 8 4 4 8 4 8 4 8 4
8 4 8 4 4 8 4 8 4 8 4 8 4 8 4 4 8 4 8 4
8 4 8 4 8 4 4 8 4 8 4 8 4 8 4 8 4 4 8 4
8 4 8 4 8 4 8 4 4 8 4 8 4 8 4 8 4 7 4 4
8 4 8 4 8 4 8 4 8 4 4 8 4 8 4 8 4 8 4 8
4 4 8 4 8 4 8 4 8 4 7 3 3 6 4 8 4 8 4 8
3 6 3 3 6 3 8 4 8 4 8 3 6 3 3 6 3 8 4 8
4 8 3 6 3 3 6 3 8 4 8 4 8 3 6 3 3 6 3 7
4 8 4 8 3 6 3 3 6 3 7 4 8 4 8 3 6 3 1 5
3 7 4 8 4 7 3 5 1 1 5 3 7 4 8 3 7 3 5 1
1 5 3 7 4 8 3 7 3 5 1 1 5 3 7 4 8 3 7 3
5 1 C  H C H C  H C  H C  H C ]
```

where C : $^{1 \times 3}$ H : $^{900 \times 350 \times 16 \times 19}$

The above values obtained are the best compromised solutions. The corresponding evaluation function value, that is, the minimized weight of the 23 stories frame structure, was $W(A_1, A_2, \cdots, A_{253}) = 321.704$ [ton] $(\theta \leq 0.005)$

6   Evaluation

In order to keep within the allowable constraints we have the enlarged the section modulus at the base of the structure(1st floor using 3 columns / Beam(H 900 × 350 × 16 × 19) in Fig.2 and Fig.3.

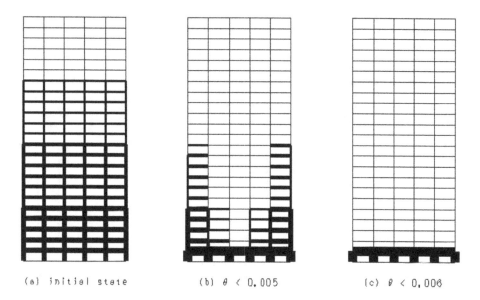

(a) initial state   (b) θ < 0.005   (c) θ < 0.006

Fig. 2. Allocation of the member subject

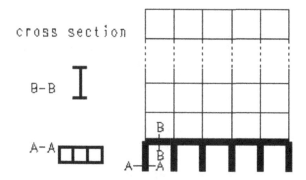

Fig, 4 column and beam of 1st story

Fig. 3. Column and beam of 1st story

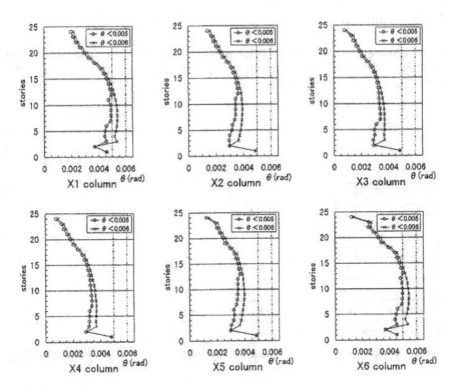

Fig. 4. Angle of relative stories displacement

In Fig.3, the constraints of the angle of relative stories displacement were satisfied by using three columns in the 1st story and increasing the beam from a height of 700[mm] to 900[mm].

After this treatment as shown in Fig.4, we can see the results of the each column's angle of relative story displacement. We can see that the results of the columns and beams show that they keep well within the constraints($\theta \leq 0.005$, $\theta \leq 0.006$). In Fig.2(a), we can see the initial state before optimization. Fig.2(b) $\theta \leq 0.005$ is the result after using our proposed method. In Fig.2(c), we have used an angle of relative story displacement of $\theta \leq 0.006$ and were able to use the lightest material in every floor and obtained a minimized W=306.886[ton] and connect with comment it has been composed composed of the member subject in which all member subjects are the lightest[4,8] except for the 1st story shown in Fig.2.

This allows the use of lighter material(small section modulus) particularly on upper floors.

In column 1 and 6 the lower floors keep within /or the limits of the constraints while quite a leeway can be observed in the upper floors.

7   Conclusion

In this paper, we formulate an OWD problem of a 23 stories frame structure for a constrained allowable stress of the dimension of each member as a statically indeterminate structure problem and solve it directly by keeping the constraints by using IGA. We discuss the efficiency of the proposed method. As a result, the number of decision (design) variables did not increase, and easily got the best compromised solution.

When a chromosome is not contained in the feasible region, it will include information about the unfeasible region's chromosome in the evaluation function in order to improve its search efficiency such as a cross-section of Square tube and H shaped steel beam.

In order to keep within the allowable constraints we have the an larged the section modulus at the based of the structure(1st floor using 3 columns / Beam(H $900 \times 350 \times 16 \times 19$).

After this treatment we applied our proposed method which allows us an increased leeway in the use of lighter materials toward the upper floors.

When the angle of relative story displacement was $\theta \leq 0.006$ and by using the same 3 columns / Beam(H $900 \times 350 \times 16 \times 19$) only on the first floor we obtained a minimized weight of W=306.886[ton].

References

1. Homma, Y., Iwatsuka, Y.: Minimum Weight Design of Elastic Frame Structure by SUMT. J. of the Society of naval Architects of Japan 143, 326–333 (1978)
2. Uchimura, H., Sato, K., Wada, A., Kokusho, S.: Steel structure improvement method focusing on lateral drift. J. Struct. Constr. Eng., AIJ 485, 73–80 (1996)
3. Goldberg, D.E., Samtani, M.P.: Engineering optimization vig geneti algorithms. In: Proc. of the 9th Conf. on Electronic Computations, ASCE, pp. 471–482 (1986)
4. Gen, M., Cheng, R.: Genetic Algorithms and Engineering Optimization. John Wiley & Sons, New York (2000)
5. Yokota, T., Wada, S., Taguchi, T., Gen, M.: Optimal weight design problem of elastic structure by GA. Comp. and Industrial Engineering (ELSEVIER) 53(2), 299–305 (2007)

A Proposal of a Genetic Algorithm for Bicriteria Fixed Charge Transportation Problem

Toshiki Shizuka and Kenichi Ida

Maebashi Institute of Technology
Maebashi 371-0861, Japan
ida@maebashi-it.ac.jp

Abstract. Transportation problem is a typical combinatorial problem. We aim at the search capacity of the solution by using a genetic algorithm with a the Bicriteria fixed Charge Transportation problem, which is an extension of the traditional transportation problem. In this paper, we improve the technique of Teramatu. In particular we propose new crossover operation for the genetic algorithm. Comparison with other methods is performed and the validity of the proposed method is shown.

Keywords: Bicriteria Fixed Charge Transportation Problem, Genetic Algorithm, Priority Level.

1 Introduction

In recent years, there has been a need for the development of algorithms which solve various combinatorial optimisation problems at high speed. Optimal solutions to large-scale combinatorial optimisation problems cannot be obtained within reasonable time because the number of possible combinations increases exponentially with problem size.

Near-optimal solutions can be obtained within reasonable time using methods such as genetic algorithms. Genetic algorithms (GA) mimic the heredity of living things and evolution and apply these principals to engineering problems. Many individual search points are considered in GA and the process of evolution drives improvement of this population so that near-optimal solutions can be obtained. The transportation problem is a typical combinatorial problem. It is a distribution system and is a fundamental network problem that has been studied by many researchers. Bicriteria fixed charge transportation problems extend the traditional problem and represent a more realistic case.

In this paper we propose a GA for bicriteria fixed charge transportation problems and compare its performance with other methods.

2 Fixed Charge Transportation Problem

There are m distribution centers and n destinations for certain goods. Transportation costs between distribution centers and destinations are known, the

T. Yoshida et al. (Eds.): AMT 2013, LNCS 8210, pp. 352–358, 2013.
© Springer International Publishing Switzerland 2013

demand at each destination is known and the supply capacity of each distribution center is known.

The fixed charge problem includes the additional consideration of a fixed charge applied when goods are conveyed from a distribution center to a destination. The problem is to determine the traffic, x_{ij}, that minimises total cost. It is formulated as follows.[2]

$$\texttt{min} \quad z_1(x) = \sum_{i=1}^{m} \sum_{j=1}^{n} (c_{ij}x_{ij} + d_{ij}g_{ij}) \tag{1}$$

$$\texttt{min} \quad z_2(x) = \sum_{i=1}^{m} \sum_{j=1}^{n} t_{ij}x_{ij} \tag{2}$$

$$\texttt{s.t.} \quad \sum_{j=1}^{m} x_{ij} = a_i \quad (i = 1, 2, \cdots, m) \tag{3}$$

$$\sum_{i=1}^{n} x_{ij} = b_j \quad (j = 1, 2, \cdots, n) \tag{4}$$

$$x_{ij} \geq 0, \forall i, j \tag{5}$$

$$\texttt{with} \quad g_{ij} = \begin{cases} 1, \text{ if } x_{ij} > 0 \\ 0, \text{ otherwise} \end{cases} \tag{6}$$

c_{ij} is the cost of carrying one unit of goods from distribution center i to destination j. d_{ij} is the fixed charge applied if $x{ij} > 0$. a_i is the supply capacity of distribution center i and b_j is the demand of destination j. It is required that supply and demand are balanced, that is, $\sum_{j=1}^{m} a_i = \sum_{i=1}^{n} b_j$.

3   Bicriteria Fixed Charge Transportation Problem

Bicriteria transportation problem that takes into account the transportation time and transportation cost is formulated as follows.

$$\texttt{min} \quad z_1(x) = \sum_{i=1}^{m} \sum_{j=1}^{n} (c_{ij}x_{ij} + d_{ij}g_{ij}) \tag{7}$$

$$\texttt{min} \quad z_2(x) = \sum_{i=1}^{m} \sum_{j=1}^{n} t_{ij}x_{ij} \tag{8}$$

$$\texttt{s.t.} \quad \sum_{j=1}^{m} x_{ij} = a_i \quad (i = 1, 2, \cdots, m) \tag{9}$$

$$\sum_{i=1}^{n} x_{ij} = b_j \quad (j = 1, 2, \cdots, n) \tag{10}$$

$$x_{ij} \geq 0, \forall i, j \tag{11}$$

$$\text{with} \quad g_{ij} = \begin{cases} 1, \text{ if } x_{ij} > 0 \\ 0, \text{ otherwise} \end{cases} \tag{12}$$

c_{ij} is the cost of carrying one unit of goods from distribution center i to destination j. d_{ij} is the fixed charge applied if $xij > 0$. t_{ij} is the time of carrying one unit of goods from distribution center i to destination j. a_i is the supply capacity of distribution center i and b_j is the demand of destination j. It is required that supply and demand are balanced, that is, $\sum_{j=1}^{m} a_i = \sum_{i=1}^{n} b_j$.

4   The Method to Propose

Without using a crossover approach of Teramatu is an existing technique, it was treating the transport problem only mutation.[2] Therefore, We propose a crossover that could correspond to this transportation problem.

4.1   Conventional Method

We used the method Teramatu using no crossover. We are shown in the following steps to send to preferentially luggage route high priority.

4.2   Preferential Ranking

The route with this cheap transportation cost and constant cost is an advantageous route by sending a load in many routes. Therefore, the method of determining the ranking of the priority, which shows the superiority, or inferiority of all routes is shown below.

1. **The priority by the maximum amount**
   Step 1: To determine the maximum amount that can be transported at the root of each of the following equation.
   $X_{ij} = \max \{a_i, b_j\}$
   $(i = 1, 2, \cdots, m)(j = 1, 2, \cdots, n)$
   Step 2: The following formula is calculated and it asks for the total cost of a transportation cost.
   $$z(x) = \sum_{i=1}^{m} \sum_{j=1}^{n} (X_{ij} x_{ij} + d_{ij})$$
   Step 3: The ranking of this priority is attached to the route in the cheap order of the total cost.

2. **The priority based on the average amount**
   Step 1: To determine the average amount that can be transported at the root of each of the following equation.

$$Y = \frac{1}{(m \times n)} \sum_{i=1}^{m} a_i$$

   Step 2: The following formula is calculated and it asks for the total cost.

$$z(x) = \sum_{i=1}^{m} \sum_{j=1}^{n} (Y x_{ij} + d_{ij})$$

   Step 3: The ranking of this priority is attached to the route in the cheap order of the total cost.

3. **The priority based on the average of the average and peak amount**
   Step 1: Determining the average of the maximum amount and average amount that can be transported at the root of each of the following equation.

$$Z_{ij} = \frac{(X_{ij} + Y)}{2}$$
$$(i = 1, 2, \cdots, m)(j = 1, 2, \cdots, n)$$

   Step 2: The following formula is calculated and it asks for the total cost.

$$z(x) = \sum_{i=1}^{m} \sum_{j=1}^{n} Z_{ij} x_{ij} + d_{ij}$$

   Step 3: The ranking of this priority is attached to the route in the cheap order of the total cost.

4. **Priority of only transportation costs**
   The ranking of this priority is attached to the route in the cheap order of the transportation cost.

5. **Priority of only fixed charge**
   The ranking of this priority is attached to the route in the cheap order of the fixed charge.

6. **Priority of only transportation time**
   The ranking of this priority is attached to the route in the cheap order of the transportation time.

7. **Priority random**
   We generate a random priority.

4.3 Mutation

The method of sending a load to the high route of preferential ranking preferentially is shown below.

Step 1: $\{i_1, i_2, \cdots, i_x\}$, $\{j_1, j_2, \cdots, j_y\}$ is chosen at random and the partial matrix W is made.
Here, $\{i_1, i_2, \cdots, i_x\}$ is the subset of $\{1, 2, \cdots, m\}$, $\{j_1, j_2, \cdots, j_y\}$ is the subset of $\{1, 2, \cdots, n\}$, and they are $2 \leq x \leq m$, $2 \leq y \leq n$.
Step 2: (3) and (4) are repeated from $k = 1$ to $x \times y$.
Step 3: The number of the line of the route of preferential ranking high to the kth is stored in i. And the number of a sequence is stored in j.
Step 4: The following formula is calculated.

$$W'_{ij} = \min\{a_{w_i}, b_{w_j}\}$$
$$a_{w_i} = a_{w_i} - W'_{ij}$$
$$b_{w_j} = b_{w_j} - W'_{ij}$$

Step 5: The partial matrix W' is returned to the portion of the gene of the original chromosome matrix.

4.4 The Proposed Corrsover

We propose a crossover method using the priority.

Step 1: Decision matrix that cross
   To determine the two parent individuals to meet the crossover probability of the number of population within. We decide to random the number of rows and columns intersect. Then, until it meets a number of rows and columns determined, to determine the respective row and column intersecting at random.

Step 2: Generation of child chromosome
   We generate a matrix child 1 and child 2 of two. Keep initialized to 0 all. We stored in the corresponding matrix child 2 of the transport volume of each row and each column was determined by the parent 1. And stores in a matrix corresponding to the transport amount of each row and column of the parent 2 determined in step1 to the child 1.

Step 3: Generation of child chromosome
   Child 1 to transport preferentially in the matrix that exists in the transportation volume of parent 1. In this case, We select randomly from the priority order that is generated. We want to transport as many as protect the constraint matrix to its high priority. Child 2 to transport preferentially in the matrix that exists in the transportation volume of parent 2. Then, to determine the transport amount of each matrix according to the priority order selected. Then, adjustment is performed to meet you also meet the constraints.

5   Numerical Experiment

In order to check the validity of the proposed method, the benchmark problem was taken up and the comparison experiment with the proposed method (proposed p-GA) and the conventional method was conducted. As the conventional methods made applicable to comparison, DCGA by Teramatu were taken up, and the object took up ran12x12, ran14x18 and ran17x17 of benchmark problems. The number of population 100. The experiment was tried 30 times on each problem. In addition, the end conditions of one trial are made into 30,000 generations. Conventional method is the mutation probability 1.0 crossover probability

Table 1. Experimental Results

	ran12×12									
	Best		Worst		Ave					
	c	t	c	t	c	t	Coverage	Number of Pareto solutions	TOPSIS	Time[s]
DCGA	2291	1590	3844	5123	2376	1590	0.74	132	0.6212	50.0
p-GA	2291	1590	3844	5123	2311.6	1590	0.77	134	0.6088	56.3
	ran14×18									
	Best		Worst		Ave					
	c	t	c	t	c	t	Coverage	Number of Pareto solutions	TOPSIS	Time[s]
DCGA	3714	1752	6875	6177	3803.7	1774.7	0.70	131	0.5666	88.9
p-GA	3714	1752	6875	6177	3767.4	1768.2	0.72	135	0.5506	91.1
	ran17×17									
	Best		Worst		Ave					
	c	t	c	t	c	t	Coverage	Number of Pareto solutions	TOPSIS	Time[s]
DCGA	1373	1757	3515	7616	1421.6	1760.2	0.85	185	0.5702	96.3
p-GA	1373	1757	3515	7616	1414.4	1759.2	0.89	195	0.5618	102.0

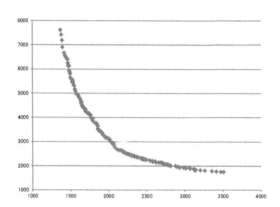

Fig. 1. conventional method:ran17x17

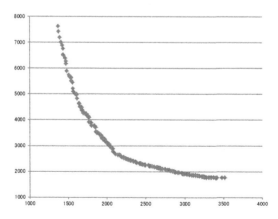

Fig. 2. Proposed method:ran17x17

0. The proposed method is 0.5 probability crossover and mutation probability. Table 1 shows the experimental results of each method. Moreover, the best solution, the worst solution, the average value, the coverage, the number of pareto solutions, the topsis, and execution time of each method for three problems are shown in Table 1. Crossover method proposed was not able to update the best solution benchmarks all problems. However, coverage and average value, number of Pareto solution, TOPSIS is superior to conventional methods in the benchmark problem of all.

6 Conclusion

In this paper, we took up the fixed charge transportation problem that extends to more realistic bicriteria fixed charge transportation problem. The proposed method was not able to update the best solution than the conventional method of numerical experiments. However, the average value is excellent benchmark problem of all. Therefore, exceeding the conventional method in the best accuracy solution when attempting 30 times. And, value coverage, the number of Pareto solution, the top law system is above than the conventional method. Therefore, The validity of the proposed method could be checked by numerical experiment. In the future, we plan to do to improve the technique so that you will be prompted to pareto solution excels in all benchmark problem.

References

1. Mitsuo, G., Cheng, R.: Genetic Algorithms and Engineering Optimization. John Wiley and Sons, Inc. (2000)
2. Kenichi, I., Hiroaki, T., Yusuke, T.C.N.: Application to fixed cherge transportation problem of genetic algorithm. The Institute of Electrical Engineers of Japan OS6_2_1–4 (2004)
3. Ida, K., Tohyama, H., Teramatu, C., Futatani, Y.: The Proposal of the Genetic Algorithm for Fixed Charge Transportation Problem. In: Proc APIEMS 2004, pp. 32, 1–5 (2004)
4. Li., Y., Gen, M., Ida, K.: Genetic Algorithms for the Fixed Charge Transportation Problem. Beijing Mathematics 4(2), 239–249 (1998)
5. Ida, K., Kimura, Y.: The Proposal of an Improved Genetic Algorithm for Floorplan Design with Non-slicing Structure. The IEICE Trans. on Fundamentals of Electro., Comm. and Comput. Sci. J87-A(3), 399–405 (2004)

The GMM Problem as One of the Estimation Methods of a Probability Density Function

Kiyoshi Tsukagoshi[1], Kenichi Ida[2], and Takao Yokota[1]

[1] Faculty of Engineering Ashikaga Institute of Technology
Ashikaga City Tochigi Prefecture Japan
[2] Faculty of Engineering Maebashi Institute of Technology
Maebashi City Gunma Prefecture Japan

Abstract. In data analysis, we must be conscious of the probability density function of population distribution. Then it is a problem why the probability density function is expressed.

The estimation of a probability density function based on a sample of independent identically distributed observations is essential in a wide range of applications. The estimation method of probability density function -- (1)a parametric method (2)a nonparametric method and (3)a semi-parametric method etc. -- it is. In this paper, GMM problem is taken up as a semi-parametric method and We use a wavelet method as a powerful new technique. Compactly supported wavelets are particularly interesting because of their natural ability to represent data with intrinsically local properties.

Keywords: GMM, wavelet, probability density function, Compactly supported, kernel.

1   Introduction

In data analysis, we must be conscious of the probability density function of population distribution.

However, the population who can express to distribution of a beautiful form which appears in the statistical textbook can hardly be found out.

Then, it is a problem why the probability density function is expressed.

The estimation of a probability density function based on a sample of independent identically distributed observations is essential in a wide range of applications.

The estimation method of probability density function –
1.   a parametric method
2.   a nonparametric method
3.   a semi-parametric method   etc.
-- it is.

In this paper, GMM problem is taken up as a semi-parametric method and We use a wavelet method as a powerful new technique.

T. Yoshida et al. (Eds.): AMT 2013, LNCS 8210, pp. 359–368, 2013.
© Springer International Publishing Switzerland 2013

Compactly supported wavelets are particularly interesting because of their natural ability to represent data with intrinsically local properties.

For this reason, the application to GMM of a wavelet is natural.

The problem of estimating density functions using data from different distributions, and a mixture of them, is considered. Maximum likelihood and Bayesian parametric techniques are summarized and various approaches using distribution-free kernel methods are expounded.

Although estimation of the probability density function is made by various methods, it comes in subsequent theoretical deployment to a deadlock in many cases.

How is the expressing method by the contaminated normal distribution expressed as one of the modes of expression of a probability density function in the form of the linear combination of the normal distribution used as the basis of the present-day statistical theory?

The importance is recognized in the field with various problems which decompose one population into some partial populations.

2   Kernel Probability Density Estimation

The method by a kernel function is one of the methods of presuming a probability density function using a normal distribution like the problem of GMM.

Both methods expressed in the form of the linear combination of a normal distribution. But the method GMM is more convenience in use.

Here, presumption of the probability density function by kernel function expression is used to make the signal for the analysis of GMM using Wavelet.

The probability density function using Gaussian kernel is expressed as follows considering $K(\bullet)$ as a standard normal distribution function.

$$\hat{f}_K(x) = \frac{1}{n}\sum_{i=1}^{n}\frac{1}{h}K\left(\frac{x - X_i}{n}\right) \tag{1}$$

$\{X_i\}_{i=1}^{n}$: sample   n:sample size   h:Bandwidth

Bandwidth h s called a smoothing parameter or window width. When Bandwidth h is a small value, the local characteristic of a density function is expressed finely, and a smoother expression is shown when Bandwidth h is a large value.

It is the same as that of the scale parameter of Wavelet. The choice of kernel Bandwidth controls the smoothness of the probability density curve.

The following graph shows the density estimate for the same data using different Bandwidths.

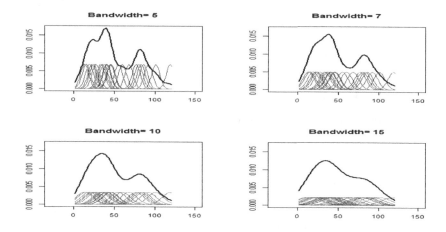

Fig. 1. Kernel function density estimation using a gauss kernel function with different bandwidth h

In Fig.1 Kernel function density estimation using a gauss kernel function with different bandwidth h for data which sample size is equal to 117930 is shown. A black line is the estimated density function.

A red line is the gauss kernel function (1/100 of 117930) which increased the size 200 times.   From now on, it will be observed in the large place of variation that the kernel function is complicated. Later, GMM analysis using which made kernel function density estimation the signal is conducted two different Bandwidth(s).

Although it is in how to decide a bandwidth with Sturges's rule, Scott's rule, and Friedman-Daiconis's rule[6].I think that deciding automatically has a problem even if it uses which rule. According to the purpose of analysis, you should decide by making Bandthwidth fixed under these rules into a standard.

3   Formulization of a Gauss System Wavelet and GMM (Gauss Mixed Model)

3.1   Formulization of GMM (Gauss Mixed Model)

GMM (Gaussian Mixture Model) is a statistics model expressed by alignment combination of the Gauss basis function, and is a technique often used by speech recognition etc. GMM consists of n normal distributions and mixed dignity ω_i of each distribution, mean μ_i, and distributed σ_i can express it.

A density function with the scale parameter which expresses a measure as position is called location scale density function. It is a location scale density function which makes a normal distribution and an average a location parameter and has standard deviation as a scale parameter.

$$f(x) = \sum_{i=1}^{n} \omega_i \frac{1}{\sqrt{2\pi}\,\sigma_i} e^{-\frac{(x-\mu_i)^2}{2\sigma^2}} \quad (i = 1, 2, \cdots, n)$$

$$\omega_i \left(\sum_{i=1}^{n} \omega_i = 1, \quad \omega_i \geq 0 \right), \quad \mu_i, \quad \sigma_i \quad (i = 1, 2, \cdots, n)$$

(2)

3.2   Wavelets Analysis

Wavelets analysis is investigating a relation with an observation signal

$\psi\ (x\ )$:   M o t h e r w a v e l e t $\psi \left(\dfrac{(x\ -\ b\ )}{a} \right)$

use a : S c a l e p a r a m e t e r   (e x p a n d s t o a t i m e b a s e d i r e c t i o n)

b : T r a n s f o r m e r (p a r a l l e l t r a n s l a t i o n t o a t i m e b a s e d i r e c t i o n)

Scale parameter: Elasticity Expansion dilation

Transformer rate: Parallel translation, move basis function Mother wavelet on a shift time-axis (it is with the Mother wavelet of a gauss system this time.)

$f(x)$ Signal function, $\psi\left(\frac{(x-b)}{a}\right)$ Wavelet transform is as follows when it is considered as a wavelet.   [2]

$$CWT(a,b) = \int_{-\infty}^{\infty} f(x) \frac{1}{\sqrt{a}} \psi\left(\frac{x-b}{a}\right) dx$$

(3)

At (3), $f(x)$ is (1) type is substituted. Moreover, let the wavelet function to be used be the following gauss of the first order(3), or a gauss wavelet function of the second order (Mexican hat) (5).

$$\psi\left(\frac{x-b}{a}\right) = -\frac{\sqrt{2}\,e^{-\frac{\left(\frac{x-b}{a}\right)^2}{2}} \left(\frac{x-b}{a}\right)}{\pi^{1/4}}$$

(4)

$$\psi\left(\frac{x-b}{a}\right) = \frac{2}{\sqrt{3}a} \pi^{-\frac{1}{4}} \left(1 - \frac{(x-b)^2}{a^2}\right) e^{-\frac{(x-b)^2}{2a^2}}$$

(5)

These become the first order differential coefficient of a gauss wavelet function, and the second order differential coefficient.

4   Analysis by Continuation Wavelet Transform

4.1   Analysis of a Normal Distribution

A normal distribution function is considered as a generality.

It is expanded and contracted by the parameter a d parallel translation of the wavelet function is carried out by the parameter b. In the frequency function of a normal distribution, since there is mean value in that the maximum is given, the differential coefficient of the first floor in there is set to 0.

$$f'(x) = -\frac{e^{-\frac{(x-\mu)^2}{2\sigma^2}} (x-\mu)}{\sqrt{2\pi}\sigma^2} = 0 \quad at \quad x = \mu$$

(6)

Moreover, the position of standard deviation $x = \mu \pm \sigma$. The differential coefficient of the second floor of the f is a position of the point of inflection used as 0.

$$f''(x) = \frac{e^{-\frac{(x-\mu)^2}{2\sigma^2}}(x^2 - 2x\mu + \mu^2 - \sigma^2)}{\sqrt{2\pi}\sigma^4} = 0 \qquad at \ x = \mu \pm \sigma \tag{7}$$

The character to be a location scale density function which makes a normal distribution and an mean a location parameter and has standard deviation as a scale parameter, The analysis of GMM is tried using the wavelet analysis which makes the scale parameter a (elastic expansion) transformer rate b (parallel translation) correspond to a basis function (Mother wavelet), and conducts signal analysis.

First, it becomes like wavelet transform, then the following type about the normal distribution of types using the primary gauss type wavelet function.

$$\int_{-\infty}^{\infty}\frac{1}{\sqrt{2\pi}}e^{-\frac{(x-\mu)^2}{2\sigma^2}}\left(-\frac{\sqrt{2}e^{-\frac{(\frac{x-b}{a})^2}{2}}(\frac{x-b}{a})}{\pi^{1/4}}\right)dx = \frac{\sqrt{2}ae^{-\frac{(b-\mu)^2}{2(a^2+\sigma^2)}}(b-\mu)}{\pi^{1/4}\sqrt{\frac{1}{a^2}+\frac{1}{\sigma^2}}(a^2+\sigma^2)} \tag{8}$$

Set $a = \sigma \ and \ b = \mu$ at (8). (8) types will be set to 0 if it becomes.

Moreover, it becomes like wavelet transform, then the following type about the normal distribution of types using the secondary gauss type wavelet function.

$$\int_{-\infty}^{\infty}\frac{1}{\sqrt{2\pi}}e^{-\frac{(x-\mu)^2}{2\sigma^2}}\left(-\frac{2e^{-\frac{(\frac{x-b}{a})^2}{2}}(-1+(\frac{x-b}{a})^2)}{\sqrt{3}\pi^{1/4}}\right)dx = \frac{2a^2e^{-\frac{(b-\mu)^2}{2(a^2+\sigma^2)}}(a^2-(b-\mu)^2+\sigma^2)}{\sqrt{3}\pi^{1/4}\sqrt{\frac{1}{a^2}+\frac{1}{\sigma^2}}(a^2+\sigma^2)^2} \tag{9}$$

In (9), it is a [small enough]. $b = \mu \pm \sigma$ It is alike, it sets and (9) types are set to 0.

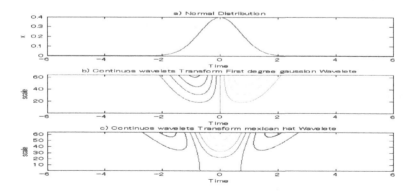

Fig. 2. Fig. 1 Wavelet transform of frequency function

a) Normal distribution curve   b)Primary gauss wavelet transform
c) Secondary gauss (Mexican hat) wavelet transform

In Fig.2, a showed the wavelet transform by the primary gauss function of a) normal distribution curve ,by b) which showed the normal distribution curve, c) The wavelet transform by a Mexican hat function was shown.

The result of Fig.2.b) -- wavelet transform is showing zero value in the point of giving mean value. And as a result, the value of a wavelet is in respect of mean value, and figure 1 c takes the local maximum. $\mu-\sigma,\mu+\sigma$ It is shown that come out and the value of a wavelet has become 0, respectively.

4.2   Analysis of GMM

A problem is a density function of observational data distribution. $f(x)$ The following forms:

$$f(x) = \omega_1 \frac{1}{\sqrt{2\pi}\,\sigma_1} e^{-\frac{(x-\mu_1)^2}{2\sigma_1^2}} + \omega_2 \frac{1}{\sqrt{2\pi}\,\sigma_2} e^{-\frac{(x-\mu_2)^2}{2\sigma_2^2}} \tag{10}$$

It comes out and expresses -- it is presuming the following parameter.

μ_i: The average of each element distribution

ω_i: The mixing rate of each element distribution

σ_i: Standard deviation of each element distribution . It carries out. Next, mixed distribution of two normal distributions is analyzed with Fig. 3.   Suppose from the left that they are the axial value (horizontal axis) which gives the contour line of 0 by a= 0 in the lowest stage of Fig.2 x_1, x_2, x_3, and x_4.

$$\begin{cases} x_1 = \mu_1-\sigma_1 & x_3 = \mu_2-\sigma_2 \\ x_2 = \mu_1+\sigma_1 & x_4 = \mu_2+\sigma_2 \end{cases} \tag{11}$$

It becomes. [4]

Give easy calculation for these. $\mu_1,\mu_2,\sigma_1,\sigma_2$. μ_1,μ_2 It is equal to the position of the x-axis of the contour line 0 of the middle of Fig. 2. Moreover ω_1,ω_2 -- it can ask by easy simultaneous equations.

$$\mu_1 = \frac{1}{2}(x_1+x_2), \quad \sigma_1 = \frac{1}{2}(x_2-x_1) \quad \mu_2 = \frac{1}{2}(x_3+x_4), \quad \sigma_2 = \frac{1}{2}(x_4-x_3) \tag{12}$$

-- it can ask by easy simultaneous equations.

It is easy to calculate dignityω_1 and ω_2 by simultaneous equations.

Specific point alpha Value of the observational data which can be set. It carries f_α out.

The following simultaneous equations are realized.

$$\begin{cases} \omega_1 + \omega_2 = 1, \\ \omega_1 \frac{1}{\sqrt{2\pi}\sigma_1} e^{-\frac{(\alpha-\mu_1)^2}{2\sigma_1^2}} + \omega_2 \frac{1}{\sqrt{2\pi}\sigma_2} e^{-\frac{(\alpha-\mu_2)^2}{2\sigma_2^2}} = f_\alpha. \end{cases} \tag{13}$$

Or to make simultaneous equations so that it may become equal to the value of two points.

A mean is determined from Fig.2b(middle). Distribution can be easily read in Fig. 3c (lower berth).

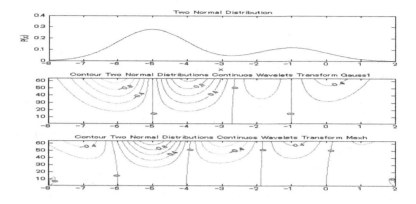

Fig. 3. Mixed part distribution of two normal distributions

5   Determination of the Scale Value for Specifying Standard Deviation

The place where the contour line of a Mexican hat wavelet becomes 0 must be written with the position which gives distribution, and must determine the value of the scale parameter a for the purpose. Therefore, a wavelet power spectrum is used as a measure of determination. The spectrum which a signal gives chooses the scale parameter which becomes the maximum, and the contour line in the position makes the transformer rate value to which the value of 0 is given the point which shows standard deviation. Signal $x(t)$ the total-calories size boiled and contained -- the -- it defines finding the integral the square.

$$E = \int_{-\infty}^{\infty} |x(t)|^2 \, dt = \|x(t)\|^2 \tag{14}$$

The relative value of the signal energy in a certain scale a and position (transformer rate) b is given from a two-dimensional wavelet energy density function.   1) and 3)

$$E(a,b) = |CWT(a,b)|^2 \tag{15}$$

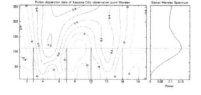

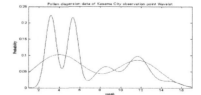

Fig. 4. Determination of standard deviation

Fig. 5. Mixed distribution of two normal distributions by wavelet power spectrum

$$\begin{cases} \omega_1 = 0.5776 & \mu_1 = \dfrac{7.0 + 2.5}{2} = 4.75 & \sigma_1 = \dfrac{7.0 - 2.5}{2} = 2.25 \\[2ex] \omega_2 = 0.4224 & \mu_2 = \dfrac{14.0 + 10.0}{2} = 12.0 & \sigma_2 = \dfrac{14.0 - 10.0}{2} = 2.00 \end{cases} \qquad (17)$$

6   The Numerical Example over Pollen Dispersion Data

The health hazard caused by cedar pollen becomes a major issue an early spring in Japan. We use the pollen dispersion data in the observation point in Kasama-City of 2004 which is downloaded from Homepage of the Ministry of the Environment. [1] The Ministry of the Environment is announcing the pollen dispersion situation every hour of from February 1st to May 31st on the homepage.

Cedar pollen problem, Hay fever is the illness which starts allergies, such as a sneeze and a runny nose, by the pollen of plants, such as a Japan cedar and a cypress, becoming a cause. It is also called seasonality allergic rhinitis.

If think about to the Kanto area is taken for an example, scattering of cedar pollen will start around in February, and scattering will decrease late in April.

And scattering of the department pollen of a cypress starts and it continues till around the end of May.

Condition, such as a sneeze, a runny nose, nasal congestion, itchiness, a feeling of a foreign substance of eyes, is in a tendency worsening in proportion to the amount of scattering of pollen. Then, We will try to divide into distribution of a Japan cedar and a cypress now.

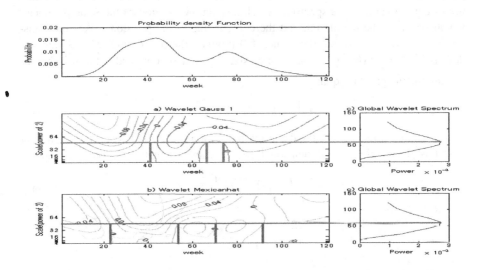

Fig. 6. Wavelet Analysis for Bandwidth=7 Kernel Estimation

The upper row of Fig. 6 shows the density function of kernel estimation, and the middle shows the analysis by a 1st order gauss wavelet, and the lower berth shows the analysis according to a Mexican hat wavelet.

The following parameter was obtained like the formula(12).

$$\begin{cases} \omega_1 = 0.6045954 & \mu_1 = 41 & \sigma_1 = 16 \\ \omega_2 = 0.203625 & \mu_2 = 75 & \sigma_2 = 10 \end{cases} \qquad (18)$$

Using this result, the result of the Kolmogorov-Smirnov test is shown below.

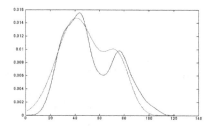
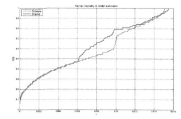

Fig. 7. GMM Estimation & Kernel Estimation   **Fig. 8.** Kolmogorov-Smirnov Test

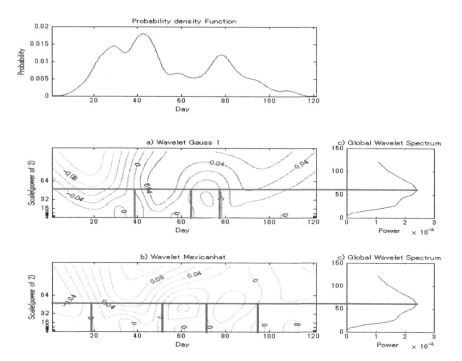

Fig. 9. Wavelet Analysis for Bandwidth=4 Kernel Estimation

The alternative hypothesis is that Kernel Density and GMM Estimation are from different continuous distributions. The result h is 1 if the test rejects the null hypothesis at the 5% significance level and 0 otherwise. Now h= 0, Asymptotic p-value of the test is 0.0691 and Test statistic 0.18.

From now on, the distribution situation of a Japan cedar and a cypress is separable. The following figure is the analysis to Kernel Density of Bandwidth=4.

If the correspondence to a detailed change is required, it is necessary to make Bandwidth small, and precise analysis is required to make a scale small.

7   Conclusion

In this research, the new solution which solves a GMM problem using continuation Wavelets transform was proposed, and the validity was verified. The feature of the solution of this research is as follows. Continuation Wavelets transform is used. It is easily analyzable using a contour line figure. After reading zero point from a graph or an output file, a solution can be calculated by easy alignment calculation. It is resetting the main part of experiment analysis to theoretical analysis as a future subject.

Although we set the parameter using the scale value which becomes the maximum wavelet power spectrum, for more suitable analysis, interactive analysis becomes more effective considering these methods as one standard. Also about the determination of bandwidth in Kernel density estimation, each rule has the feature, and it is hard to say that one method may not be recommended to used.

References

1. Addison, P.S.: The Illustrated Wavelet Transform Handbook. Institute of Physics Publishing (2002)
2. Mallat, S.: A Wavelet Tour of Signal Processing. Academic Press (1999)
3. Torrence, C., Compo, P.: A practical guide to Wavelet Analysis Bulletin of the American Meteorological Society (1998)
4. Tsukakoshi, K., Mawatari, S.: Extraction of Element Distribution of Gauss Mixture Distributions with unknown number of elements (2009)
5. Tsukakoshi, K., Mawatari, S.: Extraction of Element Distribution of Gauss Mixture Distributions By Wavelets Power Spectrum (2010)
6. Wand, M.P.: Data-Based Choice of Histogram Bin Width. The American Statistician 51(1) (February 1997)

GA for JSP with Delivery Time

Yusuke Kikuchi*, Kenichi Ida, and Mitsuo Gen

Maebashi Institute of Technology
Maebashi, 371-0861, Japan
ida@maebashi-it.ac.jp

Abstract. This paper describes a job-shop scheduling problem (JSP) of processing product subject to no delay time job. It is one objective model of the minimum delivery delay time.In this paper, the effectiveness we the numerical experiments using a benchmark problem to improve the solution accuracy and decrease execution time by adding a method to generate gene and new approach, we introduce a search method for the algorithm shorter delivery times and further there to verify.

Keywords: Job Shop Scheduling Problems, genetic algorithm, priority level.

1 Introduction

Job-shop scheduling problem (JSP) is one of the most difficult problems in combinatorial optimization problems that have been studied with many of the past researchers. JSP is designed to minimize the total work time to complete all work normally (make-span) and in reality, additional constraints and goals complicate the problem. For example, JSP and concept to delivery, late delivery (delay time) and work in process inventory generate problems.

These problems prevent issues from being considered. Delivery time is the study of Singer and Pinedo (1998), Pinedo and Singer (1999) and Asano and Ohta (2002). On the other hand,approximation method of genetic algorithm (GA) has become popular for JSP recently (Gen and Cheng, 2000).

Recently the use of a solution to the problem of job shop scheduling has been recently proposed approach to scheduling with delivery time of Abe (2009) does not consider the delivery time from Ida and Osawa (2005), execution time is very long. In this study, the effectiveness we the numerical experiments using a benchmark problem to improve the solution accuracy and decrease execution time by adding a method to generate gene and new approach, we introduce a search method for the algorithm shorter delivery times and further there to verify.

2 Delivery Time and JSP

We think a number of the machine when machining work pieces M N. In this case, it is assumed that the order of processing each job Machine, processing

---

* Maebashi Institute of Technology.

T. Yoshida et al. (Eds.): AMT 2013, LNCS 8210, pp. 369–378, 2013.
© Springer International Publishing Switzerland 2013

time on each machine and each job order is given that is technical. Processing order of each job on each machine to minimize the total work time until you have processed all the work, JSP is a problem that is to determine the schedule.

However, only one job is processed in one machine at the same time, you want to process only one job at a time, one machine is restricted, such as interruption of the work is not.

Here, we have assumed that to minimize the delivery time over each of the work at the time of each job is completed, the job end time is set in advance. We put the processing start time of o_{ij} and x_{ij} work p_{ij}, seek the processing time of the j-th operation o_{ij} job here. Also, it is possible to start at time 0 and all jobs.

D_i preset delivery of job i, and w_i and the penalty per unit time for the tardiness of job i, can be formulated as follows.

$$\min \quad \sum_{i=1}^{N} w_i \rho_i \tag{1}$$

$$\text{subject to} \quad x_{iL_i} + p_{iL_i} - \rho_i = d_i, \tag{2}$$

$$i = 1, 2, \dots, N \tag{3}$$

$$x_{ij} - x_{i,j-1} \geq p_{i,j-1}, \tag{4}$$

$$j = 1, 2, \dots, L_i, \quad i = 1, 2, \dots \tag{5}$$

$$x_{ij} - x_{st} \geq p_{st} \lor x_{st} - x_{ij} \geq p_{ij}, \tag{6}$$

$$o_{ij}, o_{st} \in E_k, \quad k = 1, 2, \dots, MST \tag{7}$$

$$x_{ij} \geq 0, \tag{8}$$

$$j = 1, 2, \dots, L_i, \quad i = 1, 2, \dots, N \text{nonnegative} \tag{9}$$

Here, E_k is a set of jobs to be processed by the machine k. Equation 1 is accomplished function, Equation 2 is a target constraint, i is an auxiliary variable representing the time of job i respectively tardiness. Equation 3 for each job order technical, Equation 4 is non-duplicating condition of jobs in each machine, Equation 5 represents the non-negative condition of production start time.

Many scheduling problem as is the case, these problems can analytically solving that involves considerable difficulty. Therefore, as a method for analysis of these meta-heuristic search due to such as GA is widely used.

3   GA for JSP

In general, if you want to apply the GA to JSP, gene expression is converted to a schedule called chromosomes. In addition, the evaluation of the JSP is done in total work time in general, but trying to express a schedule from the chromosomal gene is expressed, the Gantt chart is used.

When applied to a JSP the GA in order to reduce the time of total work, in addition to those cross-generation and initial population in genetic manipulation of normal, such as mutation, such as shift left on the Gantt chart chromosome newly generated algorithm is applied to reduce idle time such. Figure 1 shows the basic flow of GA to solve the JSP.

In this paper, we propose a method for generating Gantt charts using the standard algorithm and delivery to target a reduction in product delivery delays. In the proposed method, to generate the initial population by using the chromosome representation of Hirano. Crossover to use the weight mapping crossover(WMX) method and priority delivery. Mutation is to use of Abe. Apply the Gantt chart created by the proposed method, the algorithm for the delivery delay reduction for each chromosome.

3.1   Generation of Initial Population

If the initial population randomly generated, it is possible that early delivery job is placed at the rear of the chromosome. When considering the delivery time, it is desirable that you were lined up in order of chromosomal gene delivery time if you can to some extent. Which was to be produced by selecting a type of roulette initial population was set to select the delivery probability of each job, in this case, it is applied to generate an initial population of Figure 1. The following is an algorithm to generate an initial population considering delivery.

Generation algorithm of initial population

1. Step1: Arranged sequentially from the beginning of the gene was extracted by the extraction of non-overlapping roulette selection, to generate a chromosome.
2. Step2: Population size times the Step1, and the initial population.

3.2   Crossover

WMX can be seen as an extension of one-point crossover. Infeasible solution will be generated if the priority-based gene expression, and using the usual one-point crossover. By using WMX was fast delivery priorities and what, where, we aim to shorten the delivery delay without causing a lethal gene.

3.3   Mutation

Selected at random, cp2 cp1 two loci in the relationship of cp1 ¡ cp2 from the chromosome of interest, this mutation should be replaced when the delivery of genes cp2 g2 is less than delivery of genes g1 cp1.

This operation is applied in a mutation of Figure 1, will be processed as quickly as those early deadlines.

Mutation Algorithm

1. Step1:Randomly removed cp1, cp2 cp1 two points in the relationship of ¡cp2 from the chromosome.
2. Step2:Determine the relationship between the magnitude of the delivery of genes g1 and g2 have cp2 and cp1 chromosome.
3. Step3:Delivery of genes as early as swaps g2 from g1.

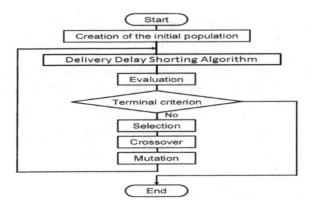

Fig. 1. GA for JSP

Further, in order to prevent entry into a local solution in addition to this, the chromosomes that are generated in the same way as the generation method of the initial population, method of generating initial group compared with the reference delivery date, a generation method that is based on penalty and delivery introduced and treated as three kinds of mutations.

These mutations are applied only to the parent population filled with mutation probability.

3.4   Selection

This section is used to select non-overlapping roulette extraction.

4   Proposed Method

4.1   Decoding Method Based on Delivery Time

Here are a placement method for the Gantt chart, which is based on the delivery date. From the beginning of the chromosome, in law arrangement from Abe and Ida have been used, but will be placed in the machine according to the order of tasks, arranged according to the delivery method of arrangement to be proposed (Figure 2).

This ensures that each job is started according to the delivery time, you will be able to prevent the overcrowding of the machine. Is prevented by the occurrence of lethal genes if all jobs if protect the delivery date is shifted left, as shown in Figure 2 does not protect the right shift.

4.2   Shorting Algorithm for Delivery Time

We propose the algorithm that aims to shorten delivery time. The algorithm shortened from Abe uses algorithms reduce aimed to shorten working hours total for JSP has been applied, in reality few problems for the purpose of shortening the only working time.

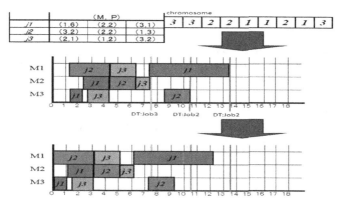

Fig. 2. Example of decording

Therefore, we propose an algorithm that aims to reduce PDTShort(Proposed Delay Time Short)Alogrithm faster delivery time.

Chart of PDTShort

1. step0:Machine reads a list relationship - Gantt chart and work that is generated from the chromosome.
2. step1:Apply a shift to the left in order or random delivery job early.
3. step2:Apply the right shift to shorten idle time.
4. step3:Algorithm is applied for shorter delivery times.
5. step4:Seek tardiness outputs a Gantt chart, compared with the best solution.Here, using the same method of generating initial population generation method to generate all the chromosomes to generate a Gantt chart, if you can not see the updates of your generation 300 returns to step 0.
6. step5:To generate a Gantt chart from chromosome to shorten the delivery time.

Right Shift for the Idle Time Reduction
Input data:Relationship list - Mechanical work, Number of jobs N, Number of machine M, Seek work o_{ij}, Work start time x_{ij}, Processing time p_{ij}.

Output data:x_{ij}, o_{ij}, p_{ij}.

1. step0:Read the Gantt chart list to apply machine relationship work, the reduction algorithm.
2. step1:Set $i = 1$,Repeat *step2* to N.
3. step2:Set $j = 1$, Repeated until M from the step4 step3.to i + 1.
4. step3:After reading the o_{ij}, reads the work machine number(Named "X" to work at this time).
5. step4:If X was the first work of the first task as a job as a machine and does not apply the right shift.I go back to *step3* to $j \leftarrow j + 1$.
6. step5:If possible, the right shift to satisfy either of the conditions of *step3*, there is idle time.Save the x_{ij} the start time of the work at that time, update the work order after a right shift o_{ij} and p_{ij} processing time work number.

7. step6:$i \leftarrow i + 1$.
8. step7:Output x_{ij}, o_{ij}, p_{ij}.

Shorting Algorithm for Delivery Time

In this paper, we first shift to the right job with no tardiness, to perform a left shift to jobs with late deliveries.

However, to apply taking into account the penalty on the priority of the algorithm applied.

Input data:Relationship list - Mechanical work, Number of jobs N, Number of machine M, Seek work o_{ij}, Work start time x_{ij}, Processing time p_{ij},Delay in delivery time ρ.

Parameters: Set S of job without delay in delivery, the number s,Set D of jobs with tardiness, the number d.

Output data:x_{ij}, o_{ij}, p_{ij}.

1. step0:Read the Gantt chart list to apply machine relationship work, the reduction algorithm.$S \leftarrow \phi, D \leftarrow \phi$.
2. step1:Set $i = 1$,Repeat *step2* to N.
3. step2Read the late delivery time of each work.Time to delivery is stored in the set S from a long job with no job tardiness.It stores the set D from the job time to delivery time is short jobs that late delivery..
4. step3:Set $i = 1$,Repeat *step4* to $s.i \leftarrow i + 1$.
5. step4:Take the job from the set S, and thereby shift right without exceeding the delivery time if possible right shift from the job close to delivery.
6. step5:Set $j = 1$,Repeat *step6* to $d.j \leftarrow j + 1$..
7. step6:Take the job from the set D, it if left-shiftable from the job Nearby work start time 0, is performed left-shift, in the range within that satisfies the constraint.
8. step7Output x_{ij}, o_{ij}, p_{ij}.

Chromosome Generation Method from the Gantt Chart

Input data:Relationship list - Mechanical work, Seek work o_{ij}, Work start time x_{ij}.

Output data:chromosome v

1. step0:Read to Relationship list - Mechanical work, $S \leftarrow \phi$.
2. step1:Set $i = 1$,Repeat *step2* to N.
3. step2:Set $j = 1$,Reads the Gantt chart chromosome of i-th, and repeated until the M *step4* from *step3*.
4. step3:To read the work start time and work x_{ij} o_{ij}, to save to chromosome v a work number to order early start time.
5. step4:$i \leftarrow i + 1$ if it is $j = M$,otherwise $j \leftarrow j + 1$.
6. step5:Output v.

5 Experiments

5.1 Condition

We were shown in the following setting conditions as well as the experimental environment in this paper.

Table 1. Delivery time calculation example (la21)

job	Σp	$F = 1.5$	$F = 1.6$
0	444	666	710
1	576	864	921
2	567	850	907
3	659	988	1054
4	603	904	964
5	451	676	721
6	471	706	753
7	423	634	676
8	440	660	704
9	717	1075	1147

Table 2. Delivery time penalty

Job number	0	1	2	3	4	5	6	7	8	9
Factor	4	4	2	2	2	2	2	2	1	1

The instance to be used for the experiment, and abz5, abz6, la16, la17, la18, la19, la20, la21, la22, la23, la24, mt10 [14] of 10 job problem 10 machine which is well known as a JSP benchmark problem.The problem setting addressed in this paper, you need a late delivery penalty and set deadlines for each job, but these settings are used the same as the experiment of Singer and Asano. Is set by the following equation: the delivery time of each job is, F is a delivery factor here. In the two reported above, and $F = 1.5 and 1.6$ a delivery factor for each problem, and evaluated the value obtained.

$$d_i = F \times \sum_{j=1}^{L_i} p_{ij}, i = 1, 2, \ldots, N \tag{10}$$

We were shown in Table 1 Delivery which was used in this experiment. Here, Σp is one obtained a total processing time for each job. Columns of $F = 1.6$ and $F = 1.5$ is what was calculated delivery time of each job, and is intended to multiply the value of the coefficient F to the value of Σp, truncating the fractional part. In Table 2, I show the penalty factor of late delivery is the first goal. The delivery delay time is obtained by multiplying this value for the delay time of each job.

In this comparison, the following parameter specifications have been used: Population size, popSize=200; Crossover probability, pC =0.50; Mutation probability,pM=0.30; Maximum generation,maxGen=10000.Termination condition to 1×10^6 individual.

5.2   Comparison with Existing Methods

In this case, it is compared with the results of Abe and Singer and Asano existing methods in order to confirm the effectiveness of the proposed method are dealing

Table 3. Experimental result 1 (F=1.5)

Instance	Opti. DT	ASA DT	S&P DT	Abe DT	pGA DT	pGA ave
abz5	69	736	109	69	70	103.6
abz6	0	0	0	0	**0**	0
la16	166	166	178	166	**166**	220.4
la17	260	573	260	260	**260**	266.2
la18	34	255	83	34	**34**	106.6
la19	21	494	76	25	29	96.2
la20	0	246	0	0	**0**	54.8
la21	0	77	16	0	**0**	11.2
la22	196	537	196	196	**196**	321.8
la23	2	466	2	2	**2**	9.2
la24	82	465	82	84	88	108.8
mt10	394	1024	394	394	438	520.8

with the same instance.However, these reports are only numbers for the late delivery.

5.3   Experimental Result

Shown in Table 3 and Table 4, the experimental results.

Here, DT is a delay in delivery time.

Singer was shown optimal solution known.

Figure 3 shows an example the value of the delivery delay time is gradually converged.

We wer carried out to explore the solution with the goal of shortening the delivery time, but, the following has been shown.

- Of the two types of delivery, the delivery time is set tight, time delivery solution was obtained in Problem 3 Problem 12 in. (Table3).
- Setting delivery time is gradual, time delivery solution was obtained in 10 issue 12 (Table4).
- Equivalent to the value of the best known, was obtained in five issues remaining 9 problems during delivery time in setting strict (Table3).
- In setting delivery is gradual, equivalent numerical and best known value was obtained in 11 issue 12 (Table4).
- If you look at the convergence status of the solution, can be seen the convergence of the solution at an early stage (Figure3).

Bold indicates what good result is obtained in this experiment, but it is the result may be a 18 issue 24 issue of The combined optimum value and known time delivery solution. In the numerical experiments of Abe, whereas it took an enormous amount of time of 2 hours 30 minutes to given problem, and the approximate solution of the equivalence almost optimal solution and optimal solution is about 3 to 5 minutes more seconds I was able to get.

Table 4. Experimental result 2 (F=1.6)

Instance	Opti. DT	ASA DT	S&P DT	Abe DT	pGA DT	pGA ave
abz5	0	0	0	0	**0**	0
abz6	0	0	0	0	**0**	0
la16	0	20	14	0	**0**	0
la17	65	129	81	65	**65**	67
la18	0	35	0	0	**0**	0
la19	0	0	0	0	**0**	0
la20	0	89	0	0	**0**	0
la21	0	0	0	0	**0**	0
la22	0	260	0	0	**0**	0
la23	0	96	0	0	**0**	0
la24	0	124	0	0	**0**	0
mt10	141	538	184	148	176	192.4

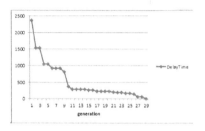

Fig. 3. Convergence process (abz6 F=1.5)

Since the search of the solution is made of capital almost 5 minutes early thing in the method of Pinedo and Asano but, it takes minutes to 1 hour 10 Depending on the problem, proposed in the problem with the goal to minimize the delivery time it was possible to confirm the effectiveness of the approach. Figure 4 shows an example of a solution that is finally obtained.

However, for that you get the best one solution is important in scheduling problems, we would like to seek a better method, taking into account the improvement of accuracy, such as Abe.

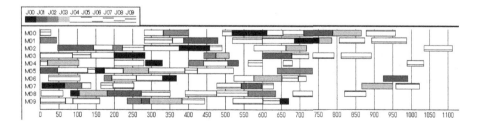

Fig. 4. Example Gant Chart (abz6 F=1.5)

6   Conclusions

In this study, the proposed algorithm of shorter lead times for delivery with job shop scheduling problem, and it was with the goal of improving the solution from the traditional model, and reduction of computation time. In this experiment, took up 12 problem by setting the 10 machine 10 job.We have set two delivery coefficients, respectively.

In a setting that is a gentle delivery, equivalent to those of the best known solution was obtained in 11 issue 12 issue of. In the second configuration, 8 issue 12 issue of was equivalent to the optimal solution known. It is considered because it is one of very important issues, since it is possible to find a solution with high accuracy in a short time, the proposed method this is effective to shorten the time in the information society in recent years. As a multi-objective scheduling problems, we would like to aim the proposed solution search algorithm of the problem more realistic as Abe future.

References

1. Asano, M., Hatanaka, K., Ohta, H.: A Heuristic Algorithm for Job-Shop Scheduling to Minimize Total Weighted Tardiness. Journal of Japan Industrial Management Association 51(3), 245–253 (2003)
2. Asano, M., Ohta, H.: A heuristic Algorithm for Job-Shop Scheduling to Minimize Total Weighted Tardiness. Computer and Industrial Engineering 42(2-4), 137–147 (2002)
3. Abe, K., Ida, K.: GA for JSP with Delivery Time and Holding Cost. Intelligent Engineering Systems through Artificial Neural Networks 19 (2009)
4. Singer, M., Pinedo, M.: A computational study of branch and bound techniques for minimizing the total weighted tardiness in job shops. IIE Trans. 30(2), 109–118 (1998)
5. Pinedo, M., Singer, M.: A shifting bottleneck heuristic for minimizing the total weighted tardiness in a job shop. Naval Res. Logist. 46(1), 1–17 (1999)
6. Nakatani, T., Ohta, T.: A Heuristic Algorithm to Minimize Total Holding Cost of Completed and Processing Production Job-Shop Scheduling Subject to No Tardy Jobs. Journal of Japan Industrial Management Association 53(6), 483–490 (2003) (in Japanese)
7. Nakatani, T., Asano, M., Ohta, T.: An Optimal Algorithm for Solving Job-Shop Scheduling to Minimize Total Holding Cost of Completed and processing Products Subject to Non- Tardy Jobs. Journal of Japan Industrial Management Association 5(2), 95–103 (2003) (in Japanese)
8. Hirano, H.: Genetic Algorithms with Cluster Averaging Method for Solving Job-Shop Scheduling Problems. Journal of Japanese Society for Artificial Intelligence 10(5), 769–777 (1995) (in Japanese)
9. Watanabe, M., Ida, K., Gen, M.: Active Solution Space and Search on Job-shop Scheduling Problem. Journal of Japanese Electrotechnical Committee 124(8), 1626–1631 (2004) (in Japanese)
10. Gen, M., Cheng, R.: Genetic Algorithms and Engineering Optimization. John Wiley and Sons, New York (2000)
11. Ida, K., Osawa, A.: Proposal of Algorithm for Shortening Idle Time on Job-shop Scheduling Problem and Its Numerical Experiments. Journal of Japan Industrial Management Association (JIMA) 56(4), 294–301 (2005) (in Japanese)
12. Gen, M., Cheng, R., Lin, L.: Network Models and Optimization: Multiobjective Genetic Algorithm Approach. Springer, New York (2008)
13. Beasley, J.E.: OR-Library: distributing test programs by electronic mail. Journal of the Operational Research Society 41(11), 1069–1072 (1990)

Advances in Multiobjective Hybrid Genetic Algorithms for Intelligent Manufacturing and Logistics Systems

Mitsuo Gen[1,2] and Kenichi Ida[3]

[1] Fuzzy Logic Systems Institute, Iizuka, Japan
[2] National Tsing Hua University, Taiwan
[3] Maebashi Institute of Technology, Maebashi, Japan

Abstract. Recently, genetic algorithms (GA) have received considerable attention regarding their potential as a combinatorial optimization for complex problems and have been successfully applied in the area of various engineering. We will survey recent advances in hybrid genetic algorithms (HGA) with local search and tuning parameters and multiobjective HGA (MO-HGA) with fitness assignments. Applications of HGA and MO-HGA will introduced for flexible job-shop scheduling problem (FJSP), reentrant flow-shop scheduling (RFS) model, and reverse logistics design model in the manufacturing and logistics systems.

Keywords: Hybrid genetic algorithms (HGA), Multiobjective HGA (MO-HGA), Reentrant flow-shop scheduling (RFS), Flexible job-shop scheduling problem (FJSP), Multiobjective reverse logistics model.

1 Introduction

Many real world applications in engineering design problems and information systems impose on more complex issues, such as complex structure, complex constraints, and multiple objectives to be handled simultaneously and make the problem intractable to the traditional approaches. Network models provide a useful way for modeling various operation management problems and are extensively used in many different types of systems: manufacturing and logistics areas. Network models and optimization for various scheduling and/or routing problems in manufacturing and logistics systems also provide a useful way as one of case studies in real world problems and are extensively used in practice [1].

Genetic algorithm (GA) has recently received a considerable attention because of its potential of being a very effective design optimization technique for solving various NP hard combinatorial optimization problems and complex information processing, manufacturing and logistics systems. Evolutionary Algorithms (EA) are stochastic optimization techniques that utilize principles of natural evolution in finding new search directions. Although, each of the evolutionary techniques, e.g. Genetic Algorithms (GA), Genetic Programming (GP) etc., have their own strengths and weaknesses for the combinatorial optimization problems. The most common algorithms studied in the literature variants of EAs [2-5].

T. Yoshida et al. (Eds.): AMT 2013, LNCS 8210, pp. 379–389, 2013.
© Springer International Publishing Switzerland 2013

Nowadays, manufacturing companies are faced with market demands for a variety of high quality products. Therefore, these companies must make their manufacturing systems more flexible, reduce costs related to production, and respond rapidly to demand fluctuations. Hence, companies need to have advanced techniques for solving the manufacturing scheduling problems, especially the developing manufacturing scheduling which is a remarkably important task in industry. However, it is very difficult to get the best schedule in manufacturing planning due to the growing complexity of the production process [6-10].

The rest of this paper is organized as follows: in Section 2, we survey a hybrid genetic algorithms (HGA) and multiobjective HGA (MO-HGA) with a recent fitness assignment mechanism. In Section 3, we introduce multiobjective flexible Job-shop Scheduling problem (MO-FJSP) with a multistage operation based GA (moGA) approach for solving FJSP. In Section 4, we introduce a reentrant flow-shop scheduling (RFS) in the hard-disk device manufacturing systems and in Section 5 a recent multiobjective reverse logistics network (mo-RLN) model by MO-HGA will be introduced and compared it with pri-awGA and LINGO.

2   Multiobjective Hybrid Genetic Algorithms

2.1   Hybrid Genetic Algorithms with Local Search and Fuzzy Logic

Genetic algorithms have proved to be a versatile and effective approach for solving combinatorial optimization problems. Nevertheless, there are many situations in which the basic GA does not perform particularly well, and various methods of have been proposed. With the hybrid approach, local optimization such as a hill-climbing or neighborhood search is applied to each newly generated offspring to move it to a local optimum before injecting it into the population. Genetic algorithms are used to perform global exploration among the population while heuristic methods are used to perform local exploitation around chromosomes. Because of the complementary properties of genetic algorithms and conventional heuristics, the hybrid approach often outperforms either method operating alone. The main idea is to use a fuzzy logic controller to compute new strategy parameter values that will be used by the genetic algorithms. A fuzzy logic controller (FLC) is comprised of four principal components:

1) a knowledge base, 2) a fuzzification interface, 3) an inference system, 4) a defuzzification interface.

The inference system is the kernel of the controller, which provides an approximate reasoning based on the knowledge base. A Hybrid Genetic Algorithms (HGA) combined with FLC routines proposed by Yun & Gen [11] and Lin & Gen [12], and the general structure of HGA in the pseudo code is described in Gen et al [1].

2.2   Multiobjective Hybrid Genetic Algorithms

Multiple objective problems arise in the design, modeling, and planning of many real complex systems in the areas of industrial production, urban transportation, capital budgeting, forest management, reservoir management, layout and landscaping of new cities, energy distribution, etc. Since the 1990s, EAs have been received considerable attention as a novel approach to multiobjective optimization problems, resulting in a fresh body of research and applications known as evolutionary multiobjective optimization (EMO) [13,27].

Without loss of generality, a multiple objective optimization problem (MOP) with q objective functions conflicting each other and m constraints can be formally represented as follows:

$$
\begin{aligned}
\max \quad & \left\{ z_1 = f_1(x), z_2 = f_2(x), \cdots, z_q = f_q(x) \right\} \\
\text{s.t.} \quad & g_i(x) \le 0, \quad i = 1, 2, \ldots, m
\end{aligned} \tag{1}
$$

We sometimes graph the MOP problem in both decision space and criterion space. S is used to denote the feasible region in the decision space and Z is used to denote the feasible region in the criterion space respectively as follows:

$$
S = \left\{ x \in R^n \,\middle|\, g_i(x) \le 0, \quad i = 1, 2, \ldots, m, x \ge 0 \right\} \tag{2}
$$

$$
Z = \left\{ z \in R^q \,\middle|\, z_1 = f_1(x), z_2 = f_2(x), \cdots, z_q = f_q(x), \quad x \in S \right\} \tag{3}
$$

here $x \in R^n$ is a vector of values of q objective functions. In the other words, Z is the set of images of all points in S. Although S is confined to the nonnegative region of R^n and Z is not necessarily confined to the nonnegative region of R^q.

There usually exists a set of solutions for the multiple objective cases which cannot be simply compared with each other. Such kind of solutions are called nondominated solutions or Pareto optimal solutions, for which no improvement in any objective function is possible without sacrificing on at least one of other objectives.

Definition 1: For a given point $z_0 \in Z$, it is nondominated if and only if there does not exist another point $z \in Z$ such that for the maximization case,

$$
\begin{aligned}
z_k > z_k^0, \quad & \text{for some } k \in \{1, 2, \ldots, q\} \\
z_l \ge z_l^0, \quad & \text{for all } l \ne k
\end{aligned}
$$

where z_0 is a dominated point in the criterion space Z with q objective functions.

2.3   Fitness Assignment Mechanisms

When applying the GAs to solve a given problem, it is necessary to refine upon each of the major components of GAs, such as encoding methods, recombination operators, fitness assignment, selection operators, and constraints handling, and so on, in order to obtain a best solution to the given problem. One of special issues in the multiobjective optimization problems is fitness assignment mechanism.

Adaptive Weight Genetic Algorithm (AWGA): Gen and Cheng proposed AWGA in which it utilizes some useful information from the current population to readjust weights to obtain a search pressure toward a positive ideal point [8]. For the examined solutions at each generation, we define two extreme points for the kth objective (maximum: z^+, minimum: z^-) as follows:

$$z_k^{max} = \max\{f_k(\mathbf{x}) \mid \mathbf{x} \in P\}, \quad k = 1, 2, \cdots, q \tag{4}$$

$$z_k^{min} = \min\{f_k(\mathbf{x}) \mid \mathbf{x} \in P\}, \quad k = 1, 2, \cdots, q \tag{5}$$

The weighted-sum objective function for a given chromosome $\mathbf{x}$ is given by the following equation:

$$eval(\mathbf{x}) = \sum_{k=1}^{q} w_k(z_k - z_k^{min}) = \sum_{k=1}^{q} \frac{z_k - z_k^{min}}{z_k^{max} - z_k^{min}} = \sum_{k=1}^{q} \frac{f_k(\mathbf{x}) - z_k^{min}}{z_k^{max} - z_k^{min}} \tag{6}$$

where w_k is adaptive weight for the kth objective function as shown in the following equation:

$$w_k = \frac{1}{z_k^{max} - z_k^{min}}, \quad k = 1, 2, \cdots, q \tag{7}$$

The equation (6) driven above is a hyperplane defined by the following extreme points (4-5) in current solutions.

Interactive Adaptive-Weight Genetic Algorithm (i-AWGA): Lin and Gen proposed an interactive adaptive-weight genetic algorithm, which is an improved adaptive-weight fitness assignment approach with the consideration of the disadvantages of weighted-sum approach and Pareto ranking-based approach [1]. They combined a penalty term to the fitness value for all of dominated solutions. Firstly, we calculate the adaptive weight $w_i = 1/ (z_i^{max} - z_i^{min})$ for each objective $i=1, 2,...,q$ by using AWGA. Afterwards, we calculate the penalty term $p(v_k)=0$, if v_k is nondominated solution in the nondominated set P. Otherwise $p(v_k')=1$ for dominated solution v_k'. Lastly, we calculate the fitness value of each chromosome by combining the i-AWGA method as follows:

$$eval(v_k) = \sum_{i=1}^{q} w_i(z_i^k - z_i^{min}) + p(v_k), \quad \forall k \in popSize \tag{8}$$

Hybrid Sampling Strategy-Based EA (HSS-EA): Zhang *et al.* proposed a hybrid sampling strategy-based evolutionary algorithm (HSS-EA) [14]. A Pareto dominating and dominated relationship-based fitness function (PDDR-FF) is proposed to evaluate the individuals. The PDDR-FF of an individual S_i is calculated by the following function:

$$eval(S_i) = q(S_i) + \frac{1}{p(S_i)+1}, \quad i = 1, 2,..., popSize \tag{9}$$

where $p()$ is the number of individuals which can be dominated by the individual S. $q()$ is the number of individuals which can dominate the individual S. The PDDR-FF can set the obvious difference values between the nondominated and dominated individuals. The general structure in the pseudo code of multiobjective hybrid genetic algorithms (MO-HGA) is described in [1]:

3   Multiobjective Flexible Job-Shop Scheduling Models

3.1   Background of FJSP Model

Flexible job-shop scheduling model is a generalization of the *job-shop scheduling problem* (JSP) [15-16] and the parallel machine environment, which provides a closer approximation to a wide range of real manufacturing systems. In particular, there are a set of parallel machines with possibly different efficiency. The *flexible job-shop scheduling problem* (FJSP) allows an operation to be performed by any machine in a work center. The FJSP model is NP-hard since it is an extension of the JSP [17].

The problem is to assign each operation to an available machine and sequence the operations assigned on each machine in order to minimize the makespan, that is, the time required to complete all jobs. The FJSP model will be formulated as a 0-1 mixed integer programming as follows:

$$\min \quad t_M = \max_{i,k} \{ c_{ik} \} \tag{10}$$

$$\min \quad W_M = \max_j \{ W_j \} \tag{11}$$

$$\min \quad W_T = \sum_{j=1}^m W_j \tag{12}$$

$$\text{s. t.} \quad c_{ik} - t_{ikj} x_{ikj} - c_{i(k-1)} \geq 0, \quad k = 2,...,K_i; \forall\, i, j \tag{13}$$

$$\sum_{j=1}^m x_{ikj} = 1, \forall\, k, i \tag{14}$$

$$x_{ikj} \in \{0,1\}, \quad \forall\, j, k, i \tag{15}$$

$$c_{ik} \geq 0, \forall\, k, i \tag{16}$$

3.2   Multistage Operation-Based GA Approach

During the past few years, the following representations for job-shop scheduling problem have been proposed [10, 13]: Parallel Machine Representation (PM-R), Parallel Job Representation (PJ-R), Kacem' approach, Priority rule-based Representation (PDR). Yang proposed a new GA-based discrete dynamic programming (GA-DDP) approach for generating static schedules in a flexible manufacturing system environment [18]. Zhang and Gen reported an effective multistage operation-based GA (moGA) approach to represent the chromosome and also proved to get better performance in solutions [19]. Originally, the encoding ideas of FJSP are followed Cheng, et al' work on a tutorial survey of GA for JSP [19, 23].

3.3   Numerical Experiments

Tables 1 and 2 illustrate the results comparing with other researcher's approaches with proposed approach in Gao, *et al* by using published for the benchmark problems [19, 20], where "ᵃ" denotes the approach "AL+CGA" proposed by Kacem *et al* [21], while "ᵇ" denotes the approach "PSO+SA" proposed by Xia and Wu (2005) [22]. The moGA with one-cut point crossover and local-search mutation solved two same data set in Kacem *et al* to compare the computational results with *popSize*: 100; p_M =0.3; p_C =0.6 [19].

Table 1. Result comparisons (8 jobs - 8machines)

	sGA	AL+CGAᵃ	PSO+SAᵇ	Proposed Approach
Makespan	7	7	7	7
W_T	53	45	44	43
Max (W_k)	7	5	6	5

Table 2. Result Comparisons (10 jobs -10 machines)

	Heuristic method (SPT)*	Classic GA*	Kacem's Approach*	moGA
t_M	16	7	7	7
W_T	59	53	45	43
W_M	16	7	6	5

4   Reentrant Flow-Shop Scheduling Models

4.1   Background of RFS Model in HDD Manufacturing

A *hard-disk devices* (HDD) manufacturing system is one of the most complicated systems depending on several constraints such as various product families with different processing time and processing flow, high flexibility machines, and one more time operation on a workstation as reentry flow of a job. Moreover, controlling processing time constraint is an important issue on the industry where requires high quality production especially in a hard-disk manufacturing system and it will be formulated as a *reentrant flow-shop scheduling (RFS) model* [24].

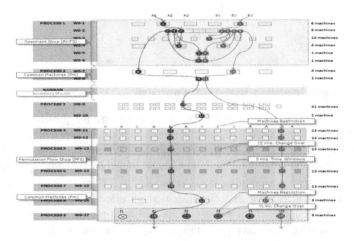

Fig. 1. The hybrid flow shop in a case of HDD manufacturing system

As show in Fig. 1, the hybrid flow shop in a real hard-disk manufacturing system, there are 9 processes with 17 workstations. Each of them has different number of machines which they also have different efficiency. Some machines might be limited by production constraints such as machine eligibility restriction and sequence dependent setup time (SDST). Moreover, the system still consists of several sub-systems for example, reentrant shop, common machines shop, and permutation shop. Unfortunately, these were located in the single system; it was very difficult to solve all by the optimization techniques [25].

4.2   Genetic Representations and Operations for RFS Problem

The operation-based encoding method is able to be applied since the problem concerns the processing step as same as the operation of a job [1]. The operation-based encoding method is able to be applied since the problem concerns the processing step as same as the operation of a job [25]. For decoding chromosomes, the feasible schedules of solving the RFS scheduling problem can be generated a schedule based on the job sequence. The two cut-point crossover and insert mutation are used for creating offspring. Refer for check and repair routine for the precedence the paper [25].

4.3   HGA with Time Window and Local Search

To develop the procedure of satisfying the time window constraint should be reasonable. The heuristic steps for checking and repairing routine for the time window constraint introduced in [25]. To improve the decoded schedule for drawing the Gantt chart, a heuristic of local search namely left-shift algorithm [24] could be combined for solving RFS problem. The left-shift procedure in the pseudo code is introduced in [25].

4.4   Numerical Experiments

Table 3. The computational results by GA, HGA and AHGA

Problem Sets			Current Practice		GA			HGA			AHGA		
			C_{max} [m]	Loss	C_{max} [m]		Loss	C_{max} [m]		Loss	C_{max} [m]		Loss
					Average	Best		Average	Best		Average	Best	
with Lot Sizes	ProcT1	7 jobs	L1	14,106	1	13,040	12,890	0	12,284	12,229	0	12,103	12,060	0
			L2	12,877	0	11,973	11,774	0	11,319	11,175	0	11,008	10,942	0
			L3	14,296	0	12,117	11,893	0	11,377	11,306	0	11,208	11,153	0
			L4	14,191	1	12,593	12,322	0	11,663	11,598	0	11,512	11,460	0
			L5	12,792	0	11,900	11,763	0	11,037	11,015	0	10,928	10,803	0
		11 jobs	L6	12,413	4	12,879	12,733	3	11,552	11,328	0	11,419	11,367	0
			L7	11,504	1	11,647	11,416	1	10,552	10,347	0	10,378	10,276	0
			L8	12,897	1	11,996	11,872	1	10,912	10,802	0	10,807	10,782	0
			L9	13,028	2	12,430	12,132	1	11,270	11,030	0	11,040	10,988	0
			L10	12,180	2	11,992	11,758	2	10,978	10,862	0	10,883	10,851	0
	ProcT2	7 jobs	L1	42,914	0	40,396	39,761	3	38,576	37,734	0	38,025	38,013	0
			L2	42,422	1	37,370	36,705	2	36,659	36,567	0	36,482	36,393	0
			L3	42,673	1	38,940	38,719	1	37,630	37,439	0	37,502	37,439	0
			L4	42,726	0	39,405	39,020	3	38,228	37,958	0	37,920	37,812	0
			L5	41,695	0	37,616	37,386	2	36,439	36,149	0	35,939	35,901	0
		11 jobs	L6	43,034	3	40,590	40,504	6	38,178	38,001	0	37,579	37,539	0
			L7	41,026	1	37,758	37,595	6	36,095	35,908	0	35,796	35,737	0
			L8	41,117	1	38,750	38,662	3	36,806	36,196	0	36,496	36,424	0
			L9	42,492	2	39,394	39,194	5	37,381	36,953	0	37,020	36,964	0
			L10	41,051	1	36,954	36,700	5	35,561	34,946	0	34,804	34,708	0

Recently Chamnanlor, *et al* proposed adaptive HGA (AHGA) for solving RFS problem with time windows in hard-disk manufacturing [24]. In the parameters of the HGA mentioned above, all the parameters except for the p_C and p_M were fixed during its search process. The computational results are shown in Table 3 in which the average values and the best solution obtained from the different computations, so it gives average of best solutions by the 10 runs of each simulation correspondingly.

5   Multiobjective Reverse Logistics Model

5.1   Background of Multiobjective Reverse Logistics Model

Reverse logistics (RL) is now the focus of attention in logistics field to realize resources recycling and low carbon society for keeping a green logistics environment. The target of the reverse logistics is the flow from recovered end-of-life products to their reusable products. Increasing the environment regulation, the reverse logistics has been emphasized by the following reasons: the economic effect resulted from the cost reduction of raw materials in manufacturing process, the propensity to consume changed to environment-friendly products, and the business strategy tried to improve a corporate image. As shown in Fig. 2, there are customer zones, several returning centers, several processing centers, and one manufacturer in this network where the reverse logistics recover products to the customers.

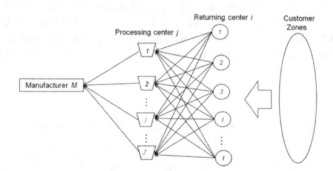

Fig. 2. Network model of revese logistics

5.2   Mathematical Model of Multiobjective Reverse Logistics

For example, in some cases, the companies may need to open more facility locations in order to decrease the total delivery tardiness and fulfill higher customer satisfaction, which may lead to a greater fixed opening cost. The multiobjective reverse logistics network model formulated as a MIP as follows:

$$\min \quad f_1 = \sum_{t=0}^{T}\left[\sum_{j=1}^{J}c_j^{op}z_j + \sum_{i=1}^{I}\sum_{j=1}^{J}c_{ij}x_{ij}(t) + \sum_{j=1}^{J}c_{jM}x_{jM}(t) + \sum_{j=1}^{J}c_j^{H}y_j^{H}(t)\right] \tag{17}$$

$$\min \quad f_2 = \sum_{t=0}^{T}\left[\sum_{i=1}^{I}\sum_{j=1}^{J}d_{ij}x_{ij}(t) + \sum_{j=1}^{J}(d_{jM}+p_j)x_{jM}(t)\right] - t_E d_M(t) \tag{18}$$

s. t.

$$\sum_{j=1}^{J}x_{ij}(t) \le r_i(t), \quad \forall i,\ t \tag{19}$$

$$\sum_{i=1}^{I}x_{ij}(t) + y_j^{H}(t-1) \le b_j z_j, \quad \forall j,\ t \tag{20}$$

$$\sum_{j=1}^{J}x_{jM}(t) \le d_M(t), \quad \forall t \tag{21}$$

$$y_j^{H}(t-1) + \sum_{i=1}^{I}x_{ij}(t) - x_{jM}(t) = y_j^{H}(t), \quad \forall j,\ t \tag{22}$$

$$x_{ij}(t),\ x_{jM}(t), y_j^{H}(t) \ge 0,\ \forall i,\ j,\ t \tag{23}$$

$$z_j \in \{0,1\},\ \forall j \tag{24}$$

Equations (17) and (18) give the objectives. Constraint (19) represents the recovered amount of end-of-life product. Constraints (20) and (21) represent the capacity of processing center and manufacturer respectively. Constraint (22) represents the inventory amount of processing center. Constraint (23) imposes the non-negativity of decision variables $x_{ij}(t)$, $x_{jM}(t)$ and $y_j^{H}(t)$. Constraint (8) assures the binary integrality of decision variables z_j. We can solve MO-RLN model by the computational software LINGO to compare.

5.3   Multiobjective Hybrid GA Approach and Numerical Experiments

To evaluate the performance of the mo-hGA [26] of revised version of MO-HGA for solving multiobjective reverse logistics network (mo-RLN) problem, mo-hGA is compared with the pri-awGA (priority-based encoding method with adaptive weight approach). Also, we compared by percentage gap of LINGO (LINDO Systems Inc.) with mo-hGA and pri-awGA. The reason for selecting pri-awGA as a basis of comparison is its similarity to mo-hGA in using a priority-based encoding method and adaptive weight approach. Six numerical examples and one case study of mo-RLN problem solved by mo-hGA and pri-awGA methods as shown in Table 4 and the improvement rate gap (%) = 100(GA-LINGO)/LINGO according to each time period is shown in the last column.

Table 4. The comparison of LINGO, pri-awGA and mo-hGA with Optimality Gaps

Problem No.	LINGO		pri-awGA	mo-hGA	Optimality Gaps (%)			
					pri-awGA		mo-hGA	
	f_1	f_2	(f_1, f_2)	(f_1, f_2)	f_1	f_2	f_1	f_2
1	20240	4046	(20300, 4054)	(20300,4054)	0.30	0.20	0.30	0.20
2	321960	6184	(323120, 6230)	(322682, 6202)	0.36	0.74	0.22	0.29
3	643928	320540	(652210,326430)	(650255,326220)	1.29	1.84	0.98	1.77
4	1564820	742250	(1613048,768424)	(1610541,757020)	3.08	3.53	2.92	1.99
5	7009356	1427258	(7501425,1520268)	(7431584,1486842)	7.02	6.52	6.02	4.17
Case study	25853321	6140542	(29004158,6924164)	(28125485,6452588)	12.19	12.76	8.79	5.08
Average					4.04	4.26	3.21	2.25

6   Conclusions

Recently, manufacturing companies are faced with global market demands for a variety of low cost products with a high quality. For responding rapidly to demand fluctuations and reducing costs related to manufacturing scheduling and logistics networks, hybrid genetic algorithms (HGA) and multiobjective HGA (MO-HGA) have received considerable attention regarding their potential for complex manufacturing and logistics problems as one of combinatorial optimization techniques. We surveyed recent advances in HGA with local search and tuning parameters by fuzzy logic and MO-HGA with a few fitness assignment mechanisms, respectively. Applications of HGA and MO-HGA introduced multiobjective flexible job-shop scheduling problem (FJSP) model, reentrant flow-shop scheduling (RFS) model, and multiobjective reverse logistics network (MO-RLN) model for the intelligent manufacturing and logistics systems.

Acknowledgement. This work was partly supported by the Grant-in-Aid for Scientific Research (C) (JSPS: No.24510219) and Taiwan NSC (NSC101-2811-E-007-004; NSC 102-2811-E-007-005).

References

1. Gen, M., Cheng, R., Lin, L.: Network Models and Optimization: Multiobjective Genetic Algorithm Approach, 710 p. Springer, London (2008)
2. Yu, Y., Gen, M.: Introduction to Evolutional Algorithms, p. 418. Springer, London (2010)
3. Gen, M.: Genetic Algorithms and their Applications. In: Pham, H. (ed.) Springer Handbook of Engineering Statistics, ch. 38, pp. 749–773. Springer (2006)
4. Gen, M., Lin, L.: Genetic Algorithms. In: Wah, B. (ed.) Wiley Encyclopedia of Computer Science and Engineering, pp. 1367–1381. John Wiley & Sons, Hoboken (2009)
5. Gen, M., Green, D., Katai, O., McKay, B., Namatame, A., Sarker, R., Zahng, B.T.: Intelligent and Evolutionary Systems. SCI, vol. 187. Springer, Heidelberg (2009)
6. Pinedo, M.: Scheduling Theory, Algorithms and Systems, 4th edn. Prentice-Hall, Upper Saddle River (2012)
7. Gen, M., Cheng, R.: Genetic Algorithms and Engineering Design, p. 432. John Wiley & Sons, New York (1997)
8. Gen, M., Cheng, R.: Genetic Algorithms and Engineering Optimization, p. 512. John Wiley & Sons, New York (2000)

9. Cheng, R., Gen, M.: Production planning and scheduling. In: Wang, J., Kusiak, A. (eds.) Handbook of Computational Intelligence in Design and Manufacturing. CRC Press LLC (2001)
10. Gen, M., Lin, L., Zhang, H.: Evolutionary techniques for optimization problems in integrated manufacturing system: State-of-the-Art-Survey. Computers & Industrial Engineering 56(3), 779–808 (2009)
11. Yun, Y., Gen, M.: Performance analysis of adapted genetic algorithm with fuzzy logic and heuristics. Fuzzy Optimization and Decision Making 2, 161–175 (2003)
12. Lin, L., Gen, M.: Auto-tuning strategy for evolutionary algorithms: balancing between exploration and exploitation. Soft Computing 13(2), 157–168 (2009)
13. Gen, M., Lin, L.: Multiobjective genetic algorithm for scheduling problems in manufacturing systems. Industrial Engineering & Management Systems 11(4), 310–330 (2012)
14. Zhang, W., Gen, M., Jo, J.B.: Hybrid sampling strategy-based multiobjective evolutionary algorithm for process planning and scheduling problem. J. of Intelligent Manufacturing (2013), doi:10.1007/s10845-013-0814-2
15. Cheng, R., Gen, M., Tsujimura, Y.: A tutorial survey of job-shop scheduling problems using genetic algorithms-I. Representation. Computers & Industrial Engineering 30(4), 983–997 (1996)
16. Cheng, R., Gen, M., Tsujimura, Y.: A tutorial survey of job-shop scheduling problems using genetic algorithms, part II: Hybrid genetic search strategies. Computers & Industrial Engineering 36(2), 343–364 (1999)
17. Garey, M.R., Johnson, D.S., Sethi, R.: The complexity of flowshop and jobshop scheduling. Mathematics of Operations Research 1, 117–129 (1976)
18. Yang, J.B.: GA-based discrete dynamic programming approach for scheduling in FMS environments. IEEE Trans. Syst., Man, and Cybernetics-Part B 31(5), 824–835 (2001)
19. Zhang, H., Gen, M.: Multistage-based genetic algorithm for flexible job-shop scheduling problem. J. of Complexity International 11, 223–232 (2005)
20. Gao, J., Gen, M., Sun, L.: Scheduling jobs and maintenances in flexible job shop with a hybrid genetic algorithm. J. of Intelligent Manufacturing 17(4), 493–507 (2006)
21. Kacem, I., Hammadi, S., Borne, P.: Pareto-optimality approach for flexible job-shop scheduling problems: Hybridization of genetic algorithms and fuzzy logic. Math. & Comp. in Simulation 60, 245–276 (2002)
22. Xia, W., Wu, Z.: An effective hybrid optimization approach for multi-objective flexible job-shop scheduling problem. Computer & Industrial Engineering 48(2), 409–425 (2005)
23. Gen, M., Gao, J., Lin, L.: Multistage-based genetic algorithm for flexible job-shop scheduling problem. Intelligent and Evolutionary Systems 187, 183–196 (2009)
24. Abe, K., Ida, K.: Genetic local search method for re-entrant flowshop problem. In: Dagli, C.H., Enke, D.L., Bryden, K.M., Ceylan, H., Gen, M. (eds.) Intelligent Engineering Systems Through Artificial Neural Networks, vol. 18, pp. 381–387. ASME Press, New York (2008)
25. Chamnanlor, C., Sethanan, K., Chien, C.F., Gen, M.: Reentrant flow-shop scheduling with time windows for hard-disk manufacturing by hybrid genetic algorithms. In: Proc. of the Asia Pacific Indus. Eng. & Management Systems, Phuket, pp. 896–907 (2012)
26. Lee, J.-E., Chung, K.-Y., Lee, K.-D., Gen, M.: A multi-objective reverse logistics network design to optimize the total costs and delivery tardiness. Multimed. Tools Appl., 19 (2013), doi:10.1007/s11042-013-1594-6
27. Gen, M., Lin, L.: Multiobjective evolutionary algorithm for manufacturing scheduling problems: state-of-the-art survey. J. of Intelligent Manufacturing, 18 (2013), doi:10.1007/s10845-013-0804-4

A Semantic Coherence Based Intelligent Search System

Weidong Liu and Xiangfeng Luo

School of Computer Engineering and Science, Shanghai University, Shanghai, China
{Liuwd,luoxf}@shu.edu.cn

Abstract. The large-scale unordered sentences are springing up on the web since the massive novel web social Medias have emerged. Although those unordered sentences have rich information, they only provide users with incoherent information service because they have loose semantic relations. Users usually expect to obtain semantic coherent information service when they are facing massive unordered sentences. Unfortunately, general web search engines are not applicable to such issue, because they only return a flat list of unordered web pages based on keywords. In this paper, we propose a novel semantic coherence based intelligent search system. The search system can provide semantic coherence based search service, which includes choosing semantic coherent sentences and ranking the sentences by a semantic coherent way. When a user enters some semantic incoherent sentences as queries, our system can return a semantic coherent paragraph as search results. The process is demonstrated by a prototypical system and experiments are conducted to validate its correctness. The results of experiments have shown that the system can distinguish semantic coherent sentences from others and rank the sentences by a semantic coherent way with higher accuracy.

Keywords: semantic coherence, intelligent search, sentence, short text.

1   Introduction

Semantic coherence is related with associated semantics and sound structure. Coherence is defined as a "continuity of senses" and "the mutual access and relevance within a configuration of concepts and relations" [1]. In the human discourse process, semantic coherence is a key problem which assumes that readers or writers routinely attempt to construct coherent meanings and connections among text constituents [2]. Similarly, human beings usually expect to obtain semantic coherent information service when they are facing massive unordered information on the web. As various novel web social Medias appear, a large volume of short messages are transmitted by sentences such as Twitter, Facebook, micro-blogs, etc. In the volume of short messages, two associated sentences belonging to one topic may be far away from each other because they are submerged in a large scale of unordered sentence set. The massive sentences are rich in semantic information, but hard to express semantic coherent information. To meet user's expectation of pursuing semantic coherence service, it is significant to enrich these semantic incoherent sentences into a semantic coherent

T. Yoshida et al. (Eds.): AMT 2013, LNCS 8210, pp. 390–400, 2013.
© Springer International Publishing Switzerland 2013

paragraph. Although general web search engines (such as Google and Bing) provide convenient service for users, most of them are unable to provide semantic coherence based intelligent search service.

In this paper, we propose a novel semantic coherence based intelligent search system. It receives semantic incoherent sentences as user queries and returns a semantic coherent paragraph as search results. Our search system can provide semantic coherence based search service which includes 1) choosing semantic coherent sentences and 2) ranking the unordered sentences in a semantic coherent way. The rest of the paper will be organized as follows: Section 2 introduces related work; Section 3 proposes an architecture of semantic coherence based intelligent search; Section 4 gives the data collection process; Section 5 proposes search service which enfolds how to choose semantic coherent sentences and rank them in a semantic coherent way; Section 6 gives a demonstration of a prototypical system and conducts experiments on it; Section 7 makes conclusions.

2   Related Work

Depending on different information collection and services, the search engines can be divided into four categories 1) keyword based search engines, 2) directory based search engines, 3) Meta search engines and 4) intelligent semantic search engines.

Keyword based search engines mainly return web pages via keyword matching such as Google and Bing. However, they depend more on the form of keywords rather than the meaning. Directory based search engines often semi-automatically assign web pages into different theme directories such as Yahoo and Open Direcotory. Manual operation and slow updating are their deficiencies [3]. Meta search engines are established on the basis of other search engines such as Dogpile, MetaCrawler [4]. They mainly pass the user query to other search engines and return their results. But it is time consuming for a user to choose needed information from the large search results.

Compared with the three categories of search engines above, intelligent semantic search engines seek to improve search accuracy by understanding user intent and web source semantics such as NAGA, Hakia, Powerset etc. [5-14]. Technologies in intelligent semantic search engines usually include contextual analysis to disambiguate queries [5], knowledge reasoning [9-11], natural language understanding [7], ontology representation [6, 8]. Some search engines highlight search engine behavior and user behavior [12-14]. However, to our best knowledge, few search systems are proposed to search sentences or short texts with the point view of semantic coherence.

3   Semantic Coherence Based Search System Architecture

In this paper, the semantic coherence based intelligent search system aims at providing semantic coherence for unordered sentences from micro-blogs. Fig 1 outlines the system's architecture. It consists of two parts, 1) data collector and 2) searching server. Data collector mainly collects data from micro-blogs and builds indexes for them.

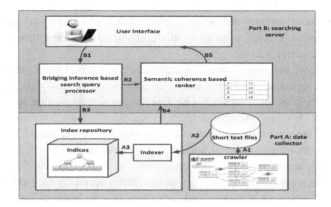

Fig. 1. Architecture of semantic coherence based intelligent search system

Searching server mainly provides semantic coherence based search service. Data collector provides data support and index structure for searching server.

1. Data collector consists of crawler, short text files and index repository.
- Web Crawler downloads micro-blog data and stores them in short text files as A1.
- Short text files store the short texts and provide data to indexer as A2.
- Index repository mainly provides an inverted index and a semantic associated index as A3.
2. Searching server consists of user interface, bridging inference based search query processor and semantic coherence based ranker.
- User interface receives some incoherent sentences as user queries and returns a semantic coherent paragraph as the search results. When a user enters a query, user interface passes the query into the bridging inference based search query processor as B1.
- Bridging inference based search query processor mainly chooses semantic coherent sentences according to user queries. It mainly uses a bridging inference based model to analyze which sentences can be activated as B3. The activated sentences can be returned to the semantic coherence based ranker as B4.
- Semantic coherence based ranker can rank the obtained unordered sentences from B2 and B4 by a semantic coherent way. It can order the unordered sentences into a semantic coherent paragraph. The paragraph is returned to user interface as B5. Its main functions include semantic coherent computing and semantic coherent ranking.

4   Data Collector

Data collector downloads short texts and builds index repository for them. The semantic coherence based intelligent search system has an obvious difference from other search systems in building index repository. Herein, we disclose how to build index repository as follows.

4.1 Index Repository

Index repository mainly includes two parts, 1) an inverted index and 2) a semantic associated index.

1. Building an inverted index

Inverted index can map keywords into sentences which are the main forms of short texts. Our system only extracts nouns to build the inverted index.

2. Building a semantic associated index

The spreading activation model assumes that semantic processing is a spreading activation among associated concepts [15]. The words or concepts are more semantic coherent if they co-occur in adjacent sentences. Our system builds a semantic associated index by building association link network ALN [10] since semantic coherent sentences can be activated by the indexed semantic associated keywords. The ALN is obtained by mining semantic associate relations from sentences in short text files.

5 Searching Server

Our search server has two core services 1) choosing semantic coherent sentences and 2) ranking the unordered sentences in a semantic coherent way. So we introduce bridging inference based search query processor and semantic coherence based ranker.

5.1 Bridging Inference Based Search Query Processor

Bridging inference is particularly central to the semantic coherence which links objects or events in sentences to narrow semantic coherency gaps [16]. Guided by bridging inference theory, bridging inference based search query processor mainly chooses semantic coherent sentences from short text files. It includes 1) sentence set representation model and 2) semantic coherent sentence activation model.

1. Sentence set representation model

Sentence set semantic associated link net (SALN) is used to represent the sentence set of user query. It is denoted by

$$SALN=<SN, SL> \qquad (1)$$

where $SN=\{kw1,kw2,kw3,\cdots,kwn\}$ denotes the keywords in a sentence set; n denotes the number of the identify keywords; SL is a set of semantic associated relations belonging to $SN \times SN$ and is denoted as

$$SL = \begin{bmatrix} 0 & w_{12} & \cdots & w_{1n} \\ w_{21} & 0 & \cdots & w_{2n} \\ \vdots & \vdots & \ddots & \vdots \\ w_{n1} & w_{n2} & \cdots & 0 \end{bmatrix} \tag{2}$$

where w_{ij} ($i{\neq}j$) denotes the weight of the existing association rule kwi→kwj in the sentences set of user query.

2. Semantic coherent sentence activation model

Given sentence set semantic associated link net (SALN), the bridging inference query processor will find out each unlinked keywords pairs in SALN. It searches the index repository to discover associated rules that directly link the unlinked keywords and activates some semantic coherent sentences that include these rules.

5.2   Semantic Coherence Based Ranker

Spreading activation model assumes that specific keywords are distributed in semantic associated net [15]. The graph-based features of semantic associate net exhibit measurable for semantic coherence. Such features are used in 1) semantic coherent computing and 2) semantic coherent ranking.

1. Semantic coherent computing method

Heuristic 1. A semantic coherent sentence set usually has high semantic compactness.

Semantic compactness suggests that the associated keywords in coherent sentence set have compact semantic relations. When the semantic relations of sentence set are compact, readers can quick obtain the main semantic relations of the sentence set.

Definition 1. Coherent compactness, CCN

CCN represents the extent of cross referencing of keywords in SALN. The measure is defined as

$$CCN = \frac{Max - \sum_{u \in N} \sum_{v \in N} d(u,v)}{Max - Min} \tag{3}$$

where $Max = N \times (N \times (N-1))$; $Min = N \times (N-1)$; $d(u, v)$ is the shortest path length of the keyword u and v; N is the number of keywords in SALN.

When CCN is high, keywords have straightforward explanations by associating others keywords with fewer steps. Low CCN makes accessing two associated keywords difficult since the keywords are far from each other.

Heuristic 2. A semantic coherent sentence set is highly semantics substantive.

Substantive semantics means that the keywords in sentence set are clustered to represent semantics. Massive single keywords often do not give substantive meaning of the sentence set, which often causes empty of matter of the sentence set.

Definition 2. Semantic coherence clustering, SCC

SCC reflects the degree to which the keywords involve the clustered semantics. The measure is defined as

$$SCC = \frac{1}{N}\sum_{1}^{N} AC_i \tag{4}$$

where AC_i denotes the clustering coefficient of keyword i; N is the number of keyword of SALN.

When SCC is high, the linked keywords highly involve the clustered semantics. The clustered keywords can express the main points of sentence set (e.g. topic keywords). Low SCC shows the keywords weakly involve the clustered semantics.

Heuristic 3. A semantic coherent sentence set usually has relative stable semantic variance.

Semantic variance often means that the overall clustered semantic change in a sentence set. A semantic coherent sentence set usually has more stable change than incoherent one.

Definition 3. Semantic coherence variance, SCV

SCV has positive correlation with standard deviation of coefficient distribution on different degree of keywords, which represents how much variation of semantic distribution around average clustering coefficient. It can be defined by

$$SCV = 1 - \exp\left(\left(\frac{-1}{n-1} \times \sum_{i}(AC_i - \overline{AC})^2\right)^{\frac{1}{2}}\right) \tag{5}$$

where AC_i means average clustering coefficient of keyword degree i; $\overline{AC}$ means the average clustering coefficient of all keywords . n is the number of different degree of keywords in SALN .

High SCV suggests that clustering coefficient is imbalanced distribution on different degree of keywords. Low SCV gives relatively stable semantic change.

2. Semantic coherent ranking method

In previous sections, semantic coherent features have been computed. For sentence set S_i with sentence order j, its coherent state is represented by coherent features $CFs(S_{ij})$.

$$CFs(S_{ij}) = \{CCN_{ij}, SCC_{ij}, SCV_{ij}\} \tag{6}$$

By analyzing these coherent features, it is found that different features have different effects on semantic coherence.

Theorem 1. Given a pairs permutations (S_{ij}, S_{ik}) of sentence set S_i, if S_{ij} performs higher semantic coherence than S_{ik}, then there is a weight vector $\vec{w}$ satisfying

$$\vec{w} \cdot CFs(S_{ij}) > \vec{w} \cdot CFs(S_{ik}) \tag{7}$$

where $\vec{w}$ is the weight vector; $CFs(S_{ij})$ denotes the coherent features of sentence set S_i with sentence order j; $CFs(S_{ik})$ denotes the coherent features of sentence set S_i with sentence order k.

So, the goal of semantic coherent computing is learning the weight vectors, which minimizes the number of violations of formula 7. We use the SVM^{light} Package to learn the weight vector [17].

6   Demonstration of a Prototypical System and Verification Experiments

In this part, we give a system demonstration to show our work and conduct some experiments to validate the correctness of semantic coherence based ranking.

6.1   Demonstration of a Prototypical System of Semantic Coherence Based Intelligent Search

Figure 2 shows the interface of the semantic coherence based intelligent search system. The interface includes four parts 1) user query box, 2) sentence set search box, 3) semantic coherent sentence list and 4) semantic coherent paragraph list.

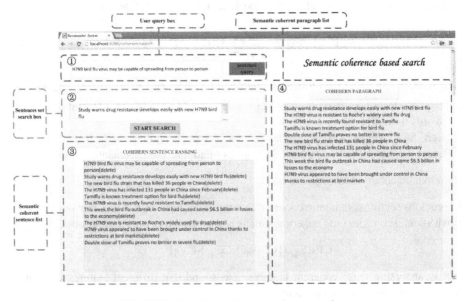

Fig. 2. User interface and system demonstration

1. User query box mainly receives the sentence that a user enters.
2. Sentence set search box displays a sentence set. The sentence set consists of all the sentences that the user enters.

3. Semantic coherent sentence list can show all the semantic coherent sentences received from bridging inference based search query processor.
4. Semantic coherent paragraph list gives a semantic coherent paragraph returned by semantic coherence based ranker.

Here, we give a simple case study to demonstrate the function of search system. Our current prototypical system includes 1627 sentences downloaded from twitter about flu virus. When the users read two sentences that "H7N9 bird flu virus may be capable of spreading from person to person" and "Study warns drug resistance develops easily with new H7N9 bird flu" in twitter, they may do not obtain the semantic coherence about the H7N9 bird flu. In this case, they can use the semantic coherence based intelligent search system to get a semantic coherence paragraph by the three steps as follows. 1) They can input these incoherent sentences into user query box and add the sentences into sentence set search box as ① and ② in fig2. 2) After they have searched, many semantic coherent sentences are returned to semantic coherent sentence list as ③. 3) The unordered sentences are organized into a semantic coherent paragraph to user as ④.

6.2   Validating the Correctness of Semantic Coherence Based Search Service

Data Set
To validate the correctness of semantic coherence based search service in our system, we downloaded 10000 news about health on reuters website (http://www.reuters.com) from March 2009 to August 2009 as benchmark data set. Table 1 gives a description of the data set. It contains 10000 texts and each text has average 15.67 sentences. We shuffle sentences of the texts and use these unordered sentence set as experiments data in the following experiments.

Table 1. Description of data set

Data base	Health news in routers
# News(benchmark data)	10000
Avg.# Sents	15.67

1. Choosing semantic coherent sentence

To verify the semantic coherence based search system can disguise the semantic coherent sentence from others, we make the system choose the most coherent sentence from different number of sentences. For a text, its writer always chooses semantic coherent sentences as best choices. Similarly, a better intelligent search system always can choose a coherent sentence from others as writers do.

We randomly select 1000 texts from data set in table 1. For each text, one sentence is removed. The remaining sentences can be regard as a question with one blank. The removed sentence is the only one right answer to the question. The right answer is mixed with other N-1 sentences which are randomly selected from other texts to form N options. Our system chooses the right answer from N options for the 1000 questions. The above process is conducted 10 times and then we calculate accuracy by formula 8.

$$\text{Accuracy} = \frac{\#\text{questions correctly choosing answer}}{\text{Total}\,\#\,\text{questions}} \tag{8}$$

Table 2 shows the average accuracy of choosing a semantic coherent sentence. For 2 options, the accuracy is 89.5%. For 5 options, the accuracy is 88.7%. When sentences increase to 100, our system can choose the semantic coherent sentence with 78.7% accuracy. It confirms that our system can choose semantic coherent sentences with high accuracy. When sentence options do not only include the right answer but also other semantic related sentences, the accuracy will decrease. The above case will occur with the higher possibility as the sentence options increase. So the average accuracy decreases with the sentence options increases in table 2.

Table 2. Average accuracy in choosing semantic coherent sentence

#Sentence options	Accuracy (%)
2	89.5
5	88.7
10	88.1
20	85.6
50	83.5
100	78.7

2. Ranking semantic coherent sentence sets

To validate the semantic coherence based search system can give correct coherent rank of the coherent sentence set. We make the system ranking the sentence set by a semantic coherent way. For a text, the original sentence set is always more semantic coherent than its permutations. A better intelligent search system will rank the original sentence order higher than its permutations.

We randomly select 500 texts as questions from data set in table 1. For each text, its sentences are permuted up to N-1 times. The N-1 permutations are mixed with original sentences set to form N sentence set options. Our system ranks the N sentence set options by the semantic coherence based rank method and records the rank of original sentence order for the 500 questions. The experiments are conducted 10 times, and then we calculate the proportion of the original sentence set in different rank range as accuracy by formula 9.

$$\text{Accuracy} = \frac{\#\text{questions ranking orginal sentence order in rank range}}{\text{Total}\,\#\,\text{questions}} \tag{9}$$

Table 3 shows the rank of the original sentence set and its proportion in all the questions. For 2 options, the proportion of the original sentence set in rank 1 are

80.3%. For 5 options, the original sentence set is rank 1 with 73.1% and rank 1-2 with 81.9%. For 50 options, the original sentence set is rank 1 with 59.2%, rank 1-2 with 70.4% and rank 1-5 with 79.1%. As the number of the options increases, the rank of the original sentence set keeps 1-5 with high proportion. It confirms that our system can rank semantic coherent sentence set with higher accuracy.

Table 3. Average proportion of sematnic coherent sentence set in different rank range

#Sentence set options	Rank range	Accuracy (%)
2	1	80.3
5	1	73.1
	1-2	81.9
10	1	63.6
	1-2	72.2
20	1	62.7
	1-2	70.1
	1-5	81.6
50	1	59.2
	1-2	70.4
	1-5	79.1

7   Conclusion

With increasing boom of micro-blogs, sentences become the main message passing forms and massive unordered sentences are emerging on the web. Although the sentences contain useful information, loose associations and unordered distribution make users hard to obtain semantic coherence from them. To help users to obtain semantic coherence based search service, we propose a novel semantic coherence based intelligent search system. Our system can receive some semantic incoherent sentences as user queries and get a semantic coherent paragraph as search results. The system can choose semantic coherent sentences and rank the sentences in a semantic coherent way. Our contributions are as follows.

1. To meet user needs of pursuing semantic coherence of unordered short texts, we develop a novel semantic coherence based intelligent search system. Our system can work on large scale unordered sentences by providing semantic coherence based search service.
2. To provide semantic coherence based search service, we integer some cognitive theories into the intelligent search system. We propose bridging inference based search query processor to choose coherent sentences and semantic coherence based ranker to rank unordered sentences into a semantic coherent paragraph.

Although the experimental data in our current search system is small, the results have exhibited great potential in providing semantic coherence based search service. We expect such system can extend into on-going works on short text automatic organization, online question-answering system and automatic text generation etc.

Acknowledgements. Research work reported in this paper was partly supported by the National Science Foundation of China under grant nos. 61071110 and the key project of shanghai municipal education commission under grant nos. 13ZZ064.

References

1. de Beaugrande, R., Dressler, W.: Introduction to text linguistics (1981)
2. Graesser, A.C., Singer, M., Trabasso, T.: Constructing inferences during narrative text comprehension. Psychological Review 101(3), 371 (1994)
3. Sullivan, D.: How search engines work. Search Engine Watch 14 (2002)
4. Zhang, J., Cheung, C.: Meta-search-engine feature analysis. Online Information Review 27(6), 433–441 (2003)
5. Kasneci, G., Suchanek, F.M., Ifrim, G., Ramanath, M., Weikum, G.: Naga: Searching and ranking knowledge. In: IEEE 24th International Conference on Data Engineering, ICDE 2008, pp. 953–962 (2008)
6. Sudeepthi, G., Anuradha, G., Babu, M.S.P.: A Survey on Semantic Web Search Engine. International Journal of Computer Science 9 (2012)
7. Dietze, H., Schroeder, M.: GoWeb: a semantic search engine for the life science web. BMC Bioinformatics 10(suppl. 10), S7 (2009)
8. Lee, D., Kwon, J., Yang, S., Lee, S.: Improvement of the Recall and the Precision for Semantic Web Services Search. In: 6th IEEE/ACIS International Conference on Computer and Information Science, ICIS 2007, pp. 763–768 (2007)
9. Luo, X., Xu, Z., Yu, J., Chen, X.: Building association link network for semantic link on web resources. IEEE Transactions on Automation Science and Engineering 8(3), 482–494 (2011)
10. Luo, X., Cai, C., Hu, Q.: Text knowledge representation model based on human concept learning. In: 2010 9th IEEE International Conference on Cognitive Informatics (ICCI), pp. 383–390 (2010)
11. Luo, X., Yu, J., Li, Q., Liu, F., Xu, Z.: Building web knowledge flows based on interactive computing with semantics. New Generation Computing 28(2), 113–120 (2010)
12. Meng, D., Huang, X.: An Interactive Intelligent Search Engine Model Research Based on User Information Preference. In: 9th International Conference on Computer Science and Informatics (2006)
13. Prakash, K.S.S., Raghavan, S.: Intelligent Search Engine: Simulation to Implementation. In: Proceedings of the iiWAS 2004, The Sixth International Conference on Information Integration and Web-based Applications Services, September 27-29 (2004)
14. Inamdar, S., Shinde, G.: An Agent Based Intelligent Search Engine System for Web mining.Research. Reflections and Innovations in Integrating ICT in Education (2008)
15. Collins, A.M., Loftus, E.F.: A spreading-activation theory of semantic processing. Psychological Review 82(6), 407–428 (1975)
16. Matsui, T.: Experimenstal pragmatics: Towards testing relevance-based predictions about anaphoric bridging inferences. Springer (2001)
17. Joachims, T.: Optimizing search engines using clickthrough data. In: Proceedings of the Eighth ACM SIGKDD International Conference on Knowledge Discovery and Data Mining, pp. 133–142 (2002)

Pyxis+: A Scalable and Adaptive Data Replication Framework

Yuwei Yang, Beihong Jin, and Sen Li

Institute of Software, Chinese Academy of Sciences, Beijing 100190, China
University of Chinese Academy of Sciences, Beijing 100190, China

Abstract. Data replication can improve the performance and availability for applications, and when it is employed by big data applications, it has to solve the challenges posed by big data applications, i.e., offering scalability and varying consistency levels. In this paper, we design and implement a data replication framework Pyxis+, whereby replication-aware applications can be developed in a rapid and convenient way. Pyxis+ allows the applications to register different consistency levels and automatically switches the consistency levels according to the change of requirements and performance. Meanwhile, on the basis of the consistency guarantees, Pyxis+ takes advantage of the consistent hashing technology to improve the scalability of data access. Simulation experimental results show that Pyxis+ can obtain relatively stable throughputs and response time by adding or removing replica managers while facing the increase of user requests.

Keywords: Replication Framework, Consistency Level, Scalability.

1 Introduction

Big data applications built on the cloud infrastructure often apply the replication technology for massive data storage. The intention lies in improving the performance of data access and enhancing the availability to tolerate component, network and hardware failures. However, the usage of replication may lead to the inconsistency of data. Meanwhile, as the CAP conjecture [1] [2] indicates, it is impossible to support consistency, availability and partition tolerance for a certain distributed service at the same time, but guaranteeing any two of the three is feasible. Therefore, these applications employing replication technology (hereinafter referred to as replication-aware applications) have to make a compromise among the consistency, availability and partition tolerance. In fact, most applications do not necessarily need strong consistency, and they prefer to ensure availability and partition tolerance by giving up strong consistency and maintaining weak consistency.

So far, there has existed different weak consistency, including eventual consistency, FIFO consistency, causal consistency and etc. Some applications only need to keep eventual consistency, which means that all the replicated data will be consistent eventually after a specific time period. But there are also many Internet-scale applications requiring stronger consistency than eventual consistency. For example, in

T. Yoshida et al. (Eds.): AMT 2013, LNCS 8210, pp. 401–412, 2013.
© Springer International Publishing Switzerland 2013

e-commerce scenarios, causal consistency is necessary to ensure that a user buys the products he has ordered. Furthermore, in some scenarios, the consistency requirements of applications may change with the variations of loads or time. Taking microblogging service as an example, there is causal consistency between publishing a microblog and commenting/transmitting the microblog. When the required response time of user requests can be satisfied, causal consistency provides better user experience. But it has to switch to eventual consistency if the required response time cannot be satisfied.

On the other hand, replication-aware big data applications expect to achieve high performance even in the face of massive data and a large number of update requests from users, and they pay much attention to dynamic failure-prone environments, owing to the fact that low performance and various failures can pose extensive negative effects on the user experience or even results in inestimable financial losses.

Unfortunately, the existing work cannot satisfy the above requirements simultaneously. In this paper, we provide a replication framework Pyxis+, where replica managers (RMs for short) manage the replicated data which can be stored in various storage systems. On the basis of Pyxis+, replication-aware applications can be developed in a rapid and convenient way. In particular, by virtue of the APIs Pyxis+ provides, replication-aware applications can customize different consistency levels for different data according to their requirements. Pyxis+ is responsible for maintaining and automatically switching consistency levels at runtime without stopping providing service. Moreover, Pyxis+ adopts a consistent hashing technique [3] to improve the scalability, and it can handle different failures besides the replica manager joining and leaving.

The rest of this paper is organized as follows. Section 2 overviews the scalable solutions of massive storage and the existing work about adaptive consistency levels. Section 3 gives the overview of Pyxis+. Section 4 presents the update/query processing in Pyxis+. Section 5 describes the adaptive and scalability mechanism. Section 6 shows the evaluation. The last section makes a conclusion of this paper.

2   Related Work

Distributed massive storage systems often employ the replication technology. Currently, one of the research focuses is to improve the scalability of such replicated storage systems. Dynamo [4] is Amazon's eventual consistent key-value store, and it adopts a variant of consistent hashing to achieve incremental scalability. In this method, it introduces the concept of "virtual nodes", that is, each physical node is assigned to multiple virtual nodes in the consistent hashing ring based on its capacity. This method takes into account heterogeneity in the physical infrastructure, and has advantages in load balancing and allowing node joining and leaving dynamically. PNUTS [5] is Yahoo!'s massively parallel and geographically distributed database system on the basis of record-level, asynchronous geographic replication. PNUTS provides novel per-record consistency guarantees, and it is proven to be highly available and scalable. [6] presents COPS, a scalable distributed key-value storage system. COPS provides causal+ consistency by tracking and explicitly checking whether causal dependencies are satisfied before exposing writes in each cluster.

Compared with the above storage systems, Pyxis+ targets applications that may access different storage systems, including key-value storage systems, distributed file systems and databases. In addition, each storage system mentioned above maintains only one consistency level, such as eventual consistency, per-record consistency or causal+ consistency. However, Pyxis+ allows applications to specify and switch between different consistency levels.

In terms of adaptive consistency, Cassandra [7] provides tunable consistency levels from the eventual consistency to the sequential consistency by configuring different numbers of replicas for read/write operations. Harmony [8] can adaptively tune the consistency level at run-time according to the application requirements. It presents an intelligent estimation model of stale reads, and then automatically adjusts the number of replicas involved in read operations to maintain a low tolerable fraction of stale reads and provide an adequate consistency level. [9] allows applications to define the consistency guarantees on the data and can automatically switch consistency guarantees at runtime. It classifies the data into three categories (A, B, and C), and treats each category differently depending on the required consistency level. The C category encompasses data under session consistency. As long as a session lasts, the system guarantees read-your-own-writes monotonicity. The A category provides serializability. Data in the B category switches between session consistency and serializability at runtime. [9] also presents different policies to adapt the consistency guarantees provided for individual data items in category B. [10] develops a conit-based continuous consistency model to capture the consistency spectrum using three application-independent metrics, numerical error, order error, and staleness. Moreover, it implements TACT, a middleware layer that enforces arbitrary consistency bounds among replicas using the metrics.

In contrast to the aforementioned researches, Pyxis+ provides two different consistency levels, that is, causal consistency and eventual consistency, and allows applications to customize and adaptively switch consistency levels according to their requirements.

3   System Overview

Pyxis+ adopts three-tier Client/Server architecture, as shown in Fig. 1. In Pyxis+, clients send messages including users' update/query requests to Front End (hereinafter, FE for short) and transmit reply messages containing request results from FE to users. FE, connected to all the RMs, is responsible for transmitting user requests from clients, and sending back request results from RMs. FE is also in charge of coordinating consistency switching at runtime.

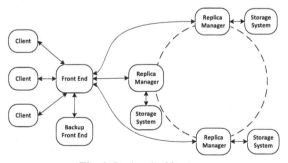

Fig. 1. Pyxis+ Architecture

In addition, several backup FEs can be deployed to improve the availability of Pyxis+. RMs are responsible for handling client requests and managing replicated data according to the consistency requirements customized by applications. In detail, all the RMs are organized into a consistent hashing ring, and thus each RM manages only a part of the data. In particular, the data managed by RMs can be stored in different kinds of storage systems, such as file systems, key/value storage systems or relational databases according to application requirements.

We assume that in Pyxis+, there is a loosely synchronized clock among all the RMs, and RMs or FEs may stop due to failures but they do not produce Byzantine behaviors. Moreover, RMs may join or leave the consistent hashing ring at any time. But two adjacent RMs in the consistent hashing ring should never fail at the same time, which means that when a failure happens to an RM, the predecessor RM and direct successor RM of the failed one will not fail during the failure handling process.

Currently, Pyxis+ supports two consistency levels, which are causal consistency and eventual consistency, and provides two kinds of consistency switching strategies: one is a customized strategy to switch consistency levels over time and the other is an adaptive switching strategy according to the service performance (here, we use the response time of client requests to measure the performance).

Pyxis+ provides several APIs (shown in Fig. 2) for its clients, of which two *register*s are used to specify the required consistency switching strategies for the data in different modules (indicated by argument

```
// Changing consistency strategies over time
void register (path, from_level, to_level, switch_time, lease_duration)
// Changing consistency strategies based on the response time of requests
void register (path, level, upper_bound, lower_bound, tolerant_duration)
// Updating the object specified by obj_id to value
boolean update(obj_id, value)
// Querying the value of the object specified by obj_id
String query(obj_id)
```

Fig. 2. Client APIs of Pyxis+

path) from the view of time and performance.

In implementation, Pyxis+ employs the TSAE (short for Time-Stamped Anti- Entropy) protocol [12] to propagate updates in a pull way, and satisfies the specifications of causal and eventual consistency by using the multi-dimensional timestamp technology, which is more available and scalable in comparison with the quorum-based replication technologies. Pyxis+ adopts a two-phase commitment protocol to realize the switches between two consistency levels, and adopts a timeout mechanism to ensure the correct execution in case of failures.

On the other hand, in order to support the high scalability required by cloud applications, we adopt the consistent hashing technology to manage data in different RMs. In specific, by using a consistent hashing function, a unique identifier is generated for each RM and object (latter refers to the replicated data to be updated/queried in the storage system) and is referred to as CHR identifier in the following. The CHR identifier of an RM is the hash value of its IP address modulo by 2^n. Thus, RMs are arranged according to their CHR identifiers in a consistent hashing ring. The CHR identifier of an object is the hash value of its *obj_id* (which is decided by the storage system) modulo by 2^n. Further, an object with CHR identifier k, indicated as Obj_k, is mapped to the RM *r* whose CHR identifier is the smallest one of the RM CHR identifiers which are large than or equal to *k*. RM *r* is treated as the primary RM of Obj_k. Meanwhile, Obj_k is also replicated on the (M-1) RMs whose positions are clockwise behind *r* in the consistent hashing ring. These (M-1) RMs are treated as the slave RMs of Obj_k, and also called successor RMs of *r*. Among the successor RMs of *r*, the RM *r'* whose position is clockwise adjacent to *r* is called as the direct successor RM of *r*, and *r* is called the predecessor RM of *r'*. Each RM maintains a finger table which keeps a part of the global routing information. Moreover, Pyxis+ allows RMs to join or leave the consistent hashing ring to adapt to varied loads on the premise of keeping the response time acceptable for users.

4   Update and Query Processing

In Pyxis+, multi-dimensional timestamp *t* is an array consisted of 256 non-negative integers, formally represented as $t = (t1, t2, t3...t256)$, where $ti \geq 0$ ($1 \leq i \leq 256$). For each RM, the last part of its IP address is used as its index in *t*.

In RM *r*, several main variables are maintained as shown in Fig. 3.

---

consistency_level: Recording the current consistency level provided by the consistent hashing ring which *r* belongs to, and its possible values are 1 and 2, representing causal and eventual consistency

rm_id : The unique identifier of *r*

pred_id: The identifier of the predecessor RM of *r*

succ_id_list: A list of the identifiers of *r*'s successor RMs

obj_id_record: Including *obj_id*, *prev* and *uid*, where *prev* and *uid* are the timestamps generated by the primary RM of the object identified by *obj_id* after the latest update operation

obj_id_map: A map of *obj_id_record*s indexed by *obj_id*

rep_ts: A timestamp, indicating the numbers of the received update operations/ACK messages

succ_rep_ts: A timestamp, recording the latest *rep_ts* of *r*'s direct successor RM

val_ts: A timestamp, indicating the numbers of the executed update operations

cid_set: A set of *cid*s, where a *cid* denotes the unique identifier of an executed update operation

wait_query_list: A list of query requests received by *r* without satisfying the execution conditions

update_record: Generated by a primary RM *p* after executing an update request from client *c*, including *client_id*, *obj_id*, *value*, *prev*, *uid*, *cid*, *rm_id*, and *counter*, where *client_id* is the unique identifier of client *c*, *obj_id* and *value* represent the identifier and value of the updated object, *prev* and *uid* are two timestamps generated by client *c* and RM *p*, *cid* is the unique identifier of the update operation, *rm_id* is the identifier of RM *p*, and *counter* is used to count the replication number of the object

ack_record: Generated by a primary RM *p* after receiving an ACK message from some client *c* for a certain update request, containing *uid*, *cid*, and *rm_id*, where *uid* is the timestamp generated by RM *p*, *cid* is the unique identifier of the corresponding update operation, and *rm_id* is the identifier of RM *p*

switch_record: Used in consistency switching process(see Section 5), consisted of *cid*, *rm_id*, and *counter*, where *cid* is the unique identifier of the switching operation, *rm_id* is the identifier of the RM generating the record, and *counter* is used to count the number of the record to be copied

op_record: An operation record, which can be *update_record*, *ack_record* or *switch_record*

op_map: A map of *op_record*s indexed by *cid*. As for the *op_map* whose index *rm_id* is *i*, its *op_record*s are sorted according to the *i*-th component of their *uid*s

op_mgt_map: A map of *op_map*s indexed by *rm_id*

Fig. 3. Main Variables in RM

In the FE, two main variables are stored for each client, see Fig. 4.

---

prev: A timestamp, used to identify the executed update operation set *s*, indicating that the follow-up update/query request from the same client depends on set *s*

rm_id: The identifier of the RM in charge of handling the requests from client *c*

Fig. 4. Main Variables in FE

RMs are responsible for processing two kinds of client requests transmitted by FE: update and query requests. In the following, we describe the update and query processing in detail.

When RM_i (indicating the RM whose *rm_id* is *i*) receives a client update request *u*, it looks up its finger table; if the primary RM of *u* is another RM *r*, it transmits *u* to *r*, otherwise, it indicates that RM_i is the primary RM of *u*, and RM_i handles *u* in Fig. 5.

---

1) RM_i determines whether the update operation has been executed, and it is deemed that *u* has been executed if one of the two conditions is satisfied: a) *u.cid* is contained in *cid_set*; b) *u.cid* exists in *op_map_i*, which means that there is an *update_record/ack_record* whose *cid* is equal to *u.cid*. If it is an *update_record*, it indicates that RM_i at least has received the same update operation; if it is an *ack_record*, the update operation is considered to have been executed. If the update operation has been carried out, RM_i sends a message to FE indicating that *u* is a repeated request, otherwise, *u* is treated as a new request, and step 2) to 6) are executed;

2) RM_i generates uid for $u : u.uid = u.prev$; $u.uid(i) = rep\_ts(i) + 1$;

3) If the consistency level is set to be eventual consistency or it is causal consistency and the condition $u.prev(i) \leq val\_ts(i)$ is satisfied, next step is to be executed, otherwise, a message indicating the failure of the execution is sent back to FE;

4) RM_i sends the operation *u* to the corresponding storage system, and waits for the execution result. If the storage system executes *u* successfully, RM_i executes step 5), otherwise, a message indicating the failure of the execution is sent back to FE;

5) A new *update_record* is generated by calling the method *make_op_record* (*u, u.uid, i, M*), where *M* is the replication number specified by applications, and then inserted into the corresponding *op_map_i*;

6) The *i*-th component of *rep_ts* is increased by 1;

7) The *i*-th component of *val_ts* is updated as that of *rep_ts*;

8) RM_i inserts *u.cid* into *cid_set*, and updates the corresponding *obj_id_record* indexed by *u.obj_id*;

9) If the consistency level is causal consistency, a message containing the *update_record* is sent to its direct successor RM, i.e., $RM_{succ\_id\_list(1)}$. If it is eventual consistency, a message indicating that the execution is successful is sent back to FE.

Fig. 5. Update Processing

In order that the update operations which have been executed by the primary RMs can be carried out by the corresponding slave RMs, the operations should be transmitted to the slave RMs as follows: each RM sends a TIMESTAMP message including its latest *rep_ts* to its predecessor RM periodically. When RM_i receives a TIMESTAMP message, it updates its *succ_rep_ts* as *rep_ts*, and compares it with the *uid*s of the update and *ack_record*s in *op_map_i* to find out the *op_record*s that have not been received by its direct successor RM *r*. Then RM_i replies a REPLICATION message containing these unreceived *op_record*s to *r*.

When RM$_i$ receives a query request q, it looks up its finger table to figure out whether the primary RM of q is another RM r, if so, it transmits q to r, otherwise, RM$_i$ is supposed to be the primary RM and it executes query processing as shown in Fig. 6.

---

1) If the consistency level is causal consistency, the query request q is sent to its direct successor RM, i.e., RM$_{succ\_id\_list(1)}$. If the consistency level is eventual consistency, q is executed locally (i.e., step 4)) or sent to one of its successor RMs;

2) When an RM r receives the query request q, if the consistency level is causal consistency, r determines whether its *pred_id* is equal to $q.rm\_id$, if so, it means that the primary RM is r's predecessor RM, and step 3) is executed, otherwise, r does not carry out further processing and throws an exception; if it is eventual consistency, step 4) is executed;

3) If the condition $q.prev(r.pred\_id) \leq r.val\_ts(r.pred\_id)$ is satisfied, step 4) is executed, otherwise, it goes to step 5);

4) The request q is sent to the corresponding storage system for execution. If q is executed successfully, a message with the query result is supposed to be received from the storage system, and local RM finds out the timestamps *prev* and *uid* with the index of *obj_id* contained by q in its *obj_id_map*. Then a message including the query result, *prev* and *uid* is sent to FE. However, if the execution of q is failed, a message indicating the failure of the execution is sent to FE;

5) The request q is inserted into *wait_query_list* to be executed when its condition is satisfied. Each RM receives a REPLICATION message periodically from its predecessor RM. After handling the message, it goes through *wait_query_list* to check out whether some of the query requests can be executed, if so, these requests are executed and removed from *wait_query_list*.

Fig. 6. Query Processing

---

5   Adaptive and Scalability Mechanism

5.1   Consistency Switching Strategy

When the consistency level in a consistent hashing ring needs to be switched, each RM is supposed to achieve a consistent state in the current consistency with FE. In Pyxis+, a two-phase commitment protocol is adopted to guarantee the correctness of runtime consistency switching. In specific, the switching process is shown in Fig. 7.

It is noted that there are some problems with the above consistency switching process. Since both FE and RMs are blocked to wait for messages at some steps, if an RM or FE crashes, FE or RMs will infinitely wait for the message from the crashed RM or FE. Therefore, a timeout mechanism is introduced to solve this kind of problem. On one hand, after sending SWITCH_REQUEST messages, FE may be blocked until it receives the replies from RMs. If it has not received all the replies in a particular time interval, FE decides to cancel consistency switching and sends GLOBAL_ABORT messages to all the RMs. On the other hand, RMs may wait for the global decision messages from FE after replying SWITCH_COMMIT messages. If a certain RM has not received the global decision message in a specific period, it should not cancel the consistency switching directly, but make sure what kind of message FE has already sent. In Pyxis+, each RM is blocked before FE failover. Since several backup FEs can be deployed, it is supposed that FEs can recover from failures in a specified duration and provide a reliable service. Moreover, we take the failover of crashed RMs into consideration. At first, the state information needs to be saved on disks. If an RM crashes without leaving any unprocessed SWITCH_REQUEST message or after sending a SWITCH_COMMIT/SWITCH_ABORT message, it can send a reply message to FE, and further decide whether to switch the consistency or not after recovering. If an RM fails after it sends a SWITCH_COMMIT message, it communicates with FE to make sure whether to carry out consistency switching after its failover.

---

1) FE sends SWITCH_REQUEST messages (including the consistency level to be switched to) to all the RMs in the consistent hashing ring;

2) When RM_i receives the SWITCH_REQUEST message, it stops handling client requests at first, then generates a *switch_record* and inserts the record into *op_map_i*. After RM_i receives all the *switch_record*s from (M-1) RMs whose positions are clockwise ahead of RM_i in the consistent hashing ring, it sends a SWITCH_ COMMIT message to FE, indicating that it is prepared to switch the consistency. However, if it has not received all the *switch_record*s in a specific time interval, it replies a SWITCH_ABORT message to FE, and cancels the consistency switching locally;

3) FE sends global decision messages to all the RMs based on their reply messages: if all the reply messages are SWITCH_COMMIT messages, FE sends GLOBAL_COMMIT messages to all the RMs; if FE receives one or more SWITCH_ABORT messages, it decides to cancel the current consistency switching process, and sends GLOBAL_ABORT messages to all the RMs;

4) The RMs which send SWITCH_COMMIT messages wait for the global decision messages from FE. If RMs receive GLOBAL_COMMIT messages, they start the consistency switching process locally; if GLOBAL_ABORT messages are received, the consistency switching process is canceled.

Fig. 7. Consistency Switching Process

In Pyxis+, the concurrent control of two or more consistency switching processes is not taken into account. FE is responsible for guaranteeing that there are not any other consistency switching processes when one process is in execution.

5.2   RM Joining and Leaving

In Pyxis+, since failures may happen to some RMs, or loads vary with time, the membership of RMs in a consistent hashing ring may change: some of them may join or leave the ring. In the following, we describe the processes of handling RM leaving, and joining. As for RM joining, there are two types of situations: one is that failover RMs rejoin the ring, and the other is that new RMs join the ring.

Aiming at ensuring that failed RMs can recover correctly, their state information must be maintained in a persistent storage. When RMs handle different *op_record*s, they not only insert them into the corresponding *op_map*s in memory, but also append them into LOG files on disks. Moreover, each RM carries out a snapshot operation once in a specified period. The snapshot operation guarantees that some failed RM can recover to the latest state before the failure happens.

Furthermore, the membership of RMs should be maintained when some RM leaves or joins the consistent hashing ring. When some RM leaves the ring, its direct successor RM is supposed to perceive its leaving. In Pyxis+, each RM sends PING messages to its predecessor RM periodically. If the PING messages sent from an RM r are timeout three times, r has to figure out whether it is itself or its predecessor RM r' that has left the ring. Therefore, r sends PING messages to FE, if the PING messages are timeout three times, it indicates that r is out of service for some reason, otherwise, r' is deemed to have left the ring, and the leaving process is shown in Fig. 8.

1) r looks up for its new predecessor RM r_{pred}, and updates its *pred_id*;

2) In order to obtain the update operations whose primary RM is r_{pred}, r sends a TIMESTAMP message to r, and r_{pred} replies a REPLICATION message to r;

3) When r receives the REPLICATION message, it executes the update operations contained in the message;

4) r stops handling the client update/query requests, and merges the *op_map$_{r'}$* and *op_map$_r$* as follows:
  a) The *op_record*s in both *op_map$_{r'}$* and *op_map$_r$* are resorted and saved into *op_map$_r$*;
  b) The *op_record*s in *op_map$_r$* are written into a new LOG file, and the original LOG files which correspond to *op_map$_{r'}$* and *op_map$_r$* are removed;

5) r synchronizes the merged *op_map$_r$* with its direct successor RM r_{succ}. In detail, r checks out the *op_record*s in *op_map$_r$*, inserts the *op_record*s satisfying specified conditions into a REPLICATION message, and sends the message to r_{succ}. As for a specific *op_record* named *op*, r compares the *uid* of *op* with *succ_rep_ts*, if *op.uid(r')* > *succ_rep_ts(r')* or *op.uid(r)* > *succ_rep_ts(r)*, then *op* is inserted into the REPLICATION message;

6) r restarts to handle client requests, and informs FE that its original predecessor r' has left the ring. When FE receives the message from r, it marks r' as left, and stops sending client requests to r'.

Fig. 8. RM Leaving Process

1) At first r sends a REJIOIN message to r_0, and the message contains the hash value q and a timestamp *sync_ts* which is equal to *r.rep_ts*;

2) When r_0 receives the REJION message from r, it looks up for r's direct successor RM r_{succ} according to q, and transmits the REJION message to r_{succ};

3) When r_{succ} receives the REJION message from r_0, it begins the data synchronization process with r. At first, r_{succ} starts the synchronization thread, and determines the *op_map*s which need to be synchronized to r. Then it reads the *op_record*s from the LOG files which correspond to the *op_map*s, and inserts some of them which satisfy particular conditions into a SYNC message. In specific, as for a certain *op_record* called *op* in a LOG file whose primary RM is r_{pri}, if *op.uid(pri)* > *sync_ts(pri)*, *op* is inserted into the message. r_{succ} goes through at most a given number of *op_record*s each time and then sends the SYNC message to r;

4) After r receives the SYNC message, it appends the *op_record*s to the corresponding *op_map*s, updates *sync_ts*, and sends a reply message containing *sync_ts* to r_{succ};

5) Receiving the reply message, r_{succ} compares its *rep_ts* with *sync_ts* contained in the message. If the difference of them (indicating the number of *op_record*s that have not been received by r) is less than or equal to a specified value *DIFF*, r_{succ} stops handling client update/query requests, and waits until the last part of the *op_record*s has been synchronized with r. Then it stops the synchronization thread, begins to handle client requests, and sends a SYNC_OVER message to r to inform the accomplishment of the synchronization process. However, if the difference of *rep_ts* and *sync_ts* is more than *DIFF*, step 3) to 5) are repeatedly executed until the synchronization is finished;

6) After r receives the SYNC_OVER message, it connects with FE, joins the ring and starts receiving and handling client requests.

Fig. 9. RM Joining Process

On the other hand, in order to guarantee the correctness of a replication service in the case of failover or new RMs joining, data synchronization needs to be carried out. For the convenience of expression, it is supposed that the RM to join the ring is r, the hash value of its IP address is q, and the RM which r communicates with in the beginning is r_o. The joining process is described in Fig. 9.

6   Evaluation

In order to evaluate the scalability of Pyxis+, we develop a micro-blogging prototype on the basis of Pyxis+ where each micro-blog is replicated three copies, and then we carry out several groups of experiments on the prototype.

Two servers (i.e., Dell PowerEdge T610) connected by a gigabit Ethernet switch are used as the experimental environment. The servers, each with a 12GB memory and two 4-core CPUs (Intel Xeon E5506, 2.13GHz) run XenServer (version 5.5), and in each XenServer, multiple virtual machines (CentOS Linux 5.4, Open JDK 6.0), each with 1GB memory and 1-core CPU, are installed. In the experiments, the number of RMs ranges from 4 to 14, and one RM and one micro-blogging service instance are deployed on one virtual machine; in addition, three FEs and several groups of clients are deployed on separate virtual machines, respectively. Experiments are designed to simulate the behaviors of micro-blogging users. Firstly, the number of users and the consistency levels required by users are set. Then, each user begins to send query/update requests, and the following request can only be sent until the execution result of the preceding request has been received for one second. Meanwhile, the average throughput and response time of query/update requests are calculated and recorded.

Fig. 10 shows the average throughput and response time of query requests in causal consistency under different numbers of users and RMs. From the results, we can see that given a specific number of RMs, the average throughput is increasing linearly with the increase of the number of users, and the average response time remains nearly the same. Moreover, given a specific number of users, the response time is decreasing with the increase of the number of RMs, which is more obvious when the number of users is 2000. Fig. 11 shows the performance of query requests in eventual consistency. From the results, similar conclusions to those of Fig. 10 are achieved. It is also seen that as for a consistency hashing ring consisting of 4 RMs, the performance arrives to the limitation when there are 1600 users, and the response time leaps beyond 700ms when the number of users increases to 2000.

Fig. 12 and Fig. 13 show the performance of update requests. Compared with the experiments of query requests, similar conclusions in terms of response time can be made. But the increase of users brings about the decrease of performance in some situations. It is observed that when the number of users is 1200 and the number of RMs is 4, or there are 1600 users and 6 RMs, the increase of users brings about the decrease of throughput. This is because when the performance arrives at the limitation, the requests from users are inserted into a queue in each RM waiting for execution, and according to the experimental procedure, each user cannot send another request until the execution result of the last request has be received for one second.

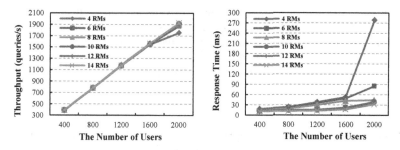

Fig. 10. Throughput and Response Time of Query Requests in Causal Consistency

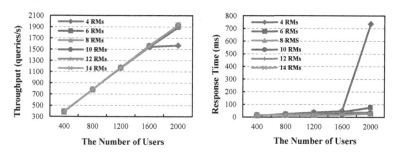

Fig. 11. Throughput and Response Time of Query Requests in Eventual Consistency

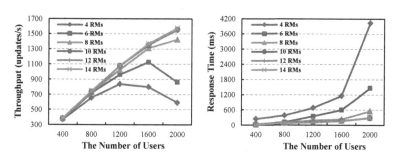

Fig. 12. Throughput and Response Time of Update Requests in Causal Consistency

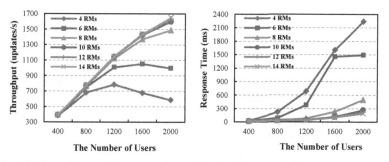

Fig. 13. Throughput and Response Time of Update Requests in Eventual Consistency

The above experimental results indicate that Pyxis+ is scalable under variant loads. On one hand, when some RMs are overloaded, inserting more RMs can increase the throughput and decrease the response time; on the other hand, when there are some underloaded RMs, removing a certain number of RMs has benefits for reducing costs without affecting the throughput and response time.

7   Conclusion

This paper presents Pyxis+, a data replication framework that provides the management of replicated data for various applications. Pyxis+ is independent of any storage systems but can put the replicated data into existing storage systems. In fact, Pyxis+ is a middleware layer that bridges the applications and the data storage systems. By now, Pyxis+ has provided multiple consistency guarantees and switching strategies, and also shown the stable performance of data access even facing the changing of user requests. Therefore, many applications can be envisioned in the future to adopt Pyxis+.

Acknowledgments. This work was supported by the National Natural Science Foundation of China under Grant No. 91124001.

References

1. Eric, A.B.: Towards Robust Distributed Systems (abstract). In: 19th Annual ACM Symposium on Principles of Distributed Computing, New York, p. 7 (2000)
2. Seth, G., Nancy, L.: Brewer's Conjecture and the Feasibility of Consistent Available Partition-tolerant Web Services. ACM SIGACT News 33, 51–59 (2002)
3. Ion, S., Robert, M., David, K.: M. Frans, K., Hari, B.: Chord: A Scalable Peer-to-peer Lookup Service for Internet Applications. In: ACM Special Interest Group on Data Communication, San Diego, pp. 149–160 (2001)
4. Giuseppe, D., Deniz, H., Madan, J., Gunavardhan, K., Avinash, L., Alex, P., Swaminathan, S., Peter, V.: Dynamo: Amazon's Highly Available Key-value Store. In: 21st ACM SIGOPS Symposium on Operating Systems Principles, New York, pp. 205–220 (2007)
5. Brian, F.C., Raghu, R., Utkarsh, S., Adam, S., Philip, B., Hans-Arno, J., Nick, P., Daniel, W., Ramana, Y.: PNUTS: Yahoo!'s Hosted Data Serving Platform. VLDB Endowment 1, 1277–1288 (2008)
6. Wyatt, L., Michael, J.F., Michael, K., David, G.A.: Don't Settle For Eventual: Scalable Causal Consistency for Wide-Area Storage with COPS. In: 23rd ACM Symposium on Operating Systems Principles, Cascais, pp. 401–416 (2011)
7. Lakshman, A., Malik, P.: Cassandra: a decentralized structured storage system. ACM SIGOPS Operating Systems Review 44, 35–40 (2010)
8. Houssem-Eddine, C., Shadi, I., Gabriel, A., Maria, S.P.: Harmony: Towards Automated Self-Adaptive Consistency in Cloud Storage. In: 2012 IEEE International Conference on Cluster Computing, Beijing, pp. 293–301 (2012)
9. Tim, K., Martin, H., Gustavo, A., Donald, K.: Consistency Rationing in the Cloud: Pay only when it matters. VLDB Endowment 2, 253–264 (2009)
10. Haifeng, Y., Amin, V.: Design and Evaluation of a Conit-Based Continuous Consistency Model for Replicated Services. ACM Transactions on Computer Systems 20, 239–282 (2002)
11. Golding, R.A., Long, D.D.E.: Modeling Replica Divergence in a Weak-consistence Protocol for Global-scale Distributed Data Bases. In: Concurrent Systems Laboratory, Computer and Information Sciences (1993)

Classifying Mass Spectral Data Using SVM and Wavelet-Based Feature Extraction

Wong Liyen[1], Maybin K. Muyeba[1], John A. Keane[2],
Zhiguo Gong[3], and Valerie Edwards-Jones[4]

[1] School of Computing, Mathematics and Digital Technology
[2] School of Computer Science, University of Manchester, UK
[3] Faculty of Science and Technology
University of Macau, China
[4] Institute for Biomedical Research into Human Movement and Health
Manchester Metropolitan University, UK
li.y.wong@stu.mmu.ac.uk, {m.muyeba,v.e.jones}@mmu.ac.uk,
jak@cs.man.ac.uk, fstzgg@umac.mo

Abstract. The paper investigates the use of support vector machines (SVM) in classifying Matrix-Assisted Laser Desorption Ionisation (MALDI) Time Of Flight (TOF) mass spectra. MALDI-TOF screening is a simple and useful technique for rapidly identifying microorganisms and classifying them into specific subtypes. MALDI-TOF data presents data analysis challenges due to its complexity and inherent data uncertainties. In addition, there are usually large mass ranges within which to identify the spectra and this may pose problems in classification. To deal with this problem, we use Wavelets to select relevant and localized features. We then search for best optimal parameters to choose an SVM kernel and apply the SVM classifier. We compare classification accuracy and dimensionality reduction between the SVM classifier and the SVM classifier with wavelet-based feature extraction. Results show that wavelet-based feature extraction improved classification accuracy by at least 10%, feature reduction by 76% and runtime by over 80%.

Keywords: SVM, wavelets, MALDI-TOF, parameter search, feature reduction.

1   Introduction

Signal data is a sequence of measurements from instruments that is either continuous or discrete and captured in intervals of time, frequency, distance, wave numbers etc. A signal of particular interest is one that is absorbed or reflected and usually measured in wavelength intervals. In recent years, there have been a number of studies on bacterial diseases and the problem of identifying species of bacteria that cause particular diseases. In particular, when signals are projected on bacterial samples, resulting ions from the compound are allowed to drift (time of flight) towards a detector. The time of flight is measured and is proportional to their mass. This data is called Matrix Assisted Laser Desorption Ionisation (MALDI) Time Of Flight (TOF) [1].

T. Yoshida et al. (Eds.): AMT 2013, LNCS 8210, pp. 413–422, 2013.
© Springer International Publishing Switzerland 2013

The number (or count) of these ions is then plotted against their mass and a spectral graph is produced. This graph shows peaks and troughs of the properties the bacterial species exhibits and is useful for identifying isolates - species, genres, types etc [2][4][6]. The signal data has lots of features, is large and has missing values. These problems motivate the approach used in this paper. Firstly, support vector machines (SVM) [22] have powerful generalisation ability for high dimensional data with missing values. Secondly, feature reduction has been known to improve classification accuracy and wavelets are a favourable choice in signal processing [5]. Wavelets are mathematical tool for decomposing data and complex functions into time and frequency components [26]. Unlike Fourier transform, wavelet transform are better suited for non-stationary signals such as MALDI-TOF mass spectra as they can distinguish different frequency signals at different times (non-stationery).

Signal classification [5] typically has two steps: feature extraction and signal classification on the reduced feature set. Our experiments are based on classification with full features using SVM, compared to classification with reduced features sets (SVM and wavelets) whilst choosing a suitable kernel classifier [25]. Some earlier work regarding initial experimentation and data mining methodology used are given in [28]

The paper is organised as follows: section 2 presents wavelets; section 3 presents mining signal data; section 4 presents experimentation and section 5 a conclusion.

2   Wavelets

Wavelets are a set of mathematical functions used to approximate data and more complex functions by dividing a signal into different frequency and time intervals called wavelets [8]. These intervals are better represented to their scales. Wavelets express a given function in terms of summation of basis functions. The wavelet basis is formed by translation and dilation of the mother wavelet. An example a mother wavelet is a Haar wavelet (fig. 1).

$$H(t)= \begin{cases} 1, & 0<t\le0.5 \\ -1, & 0.5<t\le1 \\ 0, & \text{eslewhere} \end{cases}$$

Fig. 1. Haar wavelet

The wavelet transform is performed on a continuous function, $f(t)$, and defined as

$$f(t) = \int_{-\infty}^{+\infty} h(\omega)\psi_\omega(t)dt \qquad (1)$$

where $h(\omega)$ is a weighting function and $\psi_\omega(t)$ is an othonorrnal basis function such that $\psi(2^j t - k), j, k \in Z$. By dilation and translation of the mother wavelet, we get wavelets compactly supported in their regions [9][10]. Wavelets exhibit other useful properties in addition to dilation and localization, such as smoothness, distinguishing most essential information, feature selection etc and their efficiency makes them candidates for data mining. Wavelets are designed to give excellent time resolution at high frequencies (i.e. for short durations of time at these high frequencies) and poor frequency resolution, and good frequency resolution at low frequencies and poor time resolution. Mass spectra contain noise because of contaminants and matrix material, causing varying baselines [17]. That is, to start preprocessing the data, a baseline correction is needed to remove low-frequency noise. MALDI-TOF MS spectra is recorded in signal form as (mass-to-charge ratio, millivolt signal, see figure 2), the second value shows the strength of the signal. The signal exhibits elongated (or outstanding) features above the baseline noise level and usually unevenly distributed. Feature selection (or removing noise level data) mostly focuses on selecting peaks [16] that are higher than a predetermined signal noise threshold to facilitate biomarker identification [11][12][18]. Biomarkers are measures that indicate normal biological processes, pathogenic (or organism) processes or other pharmacological responses to some therapy. Wavelets are well adapted to removing irrelevant noise level data features (Dcnoising [14]), sometimes termed smoothing.

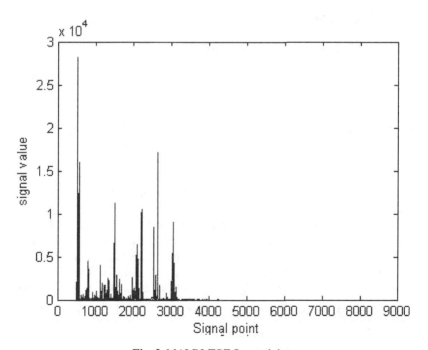

Fig. 2. MALDI-TOF Spectral data

Generally, most MALDI-TOF approaches aim to extract and quantify graph peak features accurately as these are considered to be the most interesting [7]. Other approaches use ant colony optimization to help efficiently select a set of interacting variables (features) by use of heuristic functions [16]. In addition, there are automated techniques for rapidly differentiating similarity between strains of bacteria, and in particular their taxonomic characterisation [2]. Feature selection is mainly concerned with obtaining useful features without loss of information, transforming an m-dimensional domain to a k-dimensional $k \ll m$.

Two approaches are common:

(1) Keep the largest k-coefficients, approximate rest to zero
(2) Keep the first k-coefficients, approximate rest to zero

Despite these two approaches, there is no guarantee the most important features will be obtained because there might be issues with information granulation (keep some, ignore the rest). In [5] fuzzy wavelet packets are introduced to deal with granulation of signal data into fuzzy relations (or clusters and not fuzzy sets) and reduce feature dimensionality by a fuzzy c-means approach. Information granulation by fuzzy means was presented in [15] and other fuzzy wavelet packet based feature extractions are reported in [19].

3   Mining Signal Data

Mining signal data is not new and various works exist [3]. Recently, machine learning approaches have been applied to learning MALDI-TOF data, for example use of Support Vector Machines (SVMs) and other techniques [4]. SVMs are popular classifiers that learn by examples and assign labels to objects [23][24]. They can be used for classification and regression as well as other learning tasks. SVM classifies linearly separate data by separating two clusters of data with the optimal hyperplane. The first problem, however, is that real world data is often non-linearly separable. Kernel functions provide a solution by projecting data into a higher dimensional feature space to separate by hyperplane. SVMs have been successfully applied to an increasingly wide variety of biomedical applications, for example microarray gene expression [24].

Secondly, another major problem in classifying features from signal data is handling the dimensionality problem: mapping the reduced data into a space and then classifying the result. In [5] wavelet packets are used to extract features (data/feature issues), classify and rank them (pattern evaluation) using a linear discriminant function (LDF) and others [13]. The approach is then, for a c-class problem with N labeled signal classes $X = \{(x_k, \omega_k), k = 1, 2, .., N\}$, $x_k \in \Re^n$ and $\omega_k \in C, C = \{1, 2., c\}$, a feature extraction function i.e. a mapping $f : X \rightarrow X'$ where $m \ll n$ and $X' \subseteq \Re^m$ is the reduced feature space.

To classify the feature space, we find a classifier to map the feature space into the known class labels $g : X' \rightarrow C$. To retain features that are a more accurate representation of the space for classification, the use of fuzzy clusters becomes necessary because of uncertainties in data granulation of the signal data.

To measure pattern extraction, classification metrics such as classification accuracy (rate) is applied to a number of features (or principal components) under varying discriminatory thresholds, r, for example in the fuzzy case where $0 < r < 1$ [5][19].

4   Experiments

Preliminary experiments have been done using high dimensional MALDI-TOF spectral bacteria data provided by the Medical Microbiology Department, Manchester Metropolitan University. The data consisted of 14461 features with 2 columns: the first column being mass/charge ratio and the second column being a millivolt signal also known as strength of signal. There were 14 classes of different strains of S. aureus bacteria with two (2) testing samples each i.e. total of 28 testing data. We used the LIBSVM library [21] for classification.

4.1   Data Pre-processing

The procedure for the experiment was as follows:

(1) Split data into training and testing sets
(2) Perform numerical scaling – prevent dominance in ranges
(3) Perform a Grid Search for best kernel parameters, radial (r) and degree(d)
    Parameters with best cross-validation accuracy are chosen for classifier training
(4) Predict the test data with the trained classifier

The basic procedure above is further extended with the case where we perform feature extraction with discrete wavelet transform (DWT) – denoising and decomposition with a thresholding method that only discards the portion of the data that exceeds a certain limit. Further, the signal data is then decomposed into two subsequences – the approximation coefficient and the detail coefficient. The approximation coefficients contain most of the important information (peaks of the data) and are capable of describing the underlying data characteristics [27]. The experiments only used approximation coefficients as they contain most of the peak information of the original signal. The wavelet approximations of a signal at a certain level describe generalised peak lists of the de-noised spectrum [28]. Selecting features this way reduces the dimensionality of the original data while retaining the important features [26].

Figure 3 shows the whole classification process. After data conversion and scaling, a kernel selection is performed based on particular parameter search that best produces the best cross-validation result. We used the following kernels: Linear, Radial Basis Function (RBF), Sigmoid and Polynomial. These are shown in table 1 with parameters for best model selection.

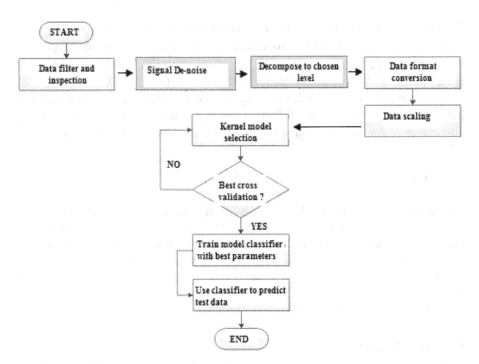

Fig. 3. SVM classification with feature selection– search for best cross validation result

Table 1. Kernelselection model parameters (Y=yes, N=no)

		Parameters			
		c Cost)	*γ* Gamma	*γ* (Radial)	*d* (degree)
Kernels	Linear	Y	N	N	N
	RBF	Y	Y	N	N
	Sigmoid	Y	Y	Y	N
	Polynomial	Y	Y	Y	Y

The experiments make use of grid search method provided in LIBSVM package. However the grid search method are time consuming and only optimize the pairs (C, γ) , therefore the experiment used, for a start, coarse grid values for the pair and later optimising the remaining parameters with finer grids.

All experiments used 5-fold cross validation. We note that Occam razor's theory mentioned in [29] suggests that smaller parameter values (in Table 1) are preferable to larger ones if they both have the same level of accuracy since building the SVM model with larger parameters takes longer. Thus our parameter search is simply based on (1) best level accuracy, (2) smallest parameter values.

4.2   Results for Experiment 1

For model selection using parameter settings in Table 1 and comparing both coarse and fine grid search, we obtained results shown in tables 2 and 3. Experiment 1 results are for signal data that does not use wavelet feature selection but only use model search for the best kernel and then performing a classification.

Table 2. Coarse Grid search cross-validation (CV%) (LIBSVM)

Kernel	c Cost)	γ Gamma	CV%
Linear	0.3125	2	76.2
RBF	1024	0.0000305	76.2
Sigmoid	1024	0.0000305	76.2
Polynomial	0.0000305	1	52.4

Table 3. Finer Grid Search cross-validation (CV%)(after MATLAB optimisation codes)

Kernel	c Cost)	γ Gamma	γ (Radial)	d (degree)	CV%
Linear	0.0078125	2	N/A	N/A	76.2
RBF	362.039	0.0000108	N/A	N/A	76.2
Sigmoid	256	0.0000305	0.0001	N/A	76.2
Polynomial	0.00195313	4	0.1	1	76.2

We found the same cross-validation accuracy of 76.2 (tables 2 and 3) across the kernels in both coarse and finer grid searches except for the polynomial kernel coarse grid which was 52.4%. Using the SVM classification on the trained data set, and computing accuracy (Equation 2), we obtained results as shown in table 4 with an equal number of support vectors (nSV).

$$Accuracy = \frac{\#Correctly\ classified\ signal\ data}{Total\ \#\ of\ signal\ data} \tag{2}$$

Table 4. Prediction accuracy –experiment 1

Kernel	Prediction accuracy	nSV	CV%	Elapsed time (s)
Linear	64.3	42	76.2	0.1412
RBF	64.3	41	76.2	0.1452
Sigmoid	64.3	42	76.2	0.1800
Polynomial	64.3	42	76.2	0.1851

Clearly as shown in table 4, linear kernel executes faster on average, given the same classification accuracy of 64.3% for all kernels. This meant that out of 28 classes (14x2), 64.3% of 28~18 classes correctly classified. It is also clear that as kernel complexity increases (e.g. linear kernel has one parameter compared to polynomial kernel with four), elapsed time also increases. Of note is the fact that cross-validation and prediction accuracy are the same. For experiment 2, we were compelled to use the linear and the RBF kernels for the reasons given above.

4.3   Results for Experiment 2

Table 4 shows experiment 1 results for cross validation, prediction accuracy, nSV and elapsed time for linear and RBK kernels using a pre-processing stage shown in figure 3.

To compare with a wavelet-based approach (experiment 2), several filters and levels were tested for the wavelet de-nosing part with two threshold levels, 50 and 70. Wavelet filters included the "bior1.1", "bior2.6", "bior3.7", "sym15" and "dmey". Results obtained are shown in table 5. The wavelet filter "bior2.6" achieved overall best performance with 75% prediction accuracy and a short elapsed time of 0.023389s. Comparing experiments 1 and 2, this represents an improvement in elapsed time of (0.14124-0.023389)/0.14124*100~83% if we compared with the best kernel "Linear kernel" from experiment 1. Prediction accuracy improved by (75-64.2857)/75*100~14.3%.

Similarly, the RBF kernel improved elapsed time by (0.1452-0.025458)/0.1452*100~82%, and accuracy by (71.4286-64.2857)/71.4286*100~10%.

Table 5. Comparison of results between experiments 1 and 2

Linear Kernel	CV%	Elapsed Time	Prediction Accuracy
Experiment 1	76.2	0.14124	64.3
Experiment 2.	76.2	0.023389	75.0
RBF kernel			
Experiment 1	76.2	0.1452	64.3
Experiment 2	76.2	0.025458	71.4

Results for experiment 2 and in table 5 confirm that classification of signal data with wavelet-based feature [26] extraction improves classification accuracy by at least 80% and runtime by at least 10%.

Table 6. File size comparison

Data Files	Exp. 1 (KB)	Exp. 2 (KB)	% Reduction
Training data	4329	987	77.2
Testing data	2917	685	76.5

Comparing data file sizes in kilobytes (KB) after data decomposition by the wavelet filter "bior2.6", table 6 shows that experiment 2 (with wavelet feature selection) had more than 77% and 76% feature reduction in the training and the testing data respectively. Both the literature in [26] and experiments shown here confirm that using SVMs with wavelet-based feature reduction improves the prediction accuracy, runtime and reduces features to be classified for MALDI-TOF signal data. These improvements would be significantly high with larger data.

5   Conclusion

The paper has presented SVM classification of MALDI-TOF signal data from bacteria colony using feature extraction by discrete wavelet transforms. Experiments tested the data with four kernels using various parameters so that a more suitable kernel was used for classification. Results showed that classification accuracy with feature selection improved accuracy by at least10%, and feature reduction by 76% and runtime by over 80%.

Further work is planned to explore other peak feature selection methods and identification of bacteria types in those peaks. Particular known denoising wavelet methods will also be used to check the cross-validation results and overall classification accuracy. Further, wavelet lifting schemes [20] may be tested with our approach. Wavelet lifting schemes are more efficient implementations of first generation wavelets and are not necessarily translates and dilates of one function, and thus do not rely on polynomial factorizations, as do Fourier transforms.

References

1. Lay, J.O.: MALDI-TOF Mass Spectrometry of Bacteria. John Wiley (2002)
2. Bundy, J., Fenselau, C.: Lectin-based Affinity Capture for MALDI-MS Analysis of Bacteria. Analy. Chem. 71(7), 1460–1463 (1999)
3. Li, T., Li, Q., Zhu, S., Ogihara, M.: A Survey on Wavelet Applications in Data Mining. SIGKDD Explorations 4(2), 49–68 (2003)
4. Bruyne, K.D., et al.: Bacterial Species Identification from MALDI-TOF Mass Spectra through Data Analysis and Machine Learning. Syst. and Appl. Microb. 34, 20–29 (2011)
5. Li, D., Pedrycz, W., Pizzi, N.J.: Fuzzy Wavelet Packet Based Feature Extraction Method and its Application to Biomedical Signal Classification. IEEE Trans. Biom. Eng. 526, 1132–1139 (2005)
6. Biotyper 2.0, http://www.bdal.com/products/software/maldi-biotyper/overview.html
7. Morris, J.S., Coombes, K.R., Koomen, J., Baggerly, K.A., Kobayashi, R.: Feature Extraction and Quantification for Mass Spectrometry in Biomedical Applications using the Mean Spectrum. Bioinformatics 21, 1764–1775 (2005)
8. Chui, C.K.: An Introduction to Wavelets. Academic Press, Boston (1992)
9. Daubechies, I.: Orthonormal Bases of Compactly Support Wavelets. Comm. Pure Appl. Math. 41, 909–996 (1988)
10. Daubechies, I.: Ten Lectures on Wavelets. Capital City Press, Montpelier (1992)
11. McDonough, R.N., Whale, A.D.: Detection of Signals in Noise, 2nd edn. Academic Press, San Diego (1995)
12. Conrad, T.O.F., Leichtle, A., Hagehülsmann, A., Diederichs, E., Baumann, S., Thiery, J., Schütte, C.: Beating the Noise: New Statistical Methods for Detecting Signals in MALDI-TOF Spectra Below Noise Level. In: Berthold, M., Glen, R.C., Fischer, I. (eds.) CompLife 2006. LNCS (LNBI), vol. 4216, pp. 119–128. Springer, Heidelberg (2006)
13. Duda, R.O., Hart, P.E., Stork, D.G.: Pattern Classification, 2nd edn. (2001)
14. Shin, H., Sampat, M.P., Koomen, J.M., Markey, M.K.: Wavelet-based Adaptive Denoising and Baseline Correction for MALDI-TOF MS. J. of Integr. Biol. 14(3), 283–295 (2010)

15. Pedrycz, W., Vukovich, G.: Feature Analysis through Information Granulation and Fuzzy Sets. Pattern Recog. 35, 825–834 (2002)
16. Resson, H.W., et al.: Peak Selection from MALDI-TOF Mass Spectra using Ant Colony Optimisation. Bioinformatics 23(5), 619–626 (2007)
17. Malyarenko, D.I., et al.: Enhancement of Sensitivity and Resolution of Surface-enhanced Laser Desorption Ionisation Time-of-flight Mass Spectrometric Records for Serum Peptides using Time-series Analysis Techniques. Clin. Chem. 51, 65–74 (2005)
18. Alexandrov, T., et al.: Biomarker Discovery in MALDI-TOF Serum Protein using Discrete Wavelet Transformation. Bioinformatics 25(5), 643–649 (2009)
19. Khushaba, R.N., Al-Jumaily, A.: Fuzzy Wavelet Packet Based Feature Extraction Method for Multifunction Myoelectric Control. J. of Biol. and Life Sci. 2(3), 186–194 (2007)
20. Sweldens, W.: Lifting Scheme: A New Philosophy in Biorthogonal Wavelet Constructions. In: SPIE Wavelet Applications in Signal and Image Processing III, vol. 2569, pp. 68–79 (1995)
21. Chih-Chung, C., Chih-Jen, L.: LIBSVM: A Library for Support Vector Machines. ACM Transactions on Intelligent Systems and Technology 2(27), 1–27 (2011)
22. Hsu, C.W., Chang, C.C., Lin, C.J.: A Practical Guide to Support Vector Classification. Bioinformatics 1(1), 1–16 (2010)
23. Boser, B.E., Guyon, I.M., Vapnik, V.N.: A Training Algorithm for Optimal Margin Classifiers. In: 5th Annual ACM Workshop on COLT, pp. 144–152 (1992)
24. Ramaswamy, R., et al.: Multiclass Cancer Diagnosis using Tumor Gene Expression Signatures. Proceedings of the National Academy of Sciences of the United States 98(26), 15149–15154 (2001)
25. Savchuk, O.Y., Hart, J.D., Sheather, S.J.: Indirect Cross-validation for Density Estimation. Amer. Stat. Ass. 105(489), 415–423 (2010)
26. Shutao, L., Chen, L., James, K.: Wavelet-based Feature Selection for Microarray Data Classification. In: Proc. Int. Joint Conference on Neur. Net. (IJCNN), pp. 5028–5033 (2006)
27. Frank-Michael, S., et al.: Support Vector Classification of Proteomic Profile Spectra Based on Feature Extraction with the Bi-orthogonal Discrete Wavelet Transform. Comp. and Visual. in Sci. 12(4), 189–199 (2009)
28. Wong, L., Muyeba, M., Keane, J.: Towards Adaptive Mining of Spectral Features. In: Proceedings of UK Workshop on Computational Intelligence, pp. 213–216 (2011)
29. Smith, M., Martinez, T.: Improving Classification Accuracy by Identifying and Removing Instances that Should Be Misclassified. In: Proc. Int. Joint Conference on Neur. Net. (IJCNN), San Jose, pp. 2690–2697 (2011)

Multi-Scale Local Spatial Binary Patterns
for Content-Based Image Retrieval

Yu Xia, Shouhong Wan, Peiquan Jin, and Lihua Yue

School of Computer Science and Technology
University of Science and Technology of China, China
xiay1989@mail.ustc.edu.cn, {wansh,jpq,llyue}@ustc.edu.cn

Abstract. Content-based image retrieval (CBIR) has been widely studied in recent years. CBIR usually employs feature descriptors to describe the concerned
characters of images, such as geometric descriptor and texture descriptor. Many
texture descriptors in texture analysis and image retrieval are based on the so-
called Local Binary Pattern (LBP) technique. However, LBP lacks of the spatial
distribution information of texture features. In this paper, we aim at improving
the traditional LBP and present a novel texture feature descriptor for CBIR
called Multi-Scale Local Spatial Binary Patterns (MLSBP). MLSBP integrates
LBP with spatial distribution information of gray-level variation direction and
gray-level variation between the referenced pixel and its neighbors. In addition,
MLSBP extracts the texture features from images on different scale levels. We
conduct experiments to compare the performance of MLSBP with five competi-
tors including LBP, Uniform LBP (ULBP), Completed LBP (CLBP), Local
Ternary Patterns (LTP), and Local Tetra Patterns (LTrP). Also three benchmark
image databases are used in the measurement, which are the Bradotz Texture
Database (DB1), the MIT VisTex Database (DB2), and the Corel 1000 Database (DB3). The experimental results show that MLSBP is superior to the competitive algorithms in terms of precision and recall.

Keywords: Content-based image retrieval, Local binary pattern, Texture
feature, Spatial distribution.

1   Introduction

Image retrieval is one of the important applications of image processing [1]. Due to
the explosive growth of digital images, there exists an urgent need for efficient image
retrieval algorithms which can search the desired images from databases. Generally,
there are three main categories of image retrieval: text-based image retrieval (TBIR),
content-based image retrieval (CBIR) and semantic-based image retrieval (SBIR) [2].
Images in TBIR systems need to be manually annotated and require much human
labor, and the annotation accuracy is subject to human perception [3]. For SBIR, researchers focus on the representation of high-level semantic of images, while low-
level features (color, texture, shape, etc.) often fail to describe the high-level semantic
concepts of images [4]. Content-based image retrieval based on the visual contents of

T. Yoshida et al. (Eds.): AMT 2013, LNCS 8210, pp. 423–432, 2013.
© Springer International Publishing Switzerland 2013

an image, such as color, texture, shape, distribution layout, etc., to represent the images. Due to the advantages of simple calculation and high efficiency, CBIR is becoming more widely used in image retrieval. The difficulty in CBIR is to find a best representation of an image for all perceptual subjectivity, due to the different view angles and illumination changes.

In this paper, we aim at improving the traditional LBP and present a novel texture feature descriptor for CBIR, which is called Multi-Scale Local Spatial Binary Patterns (MLSBP). MLSBP integrates LBP with spatial distribution information of gray-level variation direction and gray-level variation between the referenced pixel and its neighbors. In addition, MLSBP extracts the texture features from images on different scale levels. Finally, we conduct three experiments to demonstrate the performance of MLSBP and other compared methods.

The remainder of the paper is structured as follows. Section 2 introduces the related work. Section 3 gives the calculation of MLSBP. Section 4 presents the framework of proposed image retrieval algorithm and the similarity measurement. Section 5 shows our experiment results. Section 6 concludes our work and gives the directions for future studies.

2   Related Work

Due to the efficiency to image processing, texture features have been widely used in CBIR. LBP [5] has emerged as an efficient feature in texture analysis and CBIR. With the advantages of simple calculation and multi-scale characteristic, LBP performs excellent in texture image classification. Recent years, LBP has been promoted to different versions: uniform LBP (ULBP) [6], completed LBP (LBP) [7] and local ternary patterns (LTP) [8]. LBP extracted the gray-level variation pattern between center pixel and its surrounding neighbors. ULBP based on the fact that most frequent 'uniform' binary patterns correspond to primitive micro-features, such as edges, corners, and spots. Thus, ULBP reduces LBP patterns into fewer uniform patterns. At the same time, by choosing the minimum of pattern values after circular right shifting, ULBP obtained the characteristic of rotational invariance. CLBP completed LBP with magnitude information (CLBP_M) and the gray level represented by center pixels (CLBP_C). By combing CLBP_M, CLBP_C with the traditional LBP, CLBP can extract texture feature more comprehensively. LTP extended LBP with coding variation pattern by ternary result and respectively calculated the upper and lower value as the features. The local tetra patterns (LTrP) [9] based on the first-order derivatives in vertical and horizontal directions and encoded the variation with tetra codes. And through calculating high-order patterns, LTrP can extract more detail texture information. All these patterns only calculated the gray variation of pixels, but ignored the spatial distribution information of variation and direction. Thus, it is evident that the performance of these methods can be improved by extracting spatial distribution information.

3   Multi-Scale Local Spatial Binary Patterns (MLSBP)

3.1   Local Patterns

LBP was introduced in [5] for texture classification. Given a referenced center pixel, LBP is computed by comparing its gray value with its neighbors, based on Formula 3.1.

$$LBP_{P,R} = \sum_{p=1}^{P} 2^{(p-1)} \times f_{LBP}(g_p - g_c)$$

$$f_{LBP} = \begin{cases} 1, & x \geq 0 \\ 0, & else \end{cases}$$

(3.1)

where g_c is the gray-level value of the referenced pixel, g_p is the gray-level value of its neighbors, P is the number of neighbors, and R is the radius of the neighbors.

LTP was introduced in [8] for face recognition. LTP extended LBP to a three-valued code, and calculated the upper and lower value. LTP is computed by Formula 3.2.

$$LTP^{upper} = \sum_{p=1}^{P} 2^{(p-1)}, if \quad f_{LTP}(g_p - g_c) = 1$$

$$LTP^{lower} = \sum_{p=1}^{P} 2^{(p-1)}, if \quad f_{LTP}(g_p - g_c) = -1$$

(3.2)

$$f_{LTP}(x, y, t) = \begin{cases} 1, & x \geq y + t \\ 0, & |x - y| < t \\ -1, & x \leq y - t \end{cases}$$

where t is the threshold. More details about LTP can be found in [8].

3.2   Local Spatial Binary Patterns (LSBP)

LSBP we proposed extracts the spatial distribution information of gray-level variation, which is an effective supplement to LBP. LSBP is defined as Formula 3.3.

$$LSBP^{\alpha}(i, j) = (GVP_P^{\alpha} = i, GVP_Q^{\alpha} = j)$$

(3.3)

where P and Q are two pixels which are the distance **d** apart in original image, GVP_P^{α} and GVP_Q^{α} are the gray-level variation patterns (GVP) of pixel P and Q on direction α.

The GVP can be obtained by integrating the relationship results together on each direction. The relationship $f_{relation}(P, Q)$ (binary coding) between two gray values P and Q can be calculated by Formula 3.4.

$$f_{relation}(P,Q) = \begin{cases} 00, & if\ (P-Q) \geq T\ \&\ P \geq Q \\ 01, & if\ (P-Q) < T\ \&\ P \geq Q \\ 10, & if\ (Q-P) \geq T\ \&\ P < Q \\ 11, & if\ (Q-P) \geq T\ \&\ P < Q \end{cases} \qquad (3.4)$$

where T is chosen as a threshold to distinguish two pixels' gray-level values. Supposing T is small, gray-level variations are coded more precisely, which means variations with small difference can be coded as different results. This will lead to the inaccuracy of similarity measurement. On the contrary, supposing T is large, variations are coded more roughly, which means variations with big difference can be coded as the same results. This will lead to the loss of texture detail information.

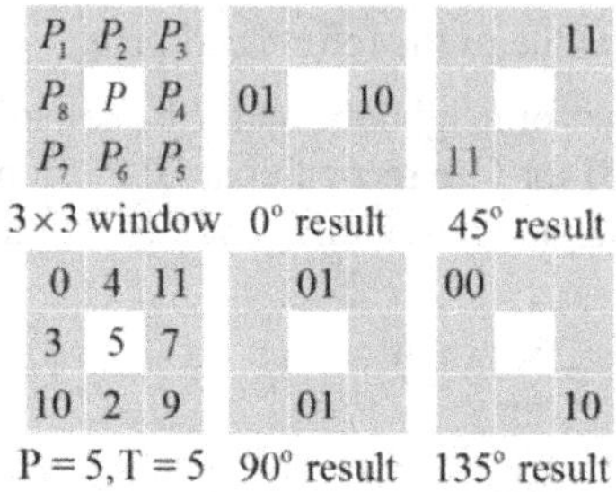

Fig. 1. Calculation of gray-level variation pattern

Fig. 1 shows an example of 9 pixels in 3×3 window, considering the center referenced pixel is P and its surrounding 8 neighbors are Q, with T is selected as 5. Based on the code results of each pixel, we can get a four binary code by integrating the two neighbors' results together on each direction to get gray-level variation pattern (GVP). The neighbors of P on $45°$ are P_3 and P_7, thus, GVP on $45°$ in the example can be coded as "1111", while $0°$ is "1001", $90°$ is "0101", $135°$ is "0010". We can get one pattern image on each direction with every pixel value in pattern image ranging from "0000" to "1111". Finally, LSBP value can be obtained based on the GVP results. Each entry (i, j) in LSBP corresponds to the number of occurrences of the pair of i and j in GVP image. To simplify the computation, LSBP (i, j) can be calculated by integrating the GVP of i and j together. For example, supposing GVP of i is "1101", j is "0110", then LSBP (i, j) can be coded as "11010110", which can be converted to decimal number "214". Through recording the occurrences of different LSBP codes (from "00000000" (0) to "11111111" (255)), LSBP histograms can be constructed to represent the spatial distribution information of texture in images. In this paper, we choose **d** as (1, 0), (0, 1), (1, 1), (-1, 1), which means the LSBP is respectively calculated with four pairs. When **d** is set as (1, 0), e.g., the pair of pixels is P and P_4. Based on the four GVP pattern images, we can obtain sixteen LSBP histograms, every four LSBP histograms on each direction.

Furthermore, we calculate the mean and standard deviation of the variations between pixels and its surrounding neighbors to reflect the gray-level variation magnitude information. Assuming the center pixel is P and the whole three pixels' values on direction α are (P_x, P, P_y). The variation Var_P^α is calculated by Formula 3.5.

$$Var_P^\alpha = |P_x - P| + |P - P_y| \tag{3.5}$$

For the pixels whose pattern values on direction α are m, the mean ($\overline{V}_m^\alpha$) and standard deviation (σ_m^α) are calculated by Formula 3.6.

$$\overline{V}_m^\alpha = \sum_{i=1}^{K} Var_{P_i}^\alpha / K, P_i \in \Phi_{MP}, \sigma_m^\alpha = \sqrt{\sum_{i=1}^{K}(Var_{P_i}^\alpha - \overline{V}_m^\alpha)^2 / K}, P_i \in \Phi_{MP} \tag{3.6}$$

where Φ_{MP} is the pixels' set whose GVP value is m on direction α, K is the number of Φ_{MP}. Through this step, we can calculate the mean and standard deviation of the 16 different patterns on $0°, 45°, 90°$, and $135°$ directions, and a $4 \times 16 \times 2 = 128$ dimensional variation magnitude vector (**VM**) was obtained to represent the gray-level variation magnitude information.

3.3   Calculation of Multi-Scale LSBP

In order to extract more detail information and contour information, multi-scale analysis is always been used in texture analysis. Due to the different combinations of P and R in Formula.(1), the LBP has the superiority of multi-scale characteristics. As Fig. 2 shows, firstly, we down-sampling the original image with 3×3 and 5×5 window. Secondly, the LSBP vectors of the sampling images were extracted. The final step is normalization of LSBP vectors to construct the MLSBP feature vector.

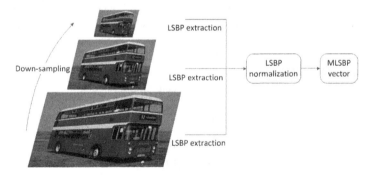

Fig. 2. Feature extraction of Multi-scale LSBP

Given an image G, it can be sampled by Formula 3.7.

$$G'(x, y) = \frac{1}{w * w} \sum_{i=-(w-1)/2}^{(w-1)/2} \sum_{j=-(w-1)/2}^{(w-1)/2} G(x+i, y+j) \tag{3.7}$$

where $w*w$ is the size of sampling window and $G(x, y)$ is one of the pixels in G. In this paper, we respectively choose w as 3 and 5. After down-sampling, the height and width of G' is reduced to $1/w$ of the height and width of G. Suppose the texture feature vectors of original image, 3×3 sampling image and 5×5 sampling image are f_1^{LSBP}, f_3^{LSBP} and f_5^{LSBP}. The MLSBP vectors can be defined by Formula 3.8.

$$f^{MLSBP} = (f_1^{LSBP}, f_3^{LSBP}, f_5^{LSBP}) \tag{3.8}$$

After the MLSBP extraction, we can use f^{MLSBP} as the feature vectors to measure the similarity between query image and images from database.

4   The Image Retrieval Algorithm Using MLSBP

4.1   Proposed Image Retrieval Framework

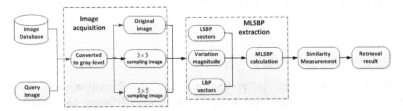

Fig. 3. The framework of proposed image retrieval algorithm

The framework of the proposed image retrieval algorithm is shown in Fig. 3. Firstly, the query image and the images in database are converted into gray images. Then we filter the original image with 3×3 and 5×5 windows. Secondly, LSBP feature vectors of $0°$, $45°$, $90°$ and $135°$, variation magnitude vectors and LBP feature vectors are extracted from the original and filtered images. After the step of calculating the MLSBP vectors, we can obtain the MLSBP extraction results. Finally, we use feature similarity to measure the similarity between the query image and the database images, and return retrieval result based on similarity calculation results.

4.2   Feature Similarity Measurement

Supposing the feature vector of the query image (Q) and images in database (DB) are represented as $f_Q(f_{Q_1}, f_{Q_2}, ..., f_{Q_L})$ and $f_{DB}(f_{DB_1}, f_{DB_2}, ..., f_{DB_L})$. L is the dimension of feature vector. This involves the selection of n top-matched images by measuring the distance between the query image and the images in database. The similarity distance to match the images is computed using Formula 4.1.

$$d = \sum_{i=1}^{L} \left| \frac{f_{DB_i} - f_{Q_i}}{1 + f_{DB_i} + f_{Q_i}} \right| \tag{4.1}$$

where f_{Q_i} and f_{DB_i} is the *ith* feature value of the feature vector. As we can know, the smaller d is, the more similar two pictures are.

5 Experiment Results

5.1 Experiment Setup

In order to prove the effectiveness of MLSBP, experiments are implemented on three benchmark databases: Bradotz Texture Database, MIT VisTex Database and CoreImage 1000 Database. The Bradotz texture database and MIT VisTex database consist of texture images, while the CoreImage 1000 database consists of natural scene images. The image retrieval algorithm we proposed is implemented by C++ and OPENCV. The performance of the proposed image retrieval algorithm is measured in terms of precision, recall, average precision (AP), and average recall (AR). For the query image Q, the indicators are defined as Formula 5.1 to 5.4.

$$\Pr ecision(I_q,n) = \frac{1}{n}\sum_{i=1}^{|DB|}\left|\delta(\phi(I_i),\phi(I_q)) \mid Rank(I_i,I_q) \leq n\right| \tag{5.1}$$

$$\operatorname{Re}call(I_q,n) = \frac{1}{N_G}\sum_{i=1}^{|DB|}\left|\delta(\phi(I_i),\phi(I_q)) \mid Rank(I_i,I_q) \leq n\right| \tag{5.2}$$

$$AP(n) = \frac{1}{|DB|}\sum_{i=1}^{|DB|}\Pr ecision(I_i,n), AR(n) = \frac{1}{|DB|}\sum_{i=1}^{|DB|}\operatorname{Re}call(I_i,n) \tag{5.3}$$

$$\delta(\phi(I_i),\phi(I_q)) = \begin{cases} 1, \phi(I_i) = \phi(I_q) \\ 0, else \end{cases} \tag{5.4}$$

where n indicates the number of retrieved images, and $|DB|$ is the size of the image database. $\phi(x)$ is the category of image x, $Rank(I_i,I_q)$ return the rank of image I_i (for the query image I_q) among all images of $|DB|$, N_G is the size of each image category in database. Given in Table.1 are the abbreviations of different methods used in the experimental discussions. In our experiments, the T used in LTP is set as 10.

Table 1. The abbreviations of different methods

Abbreviations	Methods
MLSBP	Multi-scale local spatial binary patterns
LSBP	Local spatial binary patterns
LBP	Local binary patterns
ULBP	Uniform binary patterns
CLBP	Completed local binary patterns
LTP	Local ternary patterns
LTrP	Local tetra patterns

5.2   Experiment on DB1

Database DB1 is consists of 111 different textures and used in experiment 1. These 111 textures are from Brodatz texture photographic album [10]. The size of each texture is 512×512. Each 512×512 image is divided into sixteen non-overlapping sub-images, thus creating a database of 1776 (16×111) images. Each image in the database is considered as the query image. Examples of DB1 are shown in Fig. 4. Fig. 5 illustrates the retrieval performance of the proposed method and other existing methods. T is the threshold chosen in Formula. (4). In this experiment, we set T as 10. As Fig. 5 shows, MLSBP outperforms the other existing methods.

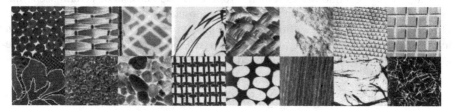

Fig. 4. Examples of texture images in DB1

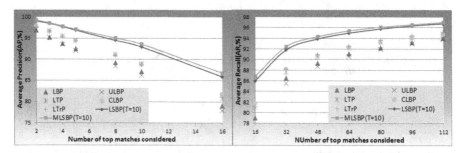

(a) AP results of existing methods on DB1   (b) AR results of existing methods on DB1

Fig. 5. Comparison of the MLSBP with other existing methods on DB1

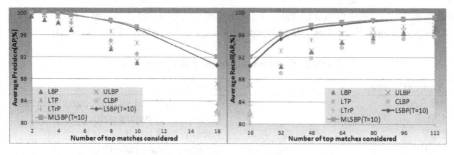

(a) AP results of existing methods on DB2   (b) AR results of existing methods on DB2

Fig. 6. Comparison of the MLSBP with other existing methods on DB2

5.3   Experiment on DB2

In this experiment, database DB2 is consists of 40 different textures collected from the MIT VisTex database [12]. The size of each texture image is 512×512, each image is divided into sixteen 128×128 non-overlapping sub-images, creating a database of 640 (16×40) images. In this experiment, we set T as 10. The retrieval result of DB2 is shown in Fig. 6.

5.4   Experiment on DB3

In experiment 3, images from the Corel database [11] have been used. This database consists of a large number of images of various contents ranging from animal images to outdoor sports and natural images. These images have been pre-classified into different categories. Each category has 100 images. For our experiment, we have chosen 1000 images to form DB3. These images are collected from ten different domains: Africans, Beach, Buildings, Buses, Dinosaurs, Elephants, Flowers, Horses, Mountains, Food. In this experiment, we set T as 20. As Fig. 7 shows, MLSBP performs better on average precision rate and average recall rate than other existing methods on DB3.

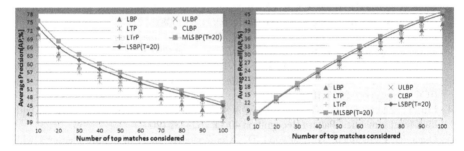

(a) AP results of existing methods on DB3   (b) AR results of existing methods on DB3

Fig. 7. Comparison of the MLSBP with other existing methods on DB3

The accurate results of AP and AR are shown in Table 2. It is evident that MLSBP outperforms other existing methods, respectively in texture images and natural scene images.

Table 2. The AP and AR results of all databases (The value in brackets is n in Formula.(12))

Methods	Average Precision(%)			Average Recall(%)		
	DB1(16)	DB2(16)	DB3(10)	DB1(112)	DB2(112)	DB3(100)
MLSBP	**86.78**	**92.06**	**75.39**	**97.00**	**99.04**	**45.81**
LSBP	85.92	90.50	72.88	96.78	98.97	44.81
LBP	79.01	82.13	70.49	94.07	96.70	41.30
ULBP	77.99	81.67	69.51	93.86	97.31	44.81
CLBP	80.86	87.05	68.90	94.78	97.80	39.48
LTP	81.72	81.92	69.55	94.70	95.73	42.71
LTrP	81.10	83.36	70.78	95.13	96.46	43.55

6   Conclusion

In this paper, we present a new texture feature descriptor referred to as MLSBP for CBIR. MLSBP not only extract the spatial distribution information of gray-level variation, but also extract the multi-scale texture feature from different scales of original image. However, MLSBP only depends on the gray image, but lose the color information of natural scene image. So the retrieval results of natural images are not every good. At the same time, the dimension of MLSBP is very large, which will influence the retrieval efficiency. In the future, we will focus our research on how to reduce the dimension of MLSBP vectors and combine MLSBP with color features to improve the retrieval accuracy.

Acknowledgements. This work was supported by the National Natural Science Foundation of China (Grant No. 61272317) and the General Program of Natural Science Foundation of AnHui of China (Grant No. 1208085MF90).

References

1. Datta, R., Joshi, D., Li, J., Wang, J.Z.: Image retrieval: Ideas, influences, and trends of the new age. ACM Computing Surveys (CSUR) 40(2), 5 (2008)
2. Liu, G.-H., Zhang, L., Hou, Y.-K., Li, Z.-Y., Yang, J.-Y.: Image retrieval based on multi-texton histogram. Pattern Recognition 43(7), 2380–2389 (2010)
3. Li, W., Duan, L., Xu, D., Tsang, I.W.-H.: Text-based image retrieval using progressive multi-instance learning. In: 2011 IEEE International Conference on Computer Vision (ICCV), pp. 2049–2055. IEEE (2011)
4. Liu, Y., Zhang, D., Lu, G., Ma, W.-Y.: A survey of content-based image retrieval with high-level semantics. Pattern Recognition 40(1), 262–282 (2007)
5. Ojala, T., Pietikäinen, M., Harwood, D.: A comparative study of texture measures with classification based on featured distributions. Pattern Recognition 29(1), 51–59 (1996)
6. Ojala, T., Pietikainen, M., Maenpaa, T.: Multiresolution gray-scale and rotation invariant texture classification with local binary patterns. IEEE Transactions on Pattern Analysis and Machine Intelligence 24(7), 971–987 (2002)
7. Guo, Z., Zhang, L., Zhang, D.: A completed modeling of local binary pattern operator for texture classification. IEEE Transactions on Image Processing 19(6), 1657–1663 (2010)
8. Tan, X., Triggs, B.: Enhanced local texture feature sets for face recognition under difficult lighting conditions. In: Zhou, S.K., Zhao, W., Tang, X., Gong, S. (eds.) AMFG 2007. LNCS, vol. 4778, pp. 168–182. Springer, Heidelberg (2007)
9. Murala, S., Maheshwari, R.P., Balasubramanian, R.: Local tetra patterns: a new feature descriptor for content-based image retrieval. IEEE Transactions on Image Processing 21(5), 2874–2886 (2012)
10. Brodatz, P.: Textures: A Photographic Album for Artists and Designers. Dover, New York (1996)
11. Corel 10000 image database,
    http://wang.ist.psu.edu/docs/related.shtml
12. MIT Vision and Modeling Group, Cambridge, Vision texture,
    http://vismod.media.mit.edu/pub/

Author Index